Essentials of Vehicle Dynamics

Essentials of Vehicle Dynamics

Joop P. Pauwelussen

AMSTERDAM • BOSTON • HEIDELBERG • LONDON
NEW YORK • OXFORD • PARIS • SAN DIEGO
SAN FRANCISCO • SINGAPORE • SYDNEY • TOKYO
Butterworth-Heinemann is an imprint of Elsevier

Butterworth-Heinemann is an imprint of Elsevier
The Boulevard, Langford Lane, Kidlington, Oxford OX5 1GB, UK
225 Wyman Street, Waltham, MA 02451, USA

ISBN: 978-0-08-100036-6

British Library Cataloguing-in-Publication Data
A catalogue record for this book is available from the British Library

Library of Congress Cataloging-in-Publication Data
A catalog record for this book is available from the Library of Congress

For information on all Butterworth-Heinemann publications visit our website at http://store.elsevier.com/

Typeset by MPS Limited, Chennai, India
www.adi-mps.com

Printed and bound in the United Kingdom

Dedication

Dedicated to my wife Petra and my children Jasper, Josien, and Joost who motivated me with their ambitions and confidence.

Contents

Preface

Teaching vehicle dynamics and control for the last 25 years, I have often struggled with the challenge of how to give students a proper understanding of the vehicle as a dynamic system. Many times, students new to the field do not currently have sufficient practice in design and experimental performance assessment, which are required for them to progress in skills and knowledge.

Fortunately, most students in automotive engineering have a minimal (and sometimes much higher) level of practical experience working on vehicles. This practical experience is usually a motivator to choose automotive engineering. However, that experience is not always matched with a sufficient level of practical knowledge of mathematics and dynamics, which is essential in vehicle dynamics and control. Lately, I have seen more and more students with a background in control or electronics who choose to specialize in automotive engineering. This should be strongly supported because future advanced vehicle chassis design requires a multidisciplinary approach and needs engineers who are able to cross borders between these disciplines.

However, these students can often be focused on a small element of the vehicle and lack a complete overview of the entire vehicle system. An overall understanding is important because this system is more complex than a linear system, which can be given any response with appropriate controllers. The tire–road contact and the interface between the vehicle and the driver especially should not be disregarded. At the end of a study, it is always asked whether the vehicle performance has been improved with respect to safety and handling, with or without the driver in the loop. Because drivers do not always respond in the way engineers expect, engineers must always be aware of the overall driver–vehicle performance assessment.

I wrote this book with the objective to address vehicle dynamics within a solid mathematical environment and to focus on the essentials in a qualitative way. Based on my experience, I strongly believe that a qualitative understanding of vehicle handling performance, with or without the driver, is the essential starting point in any research and development on chassis design, intelligent chassis management, and advanced driver support. The only way to develop this understanding is to use the appropriate mathematical tools to study dynamical systems. These systems may be highly nonlinear where the tire–road contact plays an important role. Nonlinear dynamical systems require different analysis tools than linear systems, and these tools are discussed in this book.

This book will help the reader become familiar with the essentials of vehicle dynamics, beginning with simple terms and concepts and moving to situations with greater complexity. Indeed, there may be situations that require a certain model complexity; however, by always beginning a sequence with minimal complexity and gradually increasing it, the engineer is able to explain results in physical and vehicle dynamics terms. A simple approach always improves understanding and an improved understanding makes the project simpler.

My best students always tell me, after completing their thesis project, that with their present knowledge, they could have solved their project must quicker and in a simpler way if they repeated it. This improved understanding they gained is one of the objectives of teaching.

Starting from scratch with too much complexity leads to errors in models and therefore, improper conclusions as a result of virtual prototyping (e.g., using a model approach, and more and more common in the design process). To help reader to evaluate their learning, a separate chapter of exercises is included. Many of these exercises are specially focused on the qualitative aspects of vehicle dynamics. Further, they encourage readers to justify their answers to verify their understanding.

The book is targeted toward vehicle, mechanical, and electrical engineers and engineering students who want to improve their understanding of vehicle dynamics. The content of this book can be taught within a semester. I welcome, and will be grateful for, any reports of errors (typographical and other) from my readers and thank my students who have pointed out such errors thus far. I specifically acknowledge my colleague Saskia Monsma for her critical review in this respect.

Joop Pauwelussen
Elst, The Netherlands
May 2014

Chapter | One

Introduction

Vehicle dynamics describes the behavior of a vehicle, using dynamic analysis tools. Therefore, to understand vehicle behavior, one must have a sufficient background in dynamics. These dynamics may be linear, as in case of nonextreme behavior, or nonlinear, as in a situation when tires are near saturation (i.e., when the vehicle is about to skid at front or rear tires.). Hence, the tires play a critical role in vehicle handling performance.

To improve handling comfort, the predictability of the vehicle performance from the control activities of the driver (i.e., using the steering wheel, applying the brake pedal, or the pushing the gas pedal) must be considered. The road may be flat and dry, but one should also consider cases of varying road friction or road disturbances.

In this case, the major response of the vehicle can be explained based on a linear vehicle model. The state variables, such as yaw rate (in-plane rotation of the vehicle, which is the purpose of steering wheel rotation), body slip angle (drifting, meaning the vehicle is sliding sideways), and forward speed follow from a linear set of differential equations, where we neglect roll, pitch, elastokinematic effects, etc. These effects can be added in a simple way, which will result in only slight modifications in the major handling performance. The control input from the driver causes a (rotational, translational) dynamic vehicle response, which results in inertia forces being counteracted by forces between tires and road. These forces are, in first order, proportional to tire slip. In general, tire slip describes the proportionality between local tire deformation and the longitudinal position in the tire contact area. Tire slip is related to vehicle states (yaw rate, body slip angle) or vehicle forward speed and wheel speeds, in case of braking or driving (longitudinal slip). The analysis of this linear system, with an emphasis on the vehicle (mainly tire) specific stability properties, forms the basis of vehicle handing performance and must be well understood. Any further enhancement of the model's complexity, such as adding wheel kinematics, vehicle articulations (caravan, trailer, etc.), or load transfer, will lead to an improved assessment of vehicle handling performance, but always in terms of performance modifications of the most simple dynamical vehicle system, i.e., with these effects neglected.

Essentials of Vehicle Dynamics.

The theory of linear system dynamics is well established and many tools related to state space format are available; this includes local stability analysis that refers to the eigenvalues of the linear vehicle system. Therefore, once the handling problem is formulated in (state space) mathematical terms, as follows,

$$\begin{aligned} \dot{\underline{x}} &= A.\underline{x} + D.\underline{u} \\ \underline{y} &= C.\underline{x} + D.\underline{u} \end{aligned} \tag{1.1}$$

an extensive toolbox is available to the researcher. In Eq. (1.1), $\underline{x}$ denotes the state vector (e.g., yaw rate, wheel speed), $\underline{u}$ denotes the input (e.g., steering angle, brake force), and $\underline{y}$ denotes the system output.

However, a mathematical background in system dynamics alone is not sufficient for solving vehicle dynamics problems. The experience in lecturing on vehicle dynamics shows that there is room for improvement in the mathematical background of the students, with reference to multivariate analysis, Laplace transformation, and differential equations. For this reason, we included a number of necessary commonly used tools in the appendices for further reference. These tools will help the researcher to interpret model output in physical terms. The strength of the simple linear models is the application and therefore, the interpretation to understanding real vehicle behavior. The researcher should answer questions such as:

- What is the impact of axle characteristics (force versus slip) or center of gravity position on vehicle handling performance?
- How are the axle characteristics related to kinematic design?
- How are the axle characteristics related to internal suspension compliances?
- How reliable are axle characteristics parameters and how robust are our analysis results against variations of these parameters?
- What is the impact of roll stiffness on front and rear axles on simplified model parameters?
- How can we take driving resistance (additional drive force to prevent the vehicle speed from decreasing) into account?

In addition, the contents of this book should be linked to practical experience in testing, aiming at model validation and parameter identification.

Moving to extreme vehicle behavior, a problem arises in the sense that the vehicle model becomes nonlinear. In the case of linear vehicle performance, the vehicle is either globally stable or globally unstable, with stability depending on vehicle and tire characteristics. One can analytically determine the vehicle's response for a specific driver control input and investigate the sensitivity regarding vehicle parameters. Therefore, a researcher is able to use both qualitative tools (is the model correctly described at a functional level?) and quantitative tools (does the model match experimental results?) to analyze the vehicle model in reference to experimental evidence.

For a nonlinear model, situations change principally. Nonlinear models arise if we accept that the axle characteristics depend nonlinearly on slip (i.e., when one of the axles is near saturation). A typical example of longitudinal tire behavior in terms of brake force F_x versus brake slip κ (defined in (2.19)) is shown in Figure 1.1 for various wheel loads F_z (see Section 2.4 for a more extensive treatment of longitudinal tire characteristics).

For small brake slip κ, this relationship is described as linear, with proportionality factor C_κ, between slip and tire force, as indicated in Figure 1.1. Clearly, for brake slip 0.05 or higher, this linear approximation is incorrect.

When considering safety, we must account for nonlinear model behavior. Are the driver (closed loop) and vehicle (open loop) capable of dealing with dangerous driving conditions, with or without a supporting controller?

With a stable linear model, any small disturbance (input, external circumstances) leads to a small difference in vehicle response. For a nonlinear system being originally stable, a small disturbance may result in unstable behavior, i.e., with a large difference in vehicle response. For example, with an initial condition of a vehicle approaching a stable circle, a small change could result in excessive yawing of the vehicle (i.e., stability is completely lost). Consequently, quantitative tools (i.e., calculating the response by integrating the system equations) cannot be interpreted any further in a general perspective. However, there are ways to get around this problem:

- Consider the linearization of the model around a steady-state solution (where there may be multiple solutions, in contrast to the linear model where one solution is found in general), and use the analysis tools for the linear model to find the model performance near this steady-state solution.
- Use qualitative (graphical) analysis tools specifically designed for nonlinear dynamical systems. A number of these tools are discussed in Chapter 5 and the appendixes, with distinctions made for phase plane analysis, stability and handling diagrams, the MMM method, and the "g–g" diagram.

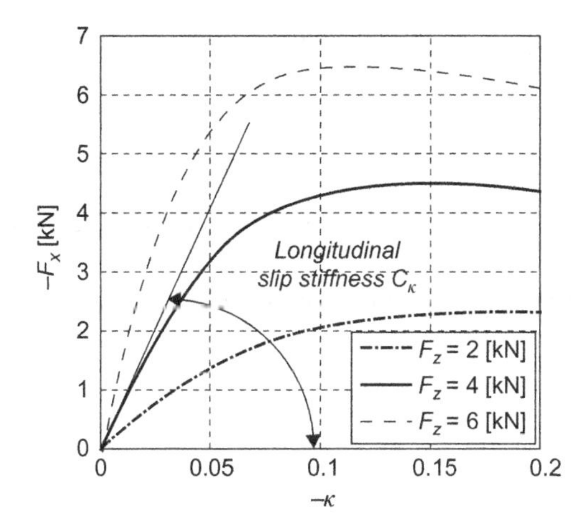

FIGURE 1.1 Longitudinal tire characteristics.

This last approach may seem to be insufficient, but remember that quantitative response only makes sense if the so-called qualitative "structural" model response is well matched. Is the order of the system correct and are trends and parameter sensitivities confirmed by the model? In other words, is the mathematical description of the model sufficient to match vehicle performance if the right parameter values are selected? For example, quadratic system performance will never be matched with sufficient accuracy to a linear model. In the same way, one must ensure that the vehicle nonlinear performance (and specifically the axle or tire performance) is well validated from experiments.

Mathematical analysis of vehicle handling always begins with the objective to understand certain (possibly actively controlled) vehicle performance, or to guarantee proper vehicle performance within certain limits. Therefore, the first priority is a good qualitative response. Moving into quantitative matching with experimental results (as many students appear to do) under certain unique circumstances only guarantees a certain performance under these unique circumstances. In other words, without further general understanding of the vehicle performance, such matching gives no evidence whatsoever on appropriate vehicle performance under arbitrary conditions. Testing and quantitative matching for all possible conditions may be an alternative of qualitative matching (and assessing the structural system properties), but this is clearly not feasible in practice.

This book is structured as follows. In Chapter 2, we will discuss fundamentals of tire behavior. The chapter follows the classical approach by first treating the free rolling tire (including rolling resistance), which is followed by discussions on purely longitudinal and lateral tire characteristics and combined slip. First, we focus on empirical tire models, which are essential elements of any vehicle handling simulation study. Second, we discuss two physical tire models: the brush model and the brush-string model. These models are not intended for use in practical simulation studies; however, they enable a deeper understanding of the physical phenomena in the tire–road contact under steady-state slip conditions.

When vehicle speed is relatively low and/or tires experience loading frequencies beyond 4 Hz (as in case of road disturbances or certain control measures), the steady-state assumption on tire performance (tire belt follows rim motions instantaneously) is no longer valid. A first step to include dynamics is to consider the tire as a first order (relaxation) system. Higher order dynamics require the belt oscillation to be incorporated in the tire model.

Chapter 3 discusses both situations in full analytical detail to allow the reader to reproduce the analytical approach. Modern tire modeling software may account for these (transient and dynamic) effects. Using such software requires an understanding of the background of the tire models used, which is what we offer to the reader.

Chapters 4 and 5 address vehicle performance. Chapter 4 discusses low-speed kinematic steering (maneuvering), which is followed by handling performance for nonzero speed in Chapter 5. Low-speed maneuvering means that tires are rolling and tire–road contact shear forces are negligibly small. The steering angle may be large and some examples of steering design are treated, showing that this force-free maneuvering can be approximated but never exactly satisfied. Chapter 4 discusses the zero lateral acceleration reference cases for the nonzero tire–road interaction forces, treated in Chapter 5.

Chapter 5 begins with a discussion of criteria for good handling performance and how it should be rated, with an emphasis on subjective and objective methodology strategies. The most basic, but still powerful, model is the single-track model (also referred to as the bicycle model), where tires are reduced to (linear or nonlinear) axles and roll behavior is neglected. In spite of its simplicity, effects such as lateral and longitudinal load transfer, alignment and compliance effects, and combined slip can be accounted for. One should be aware that the single-track model is based on axle characteristics that, in contrast to tire characteristics, depend on suspension design, which is expressed in terms of roll steer, roll camber, compliances, and aligning torque effects.

This model forces the researcher to focus on the most essential aspects of handling (either under normal driving conditions or under extreme high acceleration situations) and therefore understand the vehicle performance in terms of driver and/or control input and vehicle parameters. Straightforward extensions, such as the two-track model (distinction of left and right tires), are discussed as well.

Next, the steady-state vehicle behavior is treated in terms of understeer characteristics (response to steering input) and neutral steer point (response to external forces and moments). The concept of understeer is usually discussed in terms of linear axle characteristics, resulting in a linear relationship between steering input and vehicle lateral acceleration response in terms of the understeer gradient. The nonlinear extension is not straightforward and will be discussed in detail. We will distinguish between four definitions of understeer (and oversteer) that are identical for linear axle characteristics but are not identical for nonlinear axles. Further, we shall show that these nonlinear axle characteristics completely determine the vehicle understeer characteristics and therefore the open-loop yaw stability properties (vehicle is considered in response to steering input) and handling performance. In Chapter 6, we will show that, when the response of the driver to vehicle behavior is taken into account, the so-called closed-loop stability of the total system of driver and vehicle depends on the vehicle understeer properties as well. In addition, the vehicle response in the frequency domain is discussed, with reference to speed-dependent damping properties and (un-)damped eigenfrequencies.

As indicated earlier, nonlinear system analysis is qualitative and uses appropriate graphical assessment tools:

- *Phase plane analysis* is used to visualize solution curves near critical (steady-state) points and to support interpretation of the performance along these solution curves from a global system perspective.
- *The stability diagram* is used to visualize the type of local yaw stability in terms of axle characteristics and vehicle speed.

- *The handling diagram* is used to visualize the stable and unstable steady-state conditions in terms of axle characteristics, vehicle speed, steering angle, and curve radius.
- *The moment method (MMM) diagram* is used to visualize the vehicle potential in terms of lateral force and yaw moment (limited due to axle saturation), which basically corresponds to the phase plane representation in terms of these force and moment.
- *The "g–g" diagram* is used to link tire shear forces to vehicle lateral and longitudinal forces and therefore indicates which tire will saturate first under extreme conditions.

In Chapter 6, we discuss the vehicle–driver interface. Good handling performance cannot be assessed without considering the driver. The driver controls the vehicle by applying input signals, such as the steering wheel angle and gas or brake pedal position. Major driving tasks are guidance (e.g., following another vehicle or negotiating a curve) or stabilization (e.g., when the vehicle safety is at stake). The driver is supported in these tasks by many different types of advanced driver assistance systems. Conversely, these support systems and other onboard (infotainment) devices create an increasing number of distractions for driver.

The practical situation on the road is that the driver responds to changing vehicle and traffic conditions. That may not always be an easy task, resulting in increased workload, which, in turn, has an effect on the driver's ability to carry out driving task safely. Not only is the total closed-loop behavior relevant for the assessment of good handling performance, but the costs (effort, workload) for the driver are relevant in achieving such closed-loop performance. The assessment of driver's state is discussed in Chapter 6, with special emphasis on workload.

The vehicle–driver interface can be treated as a system, with the driver adapting to the vehicle performance. Two different cases are discussed, addressing following behavior and handling, with the final situation described in terms of path following. The driver models for both driving scenarios are special cases of the McRuer crossover model approach. In the case of following a lead vehicle, it is shown that the driver model allows us to identify the transition of the regulation phase (no safety risk) to the reaction phase (perceived increase of risk indicated by releasing the throttle) in terms of relative speed and time headway.

In the case of handling, the driver model is based on tracking a certain path at a preview distance, with a delayed steering angle response that is proportional to the observed path deviation. The relationship between the model parameters is analyzed in terms of closed-loop vehicle–driver performance, the closed-loop stability is treated, and the identification and interpretation of these parameters in terms of driver state is discussed in the final section of Chapter 6.

Chapter 7 includes exercises based on lectures and examinations at the HAN University of Applied Sciences. These exercises serve to improve the understanding of the vehicle system behavior, especially its qualitative aspects.

Chapter | Two

Fundamentals of Tire Behavior

In this chapter, attention is paid to the properties and resulting steady-state performance of tires as a vehicle component. With the tire as the prime contact between vehicle and road, the vehicle handling performance is directly related to the tire—road contact. The tires transfer the horizontal and vertical forces acting on the vehicle from steering, braking, and driving, under varying road conditions (slippery, road disturbances, etc.). Tire forces are not the only forces acting on the vehicle. Other forces acting on the vehicle could be from external disturbances (e.g., aerodynamic forces from crosswind). However, the contact between vehicle and road is by far the dominant factor in vehicle behavior and may be the difference between safe and unsafe conditions. Therefore, emphasis is put on the influence of tire properties in general and specifically in this chapter, which describes the tire steady-state behavior. Transient and dynamic tire performance will be discussed in Chapter 3.

The tire—road interface is schematically shown in Figure 2.1. The tire is a complex structure, consisting of different rubber compounds, combinations of rubberized fabric, or cords of various materials (steel, textile, etc.) that act as reinforcement elements (referred to as plies) that are embedded in the rubber with a certain orientation. The outer part of the tire is cut in a specific pattern (tread pattern design), referred to as the tire profile. The tire profile serves to guide the water away from the contact area under wet road conditions, and to adapt to the road surface in order to maintain a good contact (and therefore load transfer) between tire and road. Therefore, each tire has unique structural and geometrical design parameters. These parameters result in tire properties that, in combination with the vehicle, lead to vehicle performance. That means that the vehicle manufacturer will set up requirements for the tire manufacturer in terms of vehicle performance, which the tire manufacturer must fulfill. These requirements include many different things, such as:

- Good adherence between road and tire under all road conditions in longitudinal (braking/driving) and lateral (cornering) situations.
- Low energy dissipation (low rolling resistance).

Essentials of Vehicle Dynamics.

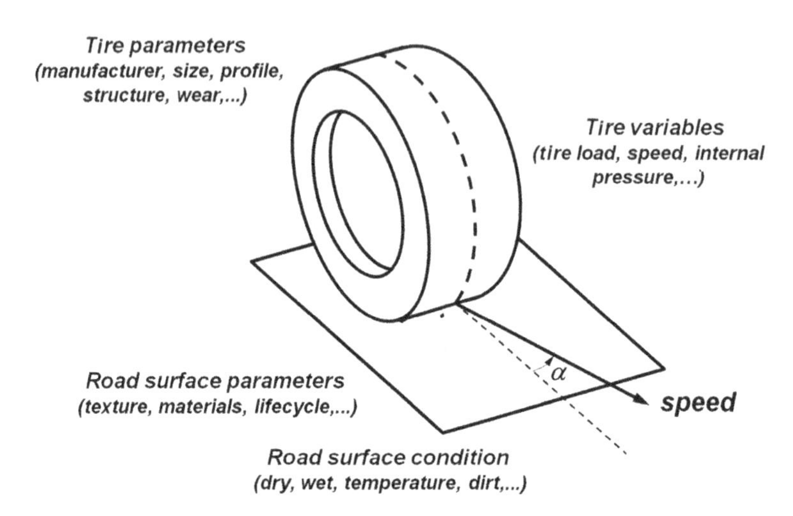

FIGURE 2.1 Tire–road interface.

- Low tire noise, which has two aspects—the effect observed inside of the vehicle and the noise emitted into the environment
  - The effect observed inside the vehicle is directly related to the vibration transfer from tire, through vehicle's suspension, toward the driver. This is a comfort issue for the driver.
  - Noise emitted into the environment is undesirable from an environmental point of view.
- Good durability and therefore, good wear resistance
  - Tire properties change with wear, which will in general lead to a higher tire stiffness in horizontal and vertical direction.
- Good comfort properties (filtering of road disturbances) and low interior noise transfer.
- Good subjective assessment, including predictability (consistency in response).

Each tire parameter has an effect on each of the tire properties, which makes the task of the tire designer a difficult one. Ultimately, this results in a compromise between these properties. Tire manufacturers are faced with the task of judging tire properties in terms of vehicle performance, and therefore must be able to understand this performance in detail for modeling and testing. In turn, the tire manufacturer determines the requirements for the component and material suppliers, i.e., for the rubber compounds, the cord materials, etc. This covers the tire parameters, but there are further considerations.

First, road has a certain structure, porosity, roughness, and thermal properties, all of which can vary. In general, the top layer of the road might be resurfaced every 5–7 years, depending on the traffic use. This means a cycle of 5–7 years for road properties. In addition, the road surface conditions may

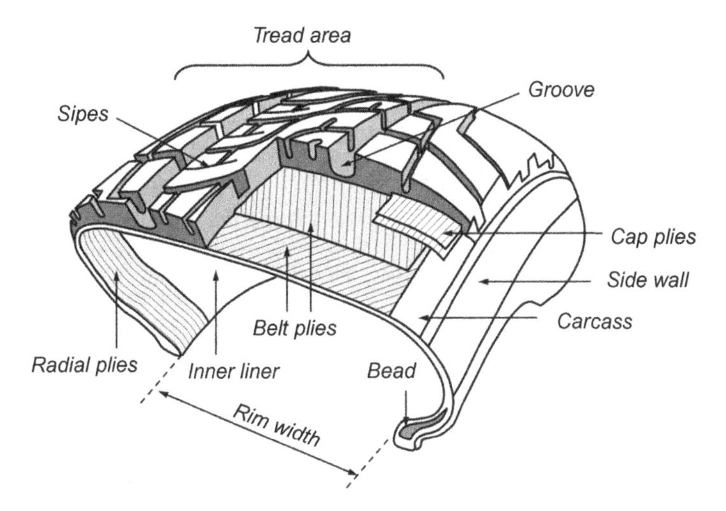

FIGURE 2.2 Schematic layout tire structure.

change due to weather conditions, day/night conditions, the traffic, and other external conditions, such as nearby housing, bridges, and viaducts.

Finally, the tire–road interface changes with the vehicle's motion. Changes in tire load will change the tire performance, which must be accounted for in the vehicle handling analysis. When the driver is cornering, the outer tires are loaded and the inner tires are unloaded. When the driver is braking or accelerating, the tire load shifts between the front and rear wheels. An increase in vehicle speed will in general lead to more critical adverse tire–road conditions. All these effects depend on the tire inner pressure.

We will take a closer look at the structure of the radial tire (Figure 2.2). The term "radial tire" refers to the radial plies, running from bead to bead, with the bead being the reinforced (with an embedded steel wire) part of the tire, connecting the tire to the rim. However, radial plies do not give the tire sufficient rigidity to fulfill the required performance under braking and cornering conditions. For that reason, the tire is surrounded by a belt with cords (steel, polyester, Kevlar, etc.) that are oriented close to the direction of travel.

The radial plies give good vertical flexibility and therefore, good ride comfort (in case of road irregularities). Cornering leads to distortion of the tire in the contact area, which evolves into deflection of rubber and extension of the cords in that area. With an almost parallel orientation of the cords in the belt, the extension of the cords is the dominant response, which means there is a large resistance (the modulus of elasticity of the cord material by far exceeds that of rubber) against this distortion and therefore, a stiff connection between vehicle and road. One could say that the different functions of the tire (i.e., having good comfort and, at the same time, good handling performance) are well covered by this distinction between radial and belt plies. The total combination of cords and plies that contributes to the tire rigidity is called the carcass.

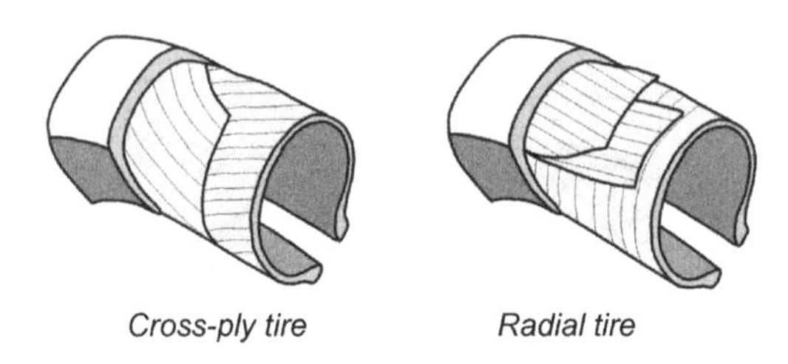

FIGURE 2.3 Schematic layout of cross-ply and radial tire.

The radial tire was patented by A.W. Savage in 1915 [43], but was not commercially successful until Michelin improved the design in the 1950s. Before that, cars were equipped with bias-ply (or cross-ply) tires with cords that run diagonally around the tire casing. For these tires, the cords have a much larger angle in relation to direction of travel (order of magnitude 40°) compared to the belt plies of a radial tire. In addition, no distinction is made between plies alongside wall and contact area. See Figure 2.3 for a schematic layout of the cross-ply and radial tire.

The cord structures (the plies) extend from contact area to sidewalls, such that deformation of the sidewalls would lead to deflections in the contact area, which would have a negative effect on wear at the shoulder of the tire. Because of the structural differences between both tires, the tread motion is reduced for the radial tire compared to the cross-ply tire, which also contributes to better fuel economy (reduced rolling resistance, see also Section 2.3). It has been shown by Moore [29] that bias-ply tires show significantly higher concentrations of shear stress, as well as normal contact pressure, at the shoulders of the tire, compared to radial tires.

The main contact between tire and road is through the tread area. Figure 2.2 indicates a tread pattern that is shaped to channel water away, with straight and s-shaped grooves that move from center of the tire to the side. We also indicated very small cuts in the pattern, referred to as sipes. These sipes are typical for winter tires and allow small motion between tread elements for rolling tires, leading to effectively larger friction on icy and snowy surfaces.

In the next section, we begin with a description of the input and output quantities of a tire. Determining what forces and moments are acting on a tire, and what input variables (such as slip, camber, and speed) these forces and moments depend on defines our language to define tire characteristics. In Section 2.2, we discuss the free rolling tire. Sections 2.3 addresses rolling resistance, with reference to all varying circumstances that can affect it. One may think of speed, additional slip such as brake or drive slip, temperature, tire pressure, tire load, etc. Sections 2.4 and 2.5 describe the tire under pure slip conditions, in case of braking/driving and cornering, respectively. In all sections, steady state behavior is assumed with the tire responding immediately to changes in slip or tire load. The phenomena in the contact area will be explained and empirical descriptions of tire characteristics will be discussed, specifically the Magic Formula description that was first introduced by Pacejka [32]. Section 2.6 discusses combined

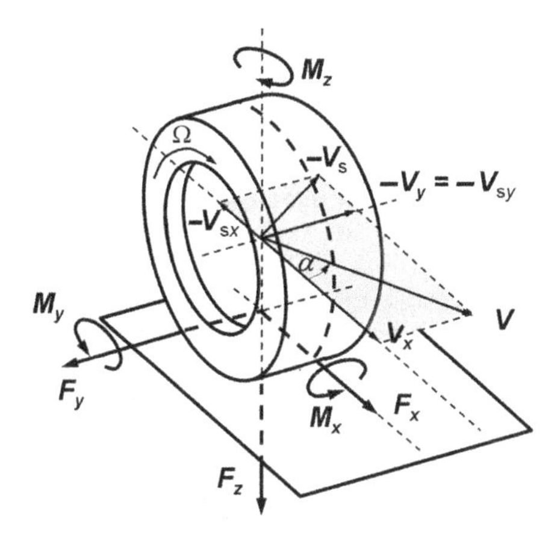

FIGURE 2.4 Forces and moments, acting on tire, speeds, and slip speeds.

slip situations (such as braking in a turn) and empirical relationships, and includes some useful simplifications.

The information covered in these sections provides the resources for anyone involved in automotive handling dynamics to resolve problems. However, for an engineer to use practical tools to describe steady-state tire behavior, the underlying physical phenomena must first be understood. For that reason, physical tire models are discussed in Section 2.7. We will treat two types of models, the brush model (which describes the local deflection in the contact area by linear springs) and the brush-string model where the belt deformation is also accounted for. These models will allow us to examine the local contact phenomena between tire and road.

2.1 TIRE INPUT AND OUTPUT QUANTITIES

A tire is schematically shown in Figure 2.4, with all the output quantities (forces and moment) and speeds indicated.

Note that the z-axis is chosen in the downward direction. There are three forces and three moments acting on the tire (*the output quantities*):

Forces		
F_x	:	Brake/drive force
F_y	:	Lateral (cornering) force
F_z	:	Tire load (to carry the vehicle weight)
Moments		
M_x	:	Overturning moment
M_y	:	Moment about the wheel axis (drive/brake torque)
M_z	:	Self-aligning moment

These forces and moments depend on a number of *input quantities*, which will be discussed in the subsequent sections:

d	:	Radial deflection—the difference between the unloaded and the loaded radius R–R_l
Ω	:	Rotational speed
γ	:	Camber angle—the angle between the normal vector to the wheel plane and the road surface (or, alternatively, the angle between the road surface normal direction and the wheel plane)
α	:	Slip angle— angle between speed direction and tire orientation in the plane parallel to the road's surface
κ	:	Longitudinal slip—the ratio of the slip speed (the difference between the rolling speed $\Omega \cdot R_e$ and the forward speed V_x in x direction) and the forward speed, where R_e is the effective rolling radius at free rolling
φ	:	Spin—the component of rotational speed in the global vertical direction, which is usually neglected except for situations when the curve radius is small (parking behavior) or for significant camber

A tire travels with a horizontal velocity V with components V_x and V_y in longitudinal and lateral direction, respectively. Due to brake or drive torque and cornering forces, slip will occur, which means that the tire slides with nonzero speed over the surface. The corresponding slip speeds V_{sx} and V_{sy} are shown in Figure 2.4 as well. Note that the slip quantities $\tan(\alpha)$ and κ, introduced previously, correspond to the negative ratios of slip speed and forward speed in x direction. The tire rolls over the surface with an angular speed Ω, leading to the rolling speed:

$$V_r = \Omega \cdot R_e \tag{2.1}$$

where R_e is the *effective rolling radius* of the free rolling tire. For a free rolling wheel (zero slip speed), the rolling speed coincides with V_x; therefore, the effective rolling radius is defined as the ratio between V_x and Ω under these conditions.

The effective rolling radius is not the same as the *loaded tire radius* R_l where the latter is defined as the vertical distance between the wheel center and the horizontal surface. A free rolling tire rotates around a point near the contact patch. For a rigid wheel on a flat horizontal surface, this point coincides with the single contact point between tire and road; here, the forward speed V_x equals the angular speed times (loaded = unloaded) radius.

For a pneumatic tire, the distance between points at the circumference of the tire and the wheel center varies from a value close to the unloaded radius just prior to entering the contact area, to the same value as the loaded radius just at the projection point of the wheel center on the contact area. At that point, the peripheral velocity of the tread (as relative to the wheel center) coincides with the horizontal velocity V of the wheel center for a free rolling tire. Moving out of the contact area, the tread regains its original length and the peripheral velocity returns to $\Omega \cdot R$, where R is the unloaded radius. Consequently, the rotational speed of the wheel with a pneumatic tire under free rolling conditions is less than that of a rigid wheel

$$R_l < R_e < R \tag{2.2}$$

This means that the center of rotation of the wheel usually lies somewhere below the surface. The effective rolling radius of a tire under free rolling conditions behaves different with varying tire load, as compared to the loaded tire radius. A loaded radius behaves almost linearly in the tire load F_z, i.e., the tire behaves as a linear spring with stiffness C_{Fz} in vertical direction. The effective rolling radius also varies with tire load, but tends to saturate for large F_z. This can be described, based on empirical fit, as follows (see Chapter 9 in Ref. [32])

$$R_e = R - d_0 \cdot \left[D \cdot \arctan\left(B \cdot \frac{\mathrm{d}}{\mathrm{d}_0} \right) + E \cdot \frac{\mathrm{d}}{\mathrm{d}_0} \right] \tag{2.3}$$

in which tire radial deflection d, tire radial deflection d_0 for nominal tire load F_{z0}, and fit parameters B, D, E, may vary according to

$3 < B < 12$	:	Note that d can be described as F_z/C_{Fz}, where C_{Fz} is the vertical tire stiffness. This means that B stretches the effective rolling radius characteristic curve along the F_z-axis. Large B (i.e., radial tire) means there will be a large slope of R_e, versus F_z at $F_z = 0$.
$0.2 < D < 0.4$	:	This value is related to the tread height, with larger values representing new tires.
$0.03 < E < 0.25$	:	This parameter describes the slope of the $R_e - F_z$ curve for large tire loads. Typically, the bias-ply tire corresponds to larger E-values.

We have selected the parameters values as given in Table 2.1 to derive the effective rolling radius as a function of the tire load.

These parameters were not derived from real experiments, but were selected to show the effect on the experimental $R_e - F_z$ relationship, see Eq. (2.3). The plots are shown in Figure 2.5. The loaded tire radius R_l versus F_z is also shown. One observes a different behavior for R_e than for R_l. Where R_l linearly decreases with increasing load, we see the effective tire radius saturate for a large load. For the radial tire, this occurs with a slope near zero for large F_z. Consequently, for a realistic range for the wheel load (varying around the nominal tire load of 4000 [N]), the radial tire shows very little variation in R_e, in contrast with the bias-ply tire, which shows a significant reduction with F_z. The initial change with F_z is strongest for the radial tire. Qualitatively, these results match those by Pacejka [32].

With increasing speed, the tire belt experiences a larger radial acceleration. As a result, the effective rolling radius will increase with increasing speed and increasing inflation pressure. The variation with speed is strongly dependent on the tire radial stiffness and therefore, on the tire carcass structure. In this respect, the bias-ply tire is more sensitive than the radial tire.

TABLE 2.1 Parameters for the $R_e - F_z$ Relationship, Eq. (2.3)

Parameter	Radial, New	Radial, Worn	Bias-Ply, New	Bias-Ply, Worn
C_{Fz} [N/m]	2×10^5	2×10^5	2×10^5	2×10^5
R [m]	0.32	0.32	0.32	0.32
F_{z0} [N]	4000	4000	4000	4000
B	10	10	3	3
D	0.4	0.2	0.4	0.2
E	0.03	0.03	0.2	0.2

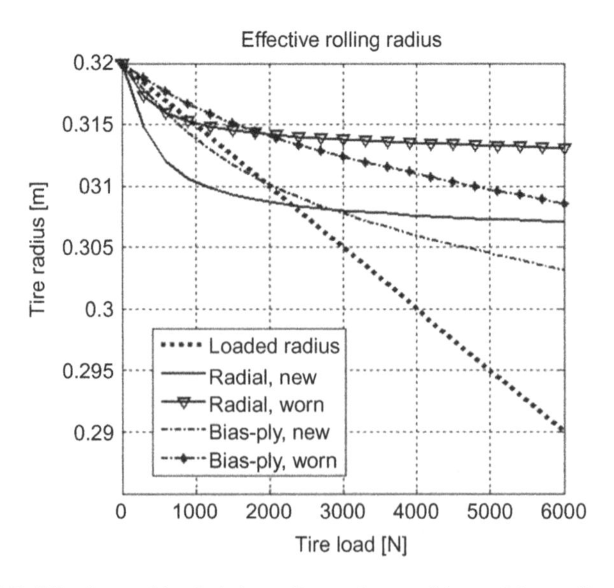

FIGURE 2.5 Effective and loaded tire radius under conditions of free rolling.

2.2 FREE ROLLING TIRE

Let us discuss the rolling tire in more detail, see Figure 2.6 (see also Ref. [29]). As the tread enters and moves through the contact area, the distance to the wheel center changes from the unloaded radius to the loaded radius, then back to the unloaded radius. Assuming complete adhesion in the contact area, i.e., no local sliding, the peripheral speed (circumferential speed with respect to the tire center) in the contact area must be equal to the forward speed of the tire and correspond to the effective rolling radius R_e. Consequently, the peripheral speed drops when entering the contact area, which suggests a negative shear stress at that point (as the rubber is being pushed into the contact zone). The opposite situation, stretching of the rubber, i.e., positive shear stress, is expected at the trailing edge of the contact area.

Let us consider the local conditions in the contact area in more detail (see Figure 2.7). With a point of the tire entering the contact area with

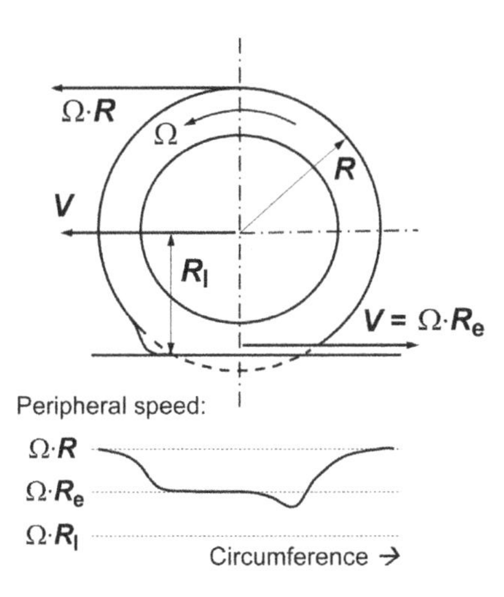

FIGURE 2.6 Free rolling tire.

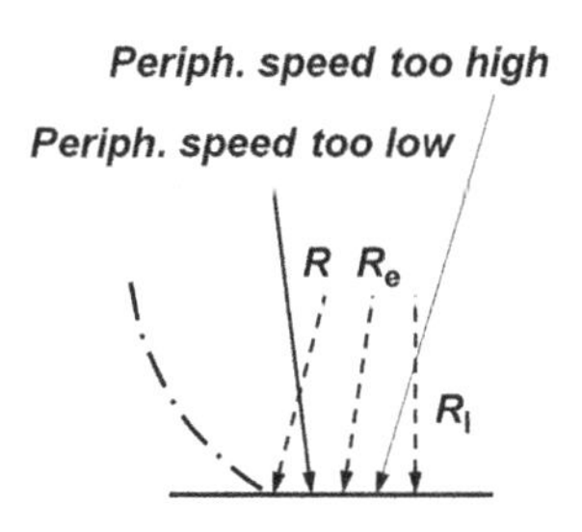

FIGURE 2.7 Behavior in contact area.

peripheral speed $\Omega \cdot R$, this speed must decrease. As we previously observed, the peripheral speed in the contact area will be equal to $\Omega \cdot R_e$ when local sliding is absent. That means that, in the front part of the contact area, points of the tire circumference should move faster, considering the distance to the wheel center (exceeding R_e). With the same points passing the center of the contact area, the distance to the wheel center equals $R_l < R_e$, which suggests that this point is moving faster than what it would be based on this distance (speed = radius × rotational speed). Consequently, the shear deformation speed in the contact area (in x direction) starts as negative and moves to be positive at the center of this area.

For the same reason, it moves back to a negative value in the last part of the contact area. The shear stress follows the shear deformation, i.e., the integral of the shear deformation speed along the contact area. This results in a shear stress pattern, as indicated in Figure 2.8. The total integral of this shear stress is equal to the rolling resistance force F_R and is negative. We also indicated the normal stress behavior between tire and road surface in Figure 2.8. The relative order of magnitude for both types of stress has no relationship to real data. The normal stress is expected to be much larger than the shear stress during free rolling. With a wheel rolling freely, i.e., without any brake or drive torque, there must be equilibrium in moment around the wheel center. Using the notations from Figure 2.8, this means that

$$F_z \cdot h = F_R \cdot R_l = f_R \cdot F_z \cdot R_l \tag{2.4}$$

for *coefficient of rolling resistance* f_R. Consequently, the resulting wheel load F_z will be slightly *in front* of the center of the contact area (the projection of the wheel center on the ground surface) with this distance h, given by

$$h = f_R \cdot R_l \tag{2.5}$$

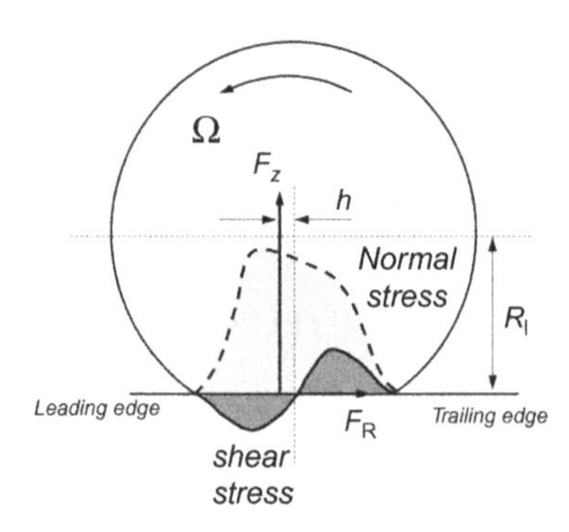

FIGURE 2.8 Shear and normal stress behavior in the contact area.

2.3 ROLLING RESISTANCE

For a rolling tire, deformation of the tire material occurs while entering the contact patch. The original (undeformed) conditions are restored when the deformed area leaves the contact patch again. This process involves energy losses, mainly due to hysteresis of the rubber material. These losses arise in the tread area, the belt, the carcass, and the sidewalls.

An overview of the various contributions in this energy loss is shown in Figure 2.9. Together, these losses correspond to the rolling resistance force f_R.

As a result, the rolling resistance is reduced for

- less hysteresis in the tire material
- less deformation of the tire.

This discussion is for a rigid flat road. For a deformable (compliant) road, such as soil, the resistance is further increased due to additional friction forces between tire and soil and the nonelastic deformation of the soil.

The rolling resistance, which is on the order of 0.01–0.05 for a rigid road or hard soil, may easily increase to 0.35 for a wet saturated soil and even higher for a soft muddy surface. In other words, a wheel on compliant soil attempts to climb out of the pit it is digging. For a concrete or tarmac road surface, f_R varies between 0.01 and 0.02.

Rolling resistance is not a fixed property of the tire. Varying conditions such as braking/driving, temperature, and speed will change the rolling resistance. Rolling resistance depends on:

- braking/driving conditions
- parasitary forces (depending on wheel alignment: toe, camber)
- temperature
- inflation pressure
- tire load
- wheel velocity
- road conditions
- tire structure, size and geometrical design (truck versus passenger car tires)
- tire aging (wear).

We discuss each of these dependencies in more detail.

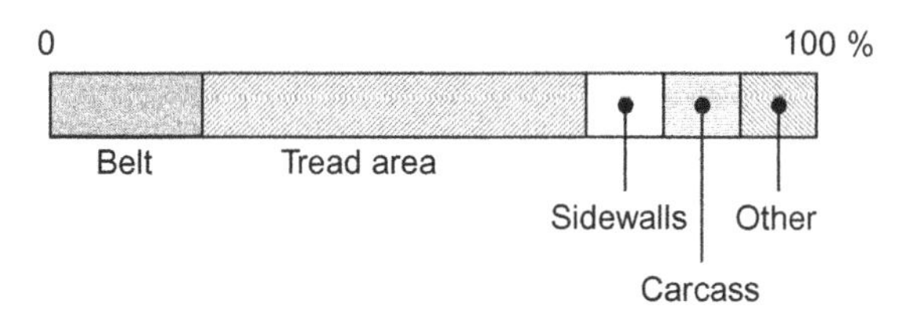

FIGURE 2.9 Contributions of tire parts to energy losses under free rolling conditions.

2.3.1 Braking/Driving Conditions

The generation of longitudinal forces is always accompanied by some sliding in part of the contact zone, as we shall see in subsequent sections. This means that more energy is lost and the rolling resistance coefficient will increase. Note that braking and traction also affect the deformation in the contact patch, which may impact rolling resistance, in addition to the occurrence of local sliding. It was shown in Ref. [10] (with reference to work of Schuring) that, during a small tractive force, the rolling resistance may decrease compared to free rolling conditions, up to a level of about 75–85% of free rolling conditions. To understand the impact of braking and driving on rolling resistance, we assume that the longitudinal force F_x linearly depends on wheel load and on longitudinal slip κ, which was introduced in Section 2.1, as follows

$$F_x(\kappa) = c_\kappa \cdot F_z \cdot \kappa \tag{2.6}$$

with

$$\kappa = \frac{\Omega \cdot R_e - V}{V} \tag{2.7}$$

assuming only longitudinal motion. This approximates a nonlinear relationship, as we shall see in the next sections. The parameter c_x is referred to as the normalized longitudinal slip stiffness. This parameter is of the order of 20 (P_{Kx1}, see Appendix 6), if we assume κ to be close to zero. For larger κ range, we estimate c_κ to be smaller, accounting for the convex shape of the F_x–κ characteristic.

Slip is negative in case of braking and positive in case of traction. The effective rolling resistance force is found from the difference between input power and effective power at the wheel–road contact

$$F_R \cdot V = M \cdot \Omega - F_x \cdot V \tag{2.8}$$

where M is the drive or brake torque and F_x is the longitudinal force between tire and road, which includes the rolling resistance force $f_R \cdot F_z$ under free rolling. Under equilibrium conditions, the moment M must be equal to contact force F_x times the loaded tire radius R_l. The variable Ω can be eliminated from Eq. (2.8) using Eq. (2.7). Replacing F_R in Eq. (2.8) with

$$F_R = f_{Rx} \cdot F_z \tag{2.9}$$

one finds

$$f_{Rx} = \frac{R_l}{R_e} \cdot c_\kappa \cdot (1 + \kappa) \cdot \kappa + f_R - c_\kappa \cdot \kappa \tag{2.10}$$

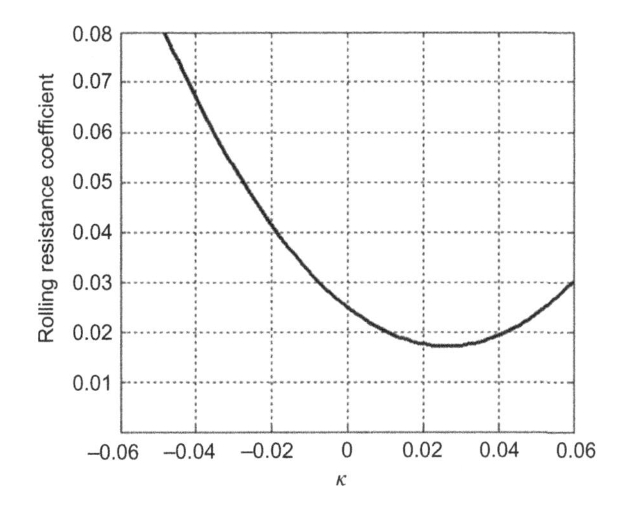

FIGURE 2.10 Rolling resistance coefficient in case of braking or driving (approximation).

In contrast to Ref. [10], we expressed the rolling resistance coefficient in terms of longitudinal slip. Genta and Morello [10] show this coefficient in terms of the normalized longitudinal force F_x/F_z, but that is a matter of substituting the value of κ in the longitudinal tire characteristics, the linearization of which is given by Eq. (2.6). We plotted this relationship for $R_l = 0.95 \times R_e$, $c_\kappa = 12$, $f_R = 0.025$ (Figure 2.10). Indeed, one observes a minimum value for positive slip (traction). It can easily be verified that this has to do with the ratio of loaded and effective tire radius and therefore, with the wheel load (see Figure 2.5).

2.3.2 Parasitary Forces: Toe and Camber

Depending on the wheel alignment, the wheel may have a small steering angle α under straight ahead driving. This will result in a small lateral force F_y, with component $F_y \cdot \sin(\alpha) \approx F_y \cdot \alpha$ in the vehicle longitudinal direction, which contributes to the rolling resistance. For small angles, the lateral force F_y can be approximated by a linear function in α, which leads to a total rolling resistance coefficient of

$$f_{R\alpha} = f_R + c_\alpha \cdot \alpha^2 \tag{2.11}$$

where c_α is the normalized cornering stiffness, which is on the order of 10–15. For a value of 15, this leads to a contribution of 0.0045 and 0.018 to $f_{R\alpha}$ for toe angle 1 and 2 [°], respectively.

In case of a camber angle γ, there are two effects. First, there is camber thrust, meaning a lateral force that can again be approximated using a linear relationship in γ, replacing Eq. (2.11) with

$$f_{R\alpha\gamma} = f_R + c_\alpha \cdot \alpha^2 + c_\gamma \cdot \alpha \cdot \gamma \tag{2.12}$$

The second effect is related to the aligning torque M_z, if present (which is generally the case). This torque, in case of camber, has a component perpendicular to the tilted wheel plane of $M_z \cdot \sin(\gamma)$, which in turn corresponds to the resistance force $M_z \cdot \sin(\gamma)/R_l$. At the same time, the rolling resistance coefficient during free rolling is reduced with a factor $\cos(\gamma)$, which leads to the following rolling resistance coefficient:

$$f_{R\alpha\gamma} = f_R \cdot \cos\gamma + \frac{M_z}{R_l \cdot F_z} \cdot \sin\gamma + c_\alpha \cdot \alpha^2 + c_\gamma \cdot \alpha \cdot \gamma \tag{2.13}$$

2.3.3 Temperature

A rolling tire has internal hysteresis losses contributing to rolling resistance. As an additional result, temperature is raised when the wheel begins to roll. This temperature raise has the following effects:

- The internal damping of rubber decreases with increasing temperature.
- The friction between road and tire decreases with temperature, resulting in a reduction of the contribution of local sliding in rolling resistance.
- The inflation pressure is increased, which reduces the tire radial deflection.

All these effects result in a reduction of the rolling resistance, and therefore, a reduction in heat dissipation, which restricts the temperature rise. Consequently, the decrease of rolling resistance tends to stabilize the temperature of the tire. From test results given in Ref. [10], in which a tire has been accelerated up to 185 [km/h] on a 2.5 m drum with the constant speed, it appears that the temperature increases up to 110 [°C] with a lag time of more than 5 [min]. This means that the temperature $T(t)$ can be well approximated by the equation:

$$\tau_{lag} \cdot \dot{T} + T = T_{saturated} \tag{2.14}$$

with a lag time τ_{lag} of about 5 [min] and $T_{saturated} = 110$ [°C]. The rolling resistance coefficient f_R decreases at the same time with the same lag time. The time histories of both temperature and rolling resistance are shown in Figure 2.11. We scaled both histories against the final tire temperature and the initial f_R-value, respectively. Note that the rubber material of a tire has very low conductivity. Therefore, sharp variations in temperature may arise through the tire wall, with the outer temperature greatly exceeding the average temperature. This outer temperature determines the contact conditions. Genta and Morello [10] indicated that the temperature was measured inside the tire body.

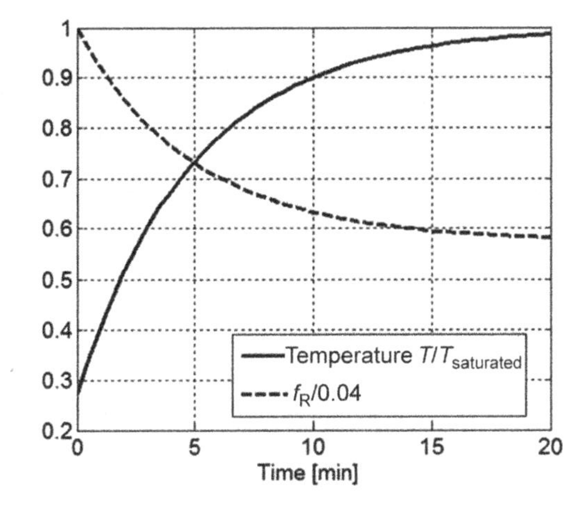

FIGURE 2.11 Variation of temperature and rolling resistance for a tire, accelerated up to 185 [km/h] (based on results by Genta and Morello [10]).

The final tire temperature depends on the wheel speed. An equilibrium value of 80 [°] tire temperature for a constant speed of 120 [km/h] is a fair value. This equilibrium value increases progressively with higher speeds.

Tests were conducted on a drum with more radial deflection compared to a flat surface. In addition, the thermal properties of the drum may affect the test results. Driving on a flat road will result in slightly smaller temperature values.

2.3.4 Forward Speed

The dependency of the rolling resistance on forward velocity V can be approximated by a higher-order formulation, with the second order being the most common one, and suggested to be a fourth order expression by Mitschke and Wallentowitz in Ref. [27], with the second-order term neglected (with the argument that this term is small compared to aerodynamic forces):

$$f_R = f_{R0} + f_{R1} \cdot \left(\frac{V}{100}\right) + f_{R4} \cdot \left(\frac{V}{100}\right)^4 ; V \text{ in } [\mathrm{km/h}] \tag{2.15}$$

The order of magnitude for the coefficients f_{R0}, f_{R1}, f_{R4} is included in Table 2.2 for nominal tire pressure, for three different types of radial (R) tires:

S	:	allowable maximum speed of 180 [km/h]
H	:	allowable maximum speed of 210 [km/h]
$M+S$	:	tires, designed for mud and snow (winter tires)

TABLE 2.2 Parameters for the R_e-F_z Relationship (2.3)

Type of Tire	f_{R0} (10^{-2})	f_{R1} (10^{-2} [h/km])	f_{R4} (10^{-2} [h^4/km^4])
SR	0.7–1.1	0.03–0.3	>0.08
HR	0.8–1.0	0.1–0.25	0.02–0.04
M+S	0.9–1.2	0.23–0.34	0.04–0.07

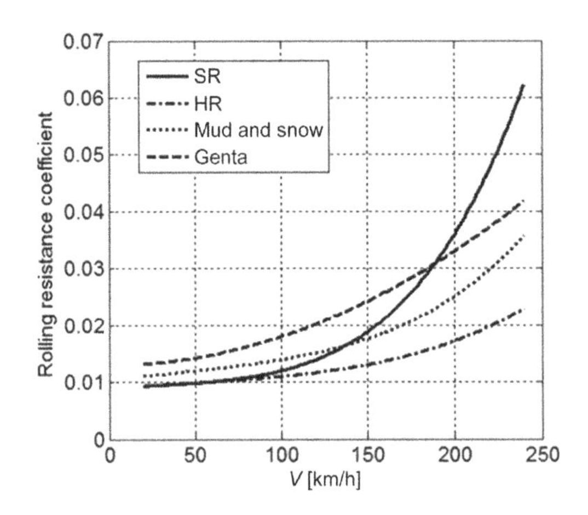

FIGURE 2.12 Rolling resistance versus speed for different types of tire and according to Eqs. (2.15) and (2.16).

From this table, a number of observations can be made:

- The range of possible values is well defined for the high performance (H) tire.
- The H-tire has the lowest value for f_{R4}, meaning that this tire is least sensitive for temperature effects.
- In contrast, the S-tire has the highest sensitivity with respect to temperature.

As mentioned, the second-order description is most commonly used

$$f_R = f_{R0} + f_{R2} \cdot \left(\frac{V}{100}\right)^2; V \text{ in } [\text{km/h}] \tag{2.16}$$

where we used the same scaling as in Eq. (2.15). It is shown in Ref. [10], by comparison of this fit with experimental results, that this expression may underestimate the behavior for high speeds.

We determined the rolling resistance coefficient according to expressions (2.15) and (2.16), where we used average values based on Table 2.2. For the second-order approximation, we selected the same parameters as in Ref. [10], $f_{R0} = 0.013$ and $f_{02} = 0.005$ [h^2/km^2]. The results are shown in Figure 2.12. As expected, the HR tire is least sensitive for speed. The SR-tire

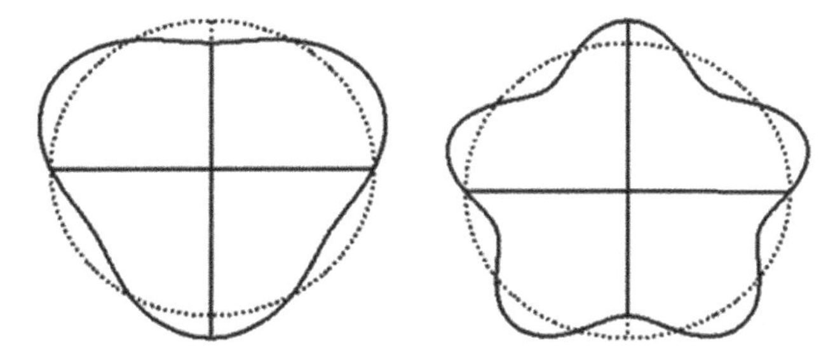

FIGURE 2.13 Some tire standing waves.

has the highest sensitivity for high speeds. For speeds that are not too high, the second-order approximation is qualitatively not very different from the higher-order fit. A sharp increase at high speed, such as for the SR-tire, is difficult to match with Eq. (2.16).

Next, it may be asked what will happen with the SR-tire when the speed is increased. A larger value of f_R means more heat dissipation. Consequently, the temperature of the tire in the contact area will increase strongly with speed. At the same time, the tire will show standing waves around the circumference, with an increasing number of modes for increasing speed. A number of these modes are shown in Figure 2.13. With increasing modes, the pressure distribution in the contact area will show more pressure concentrations, which will lead to increased heat dissipation. This self-reinforcing effect will finally destroy the tire. The speed for which the tire collapses exceeds the so-called critical speed, being the maximum speed allowed for the tire, and indicated on the tire sidewall with a speed index symbol. The references H and S, used above (for 180 and 210 [km/h], respectively) are examples of this speed index.

2.3.5 Inflation Pressure

Increasing the tire inflation pressure leads to a stiffer belt and therefore, a lower rolling resistance. On the other hand, increasing the tire load leads to more deformation and therefore, to increased rolling resistance. The critical speed increases with lower rolling resistance in these cases. An increase in temperature leads to an increased inflation pressure, which lowers the rolling resistance and corresponding heat dissipation, and therefore has a stabilizing effect regarding temperature.

Genta and Morello [10] refer to an empirical formula, suggested by the SAE, for the rolling resistance dependent on inflation pressure p_i [N/m^2], forward velocity V [m/s], and tire load F_z [N]:

$$f_R = \frac{K}{1000} \cdot \left(5.1 + \frac{5.5 \times 10^5 + 90 \cdot F_z}{p_i} + \frac{1100 + 0.0388 \cdot F_z}{p_i} \cdot V^2 \right) \quad (2.17)$$

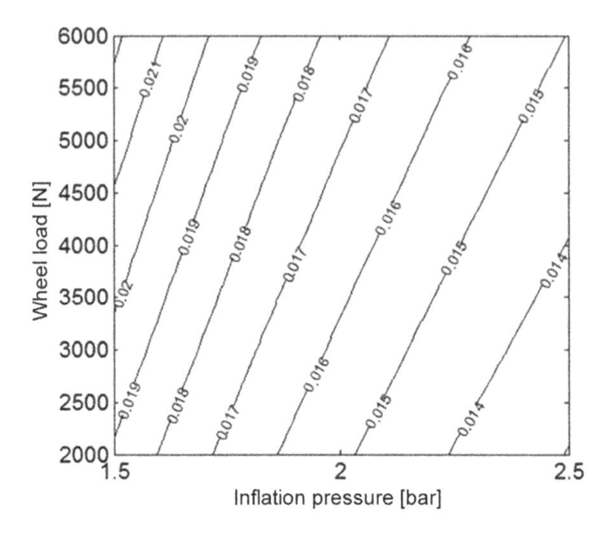

FIGURE 2.14 Rolling resistance coefficient for different inflation pressures and tire load, for 150 [km/h].

The factor K is 0.8 for radial tires and 1 for nonradial tires. We have taken a fixed speed, 150 [km/h], and determined the rolling resistance coefficient for varying inflation pressure and wheel load. Results are shown in Figure 2.14. Observe that the inflation pressure is the dominant factor in the rolling resistance coefficient. The impact of changing wheel load is small.

2.3.6 Truck Tires Versus Passenger Car Tires

For truck tires, the dependency on vehicle speed appears to be more linear, i.e., the factor f_{R4} can be neglected (see Ref. [27]). Truck tires will experience a large variation in load during normal practice. One of the performance criteria is therefore that the dependency of the rolling resistance with tire load is small, or that it shows a reduced resistance coefficient with increasing load.

Rolling resistance is important for heavy goods vehicles. About one third of the energy produced by the engine is used to compensate for the rolling resistance.

The paper by Popov et al. [40] confirms that the rolling loss (longitudinal resistance force) is almost linear in the tire load, with the slope slightly increasing with decreasing inner pressure. We determined the rolling resistance coefficients from their results (Figure 2.15). One observes a trend of reducing f_R-value for increasing wheel load and increasing inflation pressure. The F_z dependency does not correspond to Eq. (2.17). Apparently, this expression does not hold for all tires, including truck tires.

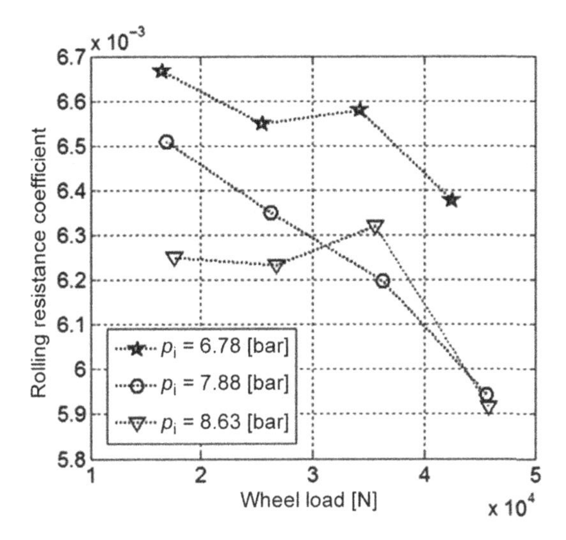

FIGURE 2.15 Rolling resistance coefficient for truck tires versus tire load and tire pressure from Ref. [40].

2.3.7 Radial Versus Bias-Ply Tires

Radial tires normally show a rolling resistance that is about 20% or more less than for bias-ply tires, and a higher value of critical speed (see Schuring [47]). This can be explained by the tire structure design, which leads to less rubber deformation energy for the radial tire compared to the bias-ply tire. This effect was increased when low rolling resistance tires were introduced in the beginning of this century, and where a significant reduction of the order of 40% was claimed with respect to conventional radial tires, i.e., ending up with half of the rolling resistance of bias-ply tires.

Other design aspects have had an impact on rolling resistance as well, such as the number and orientation of plies, the choice of rubber compounds, and the design of treads. Natural rubbers have lower damping compared to synthetic rubbers, which leads to a lower rolling resistance, however, at the cost of lower critical speed and shorter lifetime of the tire.

2.3.8 Other Effects

With a significant amount of water on the road, the tire must push away this water, which leads to a larger rolling resistance, depending on the water height h, the tire speed V, and the tire width b. This resistance will increase with speed up to the level where the full tire is floating on the water. Beyond this point, the resistance will not increase further with speed.

As reported by Gengenbach in Ref. [9], the effect of speed on the resistance force F_{RW} [N] can be expressed as:

$$F_{RW} = A(h) \cdot b \cdot V^n; \quad V \text{ in [km/h]}, \ b \text{ in [cm]} \tag{2.18}$$

with exponent n approximately equal to $n = 1.6$ if $h > 0.5$ [mm]. For $h = 0.2$ [mm], n can be approximated by $n = 2.2$. The coefficient A(h) depends on the water height h. If we express V in [m/s] and b in [m], this coefficient varies from the order of 5.5 for $h = 0.5$ [mm] to about 11.0 [$N.s^n.m^{-n-1}$] for $h = 1.0$ [mm].

Rolling resistance decreases with *wear*. Hysteresis losses occur mostly in the tread band. Hence, reducing the tread band material will result in lower resistance.

The two tire geometrical parameters having an effect on rolling resistance are:

- Tire radius
- Aspect ratio (section height/tire width).

Rolling resistance is decreased for a larger tire radius or a lower aspect ratio (low profile tires). Hence, smaller tires have a larger rolling resistance coefficient. However, such tires are usually used for lighter cars with a lower tire load and therefore lower rolling resistance force.

2.4 THE TIRE UNDER BRAKING AND DRIVING CONDITIONS

2.4.1 Braking Behavior Explained

Consider a tire under a brake torque, as indicated in Figure 2.16. The brake torque M_z must be balanced by moments due to a brake force F_x and the tire load F_z. The offset of the tire load in front of the wheel center increases with

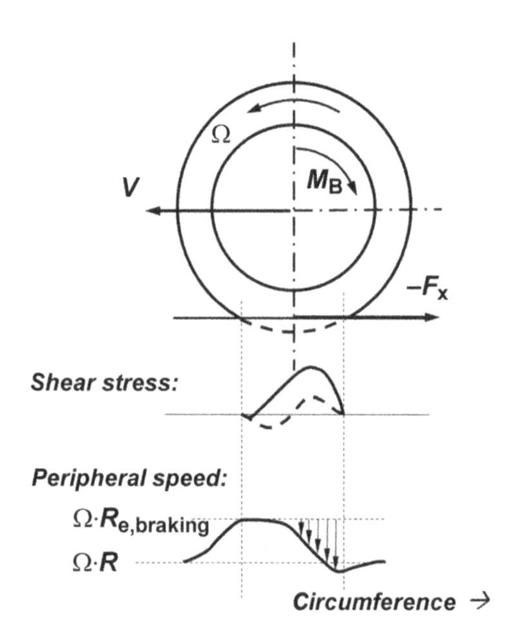

FIGURE 2.16 Braking tire.

respect to the free rolling tire. The tire will experience a slip speed of the wheel with respect to ground, reducing the angular speed and therefore increasing the effective rolling radius R_e. In the ultimate situation of a sliding nonrolling tire, this radius of rotation will become unbounded, with the center of the rotation moving to $z = \infty$. This means that, in general under braking conditions, the effective rolling radius $R_{e,\text{braking}}$ will exceed the unloaded radius. The total longitudinal shear stress in the contact area now consists of a part due to free rolling (dashed in Figure 2.16) and a superimposed shear stress caused by braking. As a result, the major part of the tire in the contact area is stretched due to the brake torque. Tread elements entering the contact area first try to adhere to the road surface, with the longitudinal deflection and therefore, the shear stress increasing linearly along the contact zone. At a certain point, the shear stress reaches the limits of friction ($\mu \cdot \sigma_z$ with local road friction μ and normal stress σ_z under Coulomb law) and the treads begin to slide. As a result, the shear stress drops down along the rear part of the contact zone. In a similar way as discussed for a free rolling tire, one arrives at a distribution of the peripheral velocity of treads (with respect to the wheel center), as shown in the bottom part of Figure 2.16.

Note that sliding begins in the rear of the contact area and extends toward the front part of the contact area for increasing brake torque, until finally sliding is apparent along the full contact area.

In case of a tire under driving conditions, the angular speed is increased and therefore, the effective rolling radius $R_{e,\text{driving}}$ is decreased. In the ultimate case of a spinning tire on the spot, the effective rolling radius has decreased to zero (no forward speed) and the point of rotation coincides with the wheel center. The drive torque must balance moments resulting from a drive force in the contact area and the tire load. The offset of the tire load in front of the wheel center is decreased with respect to the case of the free rolling tire. The shear stress is now built up from the free rolling distribution, including a triangular-shaped pattern along the contact area, and the tire tread material is experiencing a compression.

We introduce the practical longitudinal slip κ as follows

$$s_x \equiv -\kappa = \frac{V_{sx}}{V_x} = \frac{V_x - \Omega \cdot R_e}{V_x} \equiv \frac{\Omega_0 - \Omega}{\Omega_0} \tag{2.19}$$

with angular speed Ω_0 under free rolling conditions and slip speed V_{sx} of tread elements, with respect to the road surface. This slip speed is obtained from the difference between the forward speed V_x at the wheel center and the peripheral speed $\Omega \cdot R_e$.

When a driver begins braking, the wheel rotation is decelerated by the resulting brake torque and the tire brake force

$$J_{\text{wheel}} . \dot{\Omega} = -M_B - R_l \cdot F_x(\kappa) \tag{2.20}$$

with $F_x > 0$ in positive x direction (i.e., $F_x < 0$ in case of braking) and the wheel polar moment of inertia J_{wheel}. This equation is part of a larger set of equations used to solve the braking problem for a vehicle. Clearly, the forward vehicle speed (included in the preceding angular wheel velocity equation through the slip κ) is not a constant but will decrease, which is the intention of braking. The resulting forward vehicle speed follows from another equation describing the balance of the vehicle deceleration and the wheel forces

$$m \cdot \dot{V}_x = \sum_{\text{wheels}} F_x(\kappa)$$

for mass m of the vehicle, and neglecting other longitudinal forces (slope, aerodynamic drag, etc.). Note that the wheel forces and longitudinal slip are in general different for the four wheels of the vehicle.

To solve the angular wheel velocity equations for each wheel, one requires a description of F_x in terms of practical slip κ. A typical behavior of this characteristic longitudinal tire behavior is shown in Figure 2.17 for different tire loads. In the left-hand image, we plotted the absolute brake force $-F_x$ versus $-\kappa$, whereas in the right-hand image, we plotted $-\mu_x \equiv -F_x/F_z$, the *normalized tire force* (also known as the *longitudinal force coefficient* or *longitudinal friction coefficient*), for various values of the tire load.

Usually, the curves will not exactly pass the origin (due to rolling resistance and inaccuracies in the tire). Clearly, the longitudinal tire force is nearly, but not quite, proportional to the tire load. One observes a peak value and saturation value in both images for the longitudinal force coefficient indicated as μ_{xp} (*peak braking coefficient*) and μ_{xs} (the *sliding braking coefficient*, which is the limit of μ_x for pure sliding, i.e., at $\kappa = -1$). The peak value is obtained for brake slip around 0.1 and 0.15 in absolute value (10–15% slip).

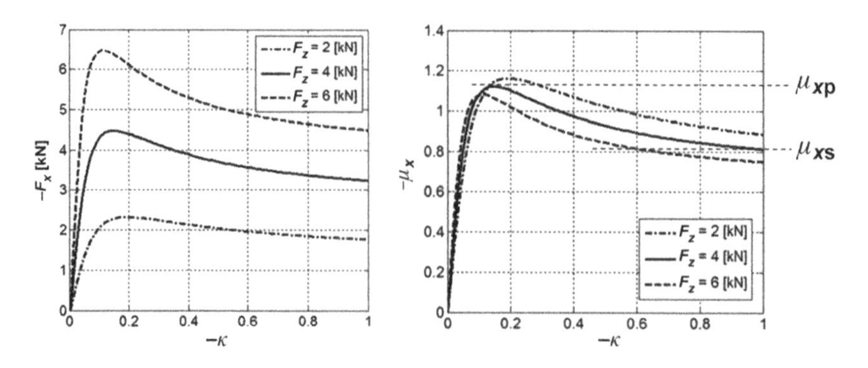

FIGURE 2.17 Brake force and normalized brake force versus brake slip κ for different wheel loads.

For small brake slip, the F_x versus κ characteristic can be approximated using a linear relationship, with the slope referred to as the *longitudinal slip stiffness* C_κ (Figure 2.18).

The peak value is the optimal value of braking, but just beyond the slip $-\kappa_0$ corresponding to this optimal value, the wheel will lock in very short time. To understand that, consider Eq. (2.20) for κ close to a value κ_1 with $|\kappa_1| > |\kappa_0|$.

Linearization around $\kappa = \kappa_1$, assuming the speed V_x reduces slowly, and considering the brake torque constant, leads in first order to

$$J_{\text{wheel}} \cdot (\dot{\Omega} - \dot{\Omega}_1) + \frac{R_1 \cdot R_e}{V_x} \cdot \frac{dF_x}{d\kappa}(\kappa_1) \cdot (\Omega - \Omega_1) = 0 \tag{2.21}$$

Because the derivative $F_x'(\kappa_1) < 0$, nontrivial solutions of this equation will blow up in time, and therefore, the system will become unstable.

This is why all new vehicles are equipped with antilock systems to prevent excessive brake slip. In the same way, one may discuss drive slip and the risk of spinning of the wheel in case of too high traction. This phenomenon can be prevented using traction control systems.

The normalized tire force μ_x (and therefore the longitudinal tire force itself) depends on the tire–road conditions:

- Road roughness. Pavement exhibits three types of roughness, micro-texture (with wavelength less than 0.5 [mm]), macro-texture (wavelength between 0.5 and 50 [mm]), and mega-texture (wavelength exceeding 50 [mm]), see Ref. [65].
- Tire tread wear.
- Wet conditions (rain, snow, ice, etc.).

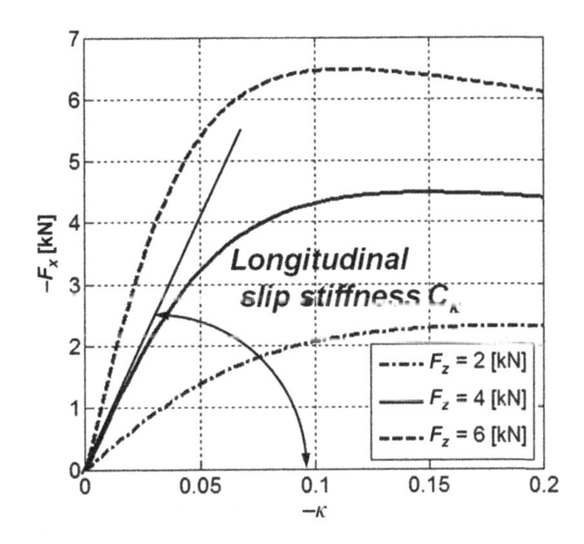

FIGURE 2.18 Longitudinal slip stiffness.

Macrotexture is related to the overall roughness of the road resulting from the number, type, and size of stone chippings, whereas microtexture is related to the roughness of the individual chippings. Idealized texture leads to sufficient drainage and significant hysteretic friction (local pressures) at the cost of tire wear. Tips should preferable be sharp to provide good friction even under wet conditions, but this may lead to abrasive wear. The existence of microtexture is due to the typical asphalt ingredients (silica, sand, quartzites).

Macrotexture and microtexture vary in time. It is known from drain asphalt that, because of many small contact zones between rubber and the ground, there is an increased polishing effect and therefore, rounded asperities, which impact the adhesive properties of the tire–road contact. Roughly speaking, one might say that macrotexture is related to a strong velocity dependence of the tire–road contact under wet road conditions, whereas microtexture is related to the slightly wet or dry-adhesive road conditions.

Under wet road conditions, the longitudinal force coefficient maximum level drops, to levels on the order of 0.6–0.8 for a wet road, to 0.4–0.5 for snow, and to levels of 0.2–0.4 for ice.

A special case is given if a significant amount of water is present on the road. To maintain contact between tire and road, the water must be evacuated. This property may be improved by adjusting the tread block pattern of the tire (longitudinal grooves, or grooves curved in an outward direction guiding the water in a radial direction away from the tire, see also Figure 2.2). With increasing speed, there is less time to remove the water and the contact zone is further reduced. Consequently, the brake force and therefore, the friction coefficient drops significantly with vehicle speed.

At a certain speed, the tire may float entirely on a film of water (*hydroplaning*), and the friction coefficient drops to very low values (<0.1). In other words, hydroplaning occurs when a tire is lifted from the road by a layer of water trapped in front of and under a tire.

One usually distinguishes between *dynamic hydroplaning* (water is not removed fast enough to prevent loss of contact) and *viscous aquaplaning* (the road is contaminated with dirt, oil, grease, leaves, etc.). Usually, regular rain will wash away the road contaminants that cause viscous aquaplaning. However, after an especially long dry period, the contaminants pile up, and a sudden rain may result in a more viscous mixture on the road, which causes unexpected, and dangerous (i.e., low friction), conditions.

Many sources exist for tire behavior under the combined effect of speed and water on the road (see Borgmann [4] and Gnadler [12]).

2.4.2 Modeling Longitudinal Tire Behavior

There are different ways to describe the slip behavior using tire models. One distinguishes between *physical models* and *empirical models*. A physical model describes the tire based on the recognized physical phenomena during braking, usually in a simplified way. Such simplified models do not aim to provide a quantitative description of the tire handling performance, but merely explain the qualitative phenomena. These phenomena will be addressed in Section 2.7. Physical models that are more complex (e.g., finite element (FE) models) are applied to derive quantitatively correct tire performance based on a detailed description of the tire structure and material properties. This means that FE models form a link between tire design and tire performance. However, FE models are very time consuming, both in CPU and preparation time.

Empirical tire models are based on a similarity approach in which experimental results are used to find parameters to tune a certain mathematical description. A well-known empirical tire model is the Magic Formula model described by Pacejka, which is often referred to as the Pacejka model [32].

The basic mathematical formula describing the longitudinal characteristics is given by the sine-version of the Magic Formula, given by

$$F_x(\kappa) = D_x \cdot \sin(C_x \cdot \arctan(B_x \cdot \kappa_x - E_x \cdot (B_x \cdot \kappa_x - \arctan(B_x \cdot \kappa_x)))) + S_{Vx} \tag{2.22}$$

with

$$\kappa_x = \kappa + S_{Hx} \tag{2.23}$$

The parameters S_{Hx} and S_{Vx} are shifts that allow the curve not to pass through the origin, which may be due to rolling resistance and tire irregularities (asymmetry). The other four parameters are:

D_x : Peak factor—determines the maximum value of F_x

C_x : Shape factor—describes whether the curves in Figure 2.17 are monotonously increasing ($0 < C_x < 1$) or include a local extreme ($C_x > 1$)

B_x : Stiffness factor—determines the slope of the curve at $\kappa_x = 0$, i.e., the longitudinal slip stiffness. This slip stiffness C_κ can easily be found to be given by

$$C_\kappa = B_x \cdot C_x \cdot D_x \tag{2.24}$$

E_x : Curvature factor—affects the behavior of the curves in Eq. (2.18) beyond the critical slip $|\kappa_0|$

Except for C_x, these factors depend on the tire load F_z. To keep the Magic Formula dimensionless, the tire load is included as its relative deviation from the *nominal tire load* F_{z0}:

$$df_z = \frac{F_z - F_{z0}}{F_{z0}} \tag{2.25}$$

TABLE 2.3 Typical Values for the Nominal Tire Load F_{z0}

Class	F_{z0} [N]
Compact class	3000
Middle class	5000
Top class	6000

The nominal tire load is related to the maximum admissible static load for the specific temperature and speed index, usually referred to as the *ETRTO value*, which is the European Tire and Rim Technical Organization value. Choosing the nominal value F_{z0} equal to 80% of this ETRTO value, a reasonable choice for F_{z0} is listed in Table 2.3.

Hence, a specific nominal tire load is related to a class of tires, with the same maximum allowable operating speed. Different nominal tire loads refer to different classes of tires, in contrast to the variation in tire load for one specific tire (due to static load variations, load transfer during cornering, etc.).

The factor D_x is related to the peak of the longitudinal force coefficient (normalized longitudinal force) and the wheel load:

$$D_x = \mu_{\text{xp}} \cdot F_z \tag{2.26}$$

Assuming pure longitudinal slip (no camber, no slip angle), this parameter μ_{xp} can be expressed in terms of F_z, as follows:

$$\mu_{\text{xp}} = (P_{Dx1} + P_{Dx2} \cdot df_z) \tag{2.27}$$

for P_{Dx1} and P_{Dx2}. Other parameters in the Magic Formula for pure longitudinal slip can be expressed as follows:

$$C_x = P_{Cx1} \tag{2.28}$$

$$B_x \cdot C_x \cdot D_x = F_z \cdot (P_{Kx1} + P_{Kx2} \cdot df_z) \cdot \exp(P_{Kx3} \cdot df_z) \tag{2.29}$$

$$E_x = (P_{Ex1} + P_{Ex2} \cdot df_z + P_{Ex3} \cdot df_z^2) \cdot (1 - P_{Ex4} \cdot \text{sign}(\kappa)) \tag{2.30}$$

$$S_{\text{H}x} = P_{\text{H}x1} + P_{\text{H}x2} \cdot df_z \tag{2.31}$$

$$S_{\text{V}x} = P_{\text{V}x1} + P_{\text{V}x2} \cdot df_z \tag{2.32}$$

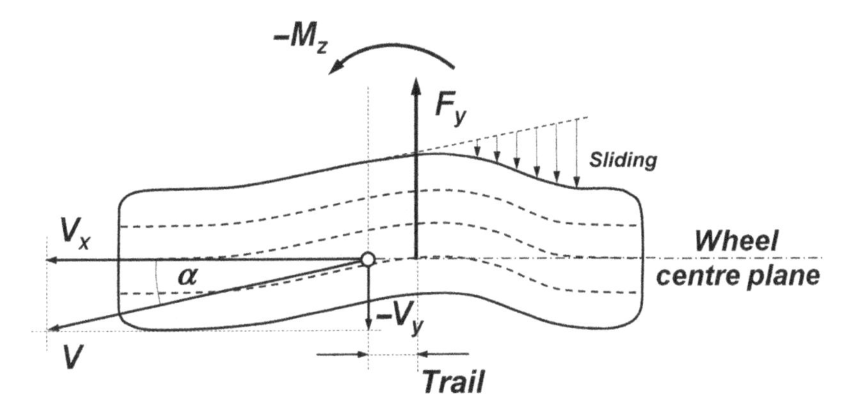

FIGURE 2.19 Tire under cornering conditions.

2.5 THE TIRE UNDER CORNERING CONDITIONS

2.5.1 Cornering Behavior Explained

Let us consider a tire under cornering conditions, as indicated in Figure 2.19 (top view), first neglecting camber. Under cornering conditions, a local velocity vector exists that is generally not parallel to the wheel center plane. This wheel center plane is defined as the symmetry plane of the tire such that forces acting in the symmetry plane do not contribute to a lateral force for the tire.

In the front part of the contact area, the treads of the tire try to follow this local speed direction, resulting in a displacement from the symmetry plane along the tire circumference within the contact area, which increases linearly from zero (just in front of the contact area) up to a situation where the induced lateral shear stress just reaches the maximum possible shear stress level, i.e., $\mu \cdot \sigma_z$ with local road friction μ and normal stress σ_z under Coulomb law.

Beyond this point, the lateral displacement reduces to zero at the trailing edge of the contact area. We discussed similar phenomena for braking and driving (traction) of the tire. Beyond the point where the shear stress first reaches $\mu \cdot \sigma_z$, the treads of the tire will slide, leading to a reduction of the shear stress in the direction of the contact area rear end. Clearly, when sliding and in the absence of longitudinal slip, the lateral shear stress will remain equal to $\mu \cdot \sigma_z$. With σ_z reducing to zero at the edges of the contact area, the friction limits for the shear stress will decrease further, and sliding is likely to extend until the contact area rear end.

Deflection of the tire is due to two separate effects:

- The deflection of the contact rubber, i.e., of the treads
- Deflection of the belt

Both compliances allow the tire to direct itself to the local speed direction, but the stiffnesses are different. In terms of physical models, one may distinguish here between the brush model and the stressed string model. Both will be discussed in more detail in Section 2.7.

We introduce the practical lateral slip as $-\tan(\alpha)$, i.e.,

$$s_y = -\tan(\alpha) = \frac{V_{sy}}{V_x} = \frac{V_y}{V_x} \tag{2.33}$$

with slip speed V_{sy}. As we will see later, the practical slip quantities correspond with a description of tire deflection in terms of deformed quantities. An alternative approach might be to express slip in terms of the undeformed coordinate system. This will result in the so-called *theoretical slip quantities*, defined as

$$\rho_x = \frac{V_{sx}}{V_r}, \quad \rho_y = \frac{V_{sy}}{V_r} \tag{2.34}$$

with

$$V_r = V_x - V_{sx} = \Omega \cdot R_e \tag{2.35}$$

We observed in Section 2.5 that, under braking, the practical slip $s_x = \kappa$ varies between -1 (locked wheel, $\Omega = 0$) and 0 ($V_{sx} = 0$). In case of a driven wheel, κ varies from 0 to $+\infty$, with the extreme case of spinning obtained if $V_x = 0$. Changing to theoretical slip values, the slip remains bounded in case of driving. It is clear that $\rho_x \rightarrow 1$ if $\kappa \rightarrow \infty$.

Using

$$\kappa = \frac{V_r - V_x}{V_x} = \frac{\Omega \cdot R_e - V_x}{V_x} \tag{2.36}$$

one easily arrives at the following relationship between practical and theoretical slip quantities

$$\rho_x = \frac{s_x}{1 - s_x}, \quad \rho_y = \frac{s_y}{1 - s_x} \tag{2.37}$$

As we observed previously, the practical brake slip s_x varies between 0 and 1, whereas under driving conditions, it is $-\infty < s_x < 0$, i.e., the practical drive slip may attain very large absolute values if the wheel spins on the spot. In contrast to the practical slip, the theoretical longitudinal slip remains bounded under driving conditions but may grow to large absolute values in case of braking when the wheel becomes locked.

Vehicle dynamics analysis requires relationships between tire lateral shear forces and slip angles at front and rear axles. This will be explained in more detail in Chapter 5. It will be shown that, by restricting to tire lateral shear forces, these four tire forces balance the centrifugal force, acting on the vehicle in local lateral direction

$$m \cdot (\dot{v}_y + V_{\text{vehicle}} \cdot r) = \sum_{\text{wheels}} F_y(\alpha) \tag{2.38}$$

with vehicle mass m, vehicle speed V_{vehicle}, vehicle lateral speed v_y at the vehicle center of gravity CoG, and yaw rate r. In addition, the moments of the four tires around the CoG must balance the total inertial moment, which can be approximated by

$$J_z \cdot \dot{r} = \sum_{\text{front wheels}} a \cdot F_y(\alpha) - \sum_{\text{rear wheels}} b \cdot F_y(\alpha) \tag{2.39}$$

with vehicle moment of inertia in vertical z-direction J_z and distances of a and b from CoG to front and rear axle, respectively. Note that, for small slip angles and steering angle δ, the vehicle speed can be approximated by V_x.

Usually, one assumes the slip angles to be identical for both front wheels and likewise for both rear wheels. Slip angles are defined by the orientation of the local velocity vector relative to the wheel symmetry plane. In terms of lateral speed and yaw rate, this leads to expressions for slip angles α_1 at front axle and α_2 at rear axle (see Chapter 5 for more details)

$$\alpha_1 \approx \tan(\alpha_1) = \delta - \frac{v_y + a \cdot r}{V_{\text{vehicle}}}, \quad \alpha_2 \approx \tan(\alpha_2) = -\frac{v_y - b \cdot r}{V_{\text{vehicle}}} \tag{2.40}$$

A typical behavior of F_y versus slip angle α is shown in Figure 2.20. Similar to the case of braking or driving, we plotted both F_y and $\mu_y \equiv F_y/F_z$, the *normalized tire force* with alternative names such as *lateral force coefficient* or *lateral friction coefficient (side force coefficient)*, for various values

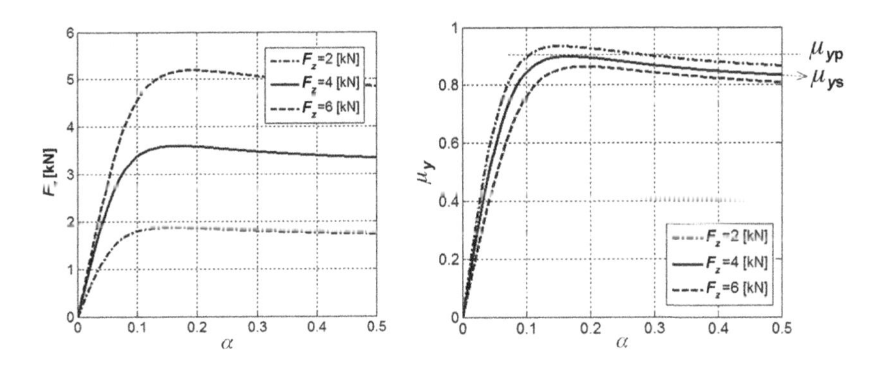

FIGURE 2.20 Cornering force and normalized cornering force versus slip angle α for different wheel loads.

of the tire load. Again, one observes the tire force to be nearly proportional to the tire load.

One observes peak values and saturation values in both images, indicated for the lateral force coefficient as the peak value μ_{yp}, and μ_{ys} as the limit of μ_y when the tire is drifting for large slip angle. The peak value is usually obtained for lateral slip near 0.1–0.2 in radians, or near 5–10 [°].

For small slip angle, the F_y versus α characteristic can be approximated using a linear relationship, with a slope that is the *lateral slip stiffness* C_α, also referred to as the cornering stiffness (Figure 2.21).

The normalized side force in Figure 2.20 indicates that the cornering stiffness tends to increase less than proportional with F_z for increasing tire load. This is shown in Figure 2.22 where the cornering force is plotted against the

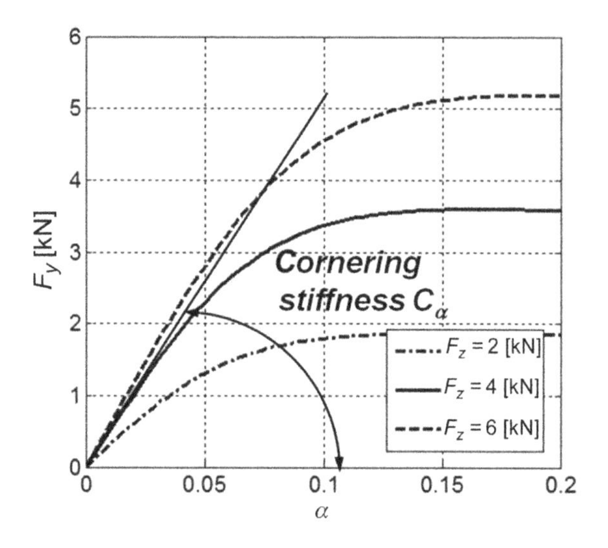

FIGURE 2.21 Cornering stiffness.

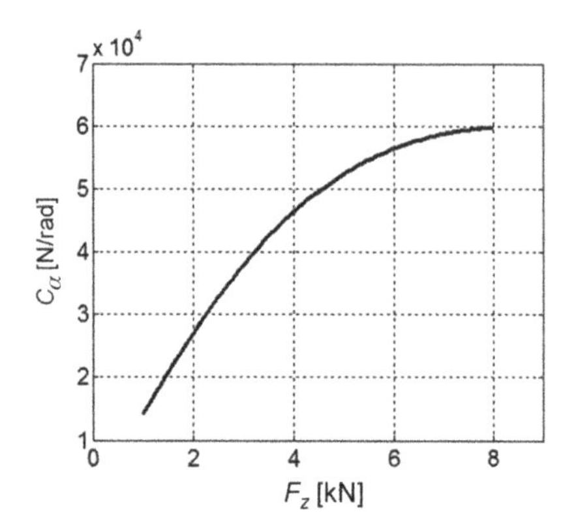

FIGURE 2.22 Cornering stiffness versus tire load.

tire load for the tire data, which is listed in Appendix 6. This nonlinear relationship is important in the sense that, during cornering, the tire load of the outer wheel will increase, whereas the inner wheel load will decrease. Because of the nonlinear dependence of cornering stiffness on tire load, the change in cornering stiffness at the outer wheel is exceeded in absolute value by the change at the inner wheel. For this reason, the average cornering stiffness for the full axle is decreased. With different roll stiffnesses at front and rear axles, this works out differently at both axles, changing the handling characteristics of the vehicle.

Lupton and Williams [22] give data for the sliding side force coefficient μ_{ys} for one specific tire but different texture depths, under wetted conditions, and for two different speeds: 50 and 80 [km/h]. Results are shown in Figure 2.23. One observes some variation in results and dependency on texture depth. Also, observe the effect of speed: increased speed lowers the friction, especially with a small texture depth (as expected).

Figure 2.19 indicates that the side force acts a small distance behind the wheel center. This distance is called the *pneumatic trail* $t_p(\alpha)$. At small slip (small α), there is almost no sliding and the adhesion part of the contact area (linearly increasing lateral deflection) extends almost over the entire contact area. This corresponds to a situation where the shear stress profile is very asymmetrical along the contact area, with a rather large pneumatic trail. With increasing slip, the sliding area increases toward the front end of the contact area. Under Coulomb law, the shear stress in the sliding area follows $\mu \cdot \sigma_z$ with road friction μ and normal stress σ_z. This normal stress is indicated in Figure 2.8. Consequently, the pneumatic trail will reduce with increasing slip. Note that the resulting vertical contact force acts slightly in front of the wheel center, meaning that the pneumatic trail may become negative for excessive sliding.

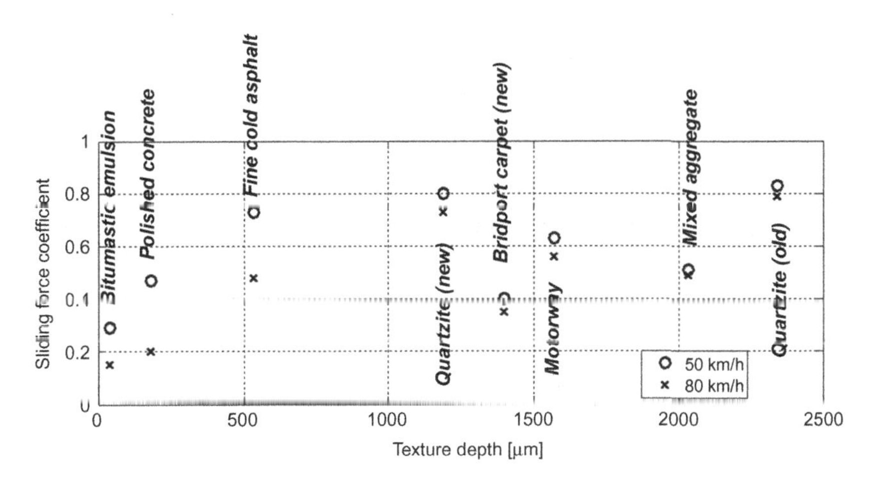

FIGURE 2.23 Side force coefficient μ_{ys} on a wetted road for different texture depths and velocities from Ref. [22].

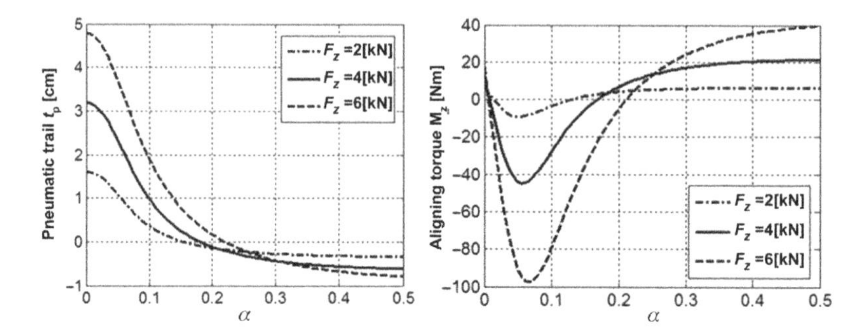

FIGURE 2.24 Pneumatic trail and aligning torque versus slip angle.

Hence, we obtain a side force $F_y(\alpha)$, starting at small values, at $\alpha = 0$, and growing to a maximum value (with the full contact area in sliding conditions, and therefore, $F_y = \mu \cdot F_z$), whereas the pneumatic trail $t_p(\alpha)$ starts at large values and reduces to small values with even negative values for excessive slip.

Pneumatic trail times side force yields the self-aligning torque (or moment) M_z, introduced in Section 2.1 (see Figure 2.4). This torque is called self-aligning because it tries to orient the tire in the speed direction. It works against the lateral deformation due to the lateral force.

For pure lateral slip (no braking or driving), the self-aligning torque can be described as follows

$$M_z(\alpha) = -t_p(\alpha) \cdot F_y(\alpha) + M_{zr}(\alpha) \tag{2.41}$$

for residual torque M_{zr}, a small torque that results from inaccuracies in the tire design that rapidly decreases in absolute value with increasing slip angle. When the tire experiences a brake or drive force in combination with lateral slip (we call that a situation of combined slip), the lateral deflection from the symmetry plane times the longitudinal force will contribute to the aligning torque, as will be discussed further in the next sections. For pure lateral slip, this contribution is omitted.

Considering the qualitative behavior of side force and pneumatic trail as described previously, we expect this aligning torque to start close to zero for $\alpha = 0$, then to grow in absolute value, but decrease again with the pneumatic trail for increasing slip.

We plotted the pneumatic trail and the aligning torque in Figure 2.24 for varying tire loads and for the tire data from Appendix 6. Observe a sharp decay in both trail and aligning torque for decreasing tire load, which is not shown for the side force (Figure 2.20). Pneumatic trail and aligning torque are much more sensitive to the tire load than the side force. The same conclusion holds with respect to road friction.

In Figure 2.25, we plotted the relative values of the aligning torque and side force (both scaled by their maximum value) in one image for

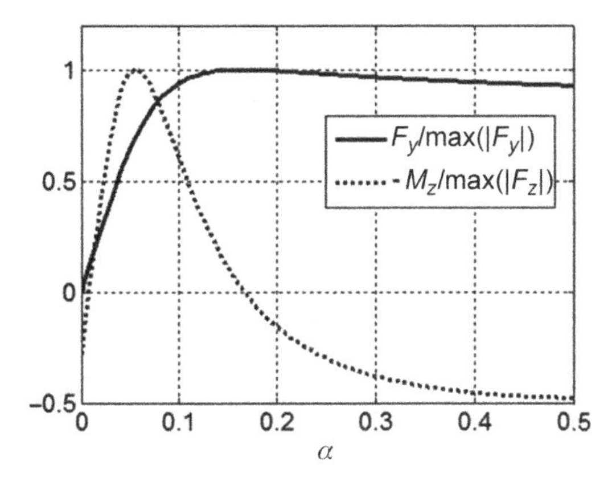

FIGURE 2.25 Scaled side force and aligning torque for $F_z = 4$ [kN].

$F_z = 4$ [kN]. Observe that the aligning torque passes its maximum at a slip angle value that is smaller than at the maximum of the side force (i.e., where the tire begins to slide). This is valuable information for the driver. The torque from the combined effect of mechanical trail (caster) and pneumatic trail is felt by the driver through the steering wheel. That means that the driver will experience a reduction in the feedback torque from the steering wheel because of the reduction of the aligning torque in absolute value. This should warn the driver that he or she is approaching a situation with an increased risk of skidding of the front axle.

2.5.2 Modeling Lateral Tire Behavior

Just like the longitudinal force description in Section 2.4.2, the lateral force can be described by the Magic Formula as follows

$$F_y(\alpha) = D_y \cdot \sin(C_y \cdot \arctan(B_y \cdot \alpha_y - E_y \cdot (B_y \cdot \alpha_y - \arctan(B_y \cdot \alpha_y)))) + S_{Vy} \tag{2.42}$$

with

$$\alpha_y = \alpha + S_{Hy} \tag{2.43}$$

and with shifts S_{Hy} and S_{Vy}. The other parameters again depend on the tire load F_z but also on the camber angle, except for C_y. Again, the relative deviation df_z of the tire load from the nominal tire load is used. According to Pacejka [32], the coefficient D_y (peak factor) can be expressed in terms of tire load F_z and camber angle γ, as follows:

$$D_y = \mu_y \cdot F_z = \frac{P_{Dy1} + P_{Dy2} \cdot df_z}{1 + P_{Dy3} \cdot \gamma^2} \cdot F_z \tag{2.44}$$

for some parameters P_{Dyi}, where the effect of decaying friction with slip speed has been neglected. Other coefficients (shape factor C_y, stiffness factor B_y, curvature factor E_y, and the shifts S_{Hy} and S_{Vy}) are found from

$$C_y = P_{Cy1} \tag{2.45}$$

$$\begin{aligned} B_y \cdot C_y \cdot D_y \equiv K_{y\alpha} = F_{z0} \cdot \frac{P_{Ky1}}{1 + P_{Ky3} \cdot \gamma^2} \\ \cdot \sin\left[P_{Ky4} \cdot \arctan\left(\frac{F_z}{F_{z0} \cdot (P_{Ky2} + P_{Ky5}) \cdot \gamma^2}\right)\right] \end{aligned} \tag{2.46}$$

$$E_y = (P_{Ey1} + P_{Ey2} \cdot df_z) \cdot \{1 + P_{Ey5} \cdot \gamma^2 - (P_{Ey3} + P_{Ey4} \cdot \gamma) \cdot \text{sign}(\alpha_y)\} \tag{2.47}$$

$$S_{Hy} = (P_{\text{Hy1}} + P_{\text{Hy2}} \cdot df_z) + \frac{F_z \cdot \gamma \cdot (P_{Ky6} - P_{\text{Vy3}} + (P_{Ky7} - P_{\text{Vy4}}) \cdot df_z)}{K_{y\alpha}} \tag{2.48}$$

$$S_{\text{Vy}} = F_z \cdot (P_{\text{Vy1}} + P_{\text{Vy2}} \cdot df_z) + F_z \cdot (P_{\text{Vy3}} + P_{\text{Vy4}} \cdot df_z) \cdot \gamma \tag{2.49}$$

where we assumed $\sin(\gamma) \approx \gamma$, and introduced certain parameters P_{jyi} with j indicating the specific factor (shape, stiffness, etc.) and $i = 1, 2, 3, \ldots$ being used to distinguish between the different Magic Formula parameters. The notation of the coefficients is taken directly from Ref. [32]. There are some differences from Ref. [32], in the sense that scaling factors are omitted and spin is assumed to be small (see Section 2.1). Furthermore, the equations in Ref. [32] also account for zero or very small values of velocity or tire load. These cases are not considered here.

The pneumatic trail, as shown in Figure 2.24, suggests a cosine version of the Magic Formula. According to Pacejka [32], the pneumatic trail can be expressed as follows in terms of slip angle α:

$$t_{\text{p}}(\alpha) = D_{\text{t}} \cdot \cos(C_{\text{t}} \cdot \arctan(B_{\text{t}} \cdot \alpha_{\text{t}} - E_{\text{t}} \cdot (B_{\text{t}} \cdot \alpha_{\text{t}} - \arctan(B_{\text{t}} \cdot \alpha_{\text{t}})))) \cdot \cos(\alpha_{\text{t}}) \tag{2.50}$$

with

$$\alpha_t = \alpha + S_{\text{H}t} \tag{2.51}$$

for some shift S_{Ht}. As in the expression for F_y, the other parameters depend on the tire load F_z and camber angle γ, except for C_{t}. The factor $\cos(\alpha)$ is introduced to account for large slip behavior. In the original equation in Ref. [32], this factor is slightly different to cover situations with very small velocity of the wheel contact center. In this chapter, we will not discuss this case further, and we refer reader to Ref. [32] for more details. According to Ref. [32], the various coefficients in Eq. (2.50) are found from

$$B_t = (Q_{Bz1} + Q_{Bz2} \cdot df_z + Q_{Bz3} \cdot df_z^2) \cdot (1 + Q_{Bz5} \cdot |\gamma| + Q_{Bz6} \cdot \gamma^2) \quad (2.52)$$

$$C_t = Q_{Cz1} \quad (2.53)$$

$$D_t = \frac{F_z \cdot R_0}{F_{z0}} \cdot (Q_{Dz1} + Q_{Dz2} \cdot df_z) \cdot (1 + Q_{Dz3} \cdot |\gamma| + Q_{Dz4} \cdot \gamma^2) \cdot \text{sign}(V_{Cx}) \quad (2.54)$$

$$E_t = (Q_{Ez1} + Q_{Ez2} \cdot df_z + Q_{Ez3} \cdot df_z^2) \cdot \{1 + 2 \cdot (Q_{Ez4} + Q_{Ez5} \cdot \gamma) \cdot \arctan(B_t \cdot C_t \cdot \alpha_t)/\pi\} \quad (2.55)$$

$$S_{Ht} = Q_{Hz1} + Q_{Hz2} \cdot df_z + (Q_{Hz3} + Q_{Hz4} \cdot df_z) \cdot \gamma \quad (2.56)$$

for unloaded tire radius R_0, and longitudinal speed V_{Cx} of the wheel contact center. The parameters Q_{Cz1}, Q_{Bz1}, ... are tire dependent (see Appendix 6 for a specific set of data). The residual torque in Eq. (2.41) can be expressed as follows

$$M_{zr} = D_r \cdot \cos(\arctan(B_r \cdot \alpha_r)) \quad (2.57)$$

with

$$\alpha_r = \alpha + S_{Hy} + \frac{S_{Vy}}{K_{y\alpha}} \quad (2.58)$$

$$B_r = Q_{Bz9} + Q_{Bz10} \cdot B_y \cdot C_y \quad (2.59)$$

$$D_r = F_z \cdot R_0 \cdot \{Q_{Dz6} + Q_{Dz7} \cdot df_z + (Q_{Dz8} + Q_{Dz9} \cdot df_z).\gamma + (Q_{Dz10} + Q_{Dz11} \cdot df_z) \cdot \gamma \cdot |\gamma|\} \cdot \cos\alpha \cdot \text{sign}(V_{Cx}) \quad (2.60)$$

for empirical factors Q_{jzi}, $j = B, D$, and $i = 1, 2, 3, \ldots$

Thus far, we have not discussed the effect of camber on the lateral tire characteristics. With only a camber angle and no slip angle, the tire tries to follow an almost circular track that is determined by the local shape of the tire cross section.

The direction of motion of the wheel is forced by the vehicle velocity vector. For example, the wheel may be moving forward in a straight path. As a result, local shear stresses arise in the contact area and build up a camber force (Figure 2.26). For a motorcycle, the camber force is the major force between tire and road that prevents the tire from sliding.

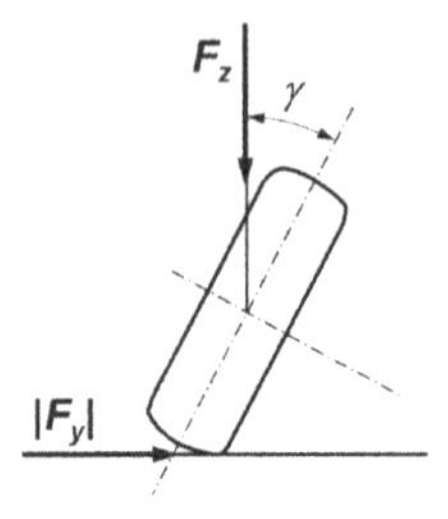

FIGURE 2.26 Camber-induced side force.

In the linear range, the side force can be expressed in terms of slip angle and camber angle, as follows

$$F_y(\alpha) = C_\alpha \cdot \alpha + C_\gamma \cdot \gamma \tag{2.61}$$

with cornering stiffness C_α and *camber stiffness* C_γ, defined as

$$C_\alpha = \frac{\partial F_y}{\partial \alpha}(\alpha = 0, \gamma = 0) \tag{2.62}$$

$$C_y = \frac{\partial F_y}{\partial \gamma}(\alpha = 0, \gamma = 0) \tag{2.63}$$

The ratio of camber stiffness and tire load is referred to as the normalized camber stiffness (also denoted as the *camber thrust coefficient*). For a passenger car tire, this value can be of the order of 1, to be compared to the value of C_α/F_z, which can range between 10 and 15 (see Figure 2.22). Hence, the camber stiffness is of the order of 10% or less, compared to the cornering stiffness.

We determined lateral force, pneumatic trail, and aligning torque for a fixed wheel load of 4000 [N] and for camber angle values of −5, 0, and 5 [°]. The results for F_y and for trail and aligning torque are shown in Figures 2.27 and 2.28, respectively.

2.6 COMBINED CORNERING AND BRAKING/DRIVING

2.6.1 Combined Slip

The discussion in the preceding sections deals with pure slip, i.e., with situations where the car is either cornering, or braking/driving. When a drive or brake torque is applied during cornering, the total horizontal force is not acting purely in the longitudinal or lateral direction, and the cornering force potential is therefore reduced. According to Figures 2.17 and 2.20, the maximum longitudinal and lateral forces for a certain wheel load F_z are given by

$$F_{x,\max} = \mu_{xp} \cdot F_z \tag{2.64}$$

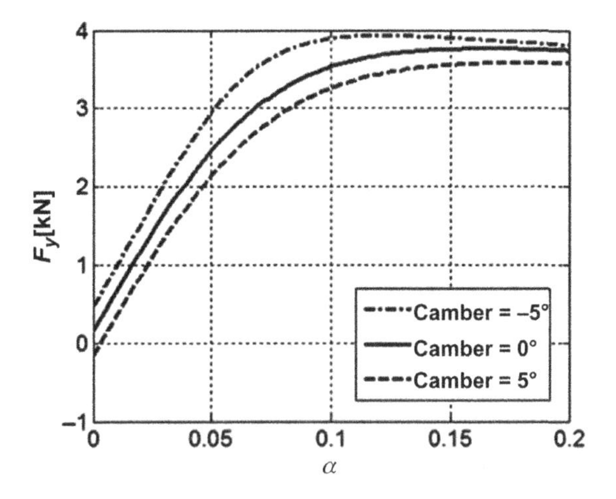

FIGURE 2.27 Side force vs. slip angle for different camber angle value and wheel load $F_z = 4000$ [N].

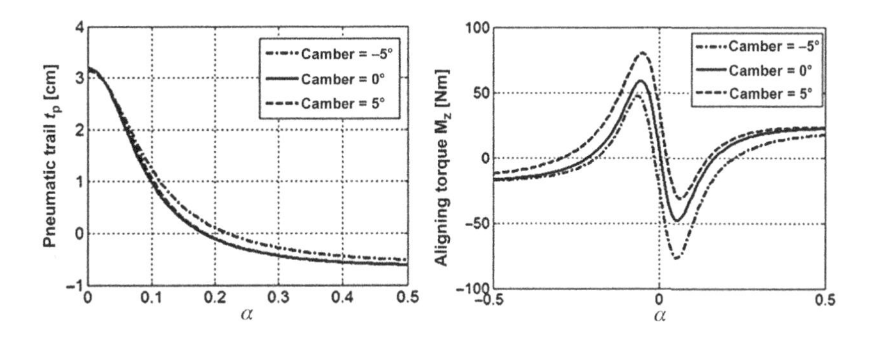

FIGURE 2.28 Pneumatic trail and aligning torque vs. slip angle for different camber angle value and wheel load $F_z = 4000$ [N].

$$F_{y,\max} = \mu_{yp} \cdot F_z \tag{2.65}$$

Clearly, the peak friction coefficients are related to the road friction, with lower road friction leading to a proportional decrease of μ_{xp} and μ_{yp}. This suggests that the envelope of the general normalized shear force under combined slip between tire and road is a closed curve, with a maximum value μ_{xp} along the x-axis (pure slip in longitudinal direction) and a maximum value μ_{yp} along the y-axis (pure slip in lateral direction). In general, the drive force will be smaller than the brake force because it is bounded by the engine power. Furthermore, this enveloping curve will depend on speed on a wet road, especially in case of a significant amount of water (aquaplaning). A tire is not a rotational symmetric object, which explains the difference in size between μ_{xp} and μ_{yp}. This enveloping curve will be different for different tires.

Let us assume the enveloping curve to be well approximated by an ellipse, as indicated by the outer curve in Figure 2.29. The right-hand part of this figure corresponds to driving ($F_x > 0$), whereas the left-hand side of this figure corresponds to braking. The outer ellipse describes the maximum shear force, which can be applied for a certain road friction and wheel load. The figure shows clearly (point A) that the side force in case of such a maximum shear force, under presence of a drive force $F_x = \mu_x \cdot F_z$, will be less than $\mu_{yp} \cdot F_z$. Likewise, applying a side force while braking or driving will reduce the longitudinal force, i.e., the braking or driving potential of the tire.

The internal ellipses are approximations for the shear force for constant slip angle α and for varying longitudinal slip. We shall see later, when we plot F_x versus F_y for fixed slip angle, based on test data, that this approximation is rather rough. These elliptic approximations were used by Genta and Morello [10] to estimate the cornering stiffness under conditions of combined slip. In absence of a brake or drive force, the side force is indicated by $F_{y0} = \mu_{y0} \cdot F_z$ (i.e., in case of pure side slip). With maximum brake or drive force, the longitudinal friction coefficient is assumed equal to μ_{xp}.

The internal ellipses in Figure 2.29 can therefore be described by the following relationship between F_x and F_y:

$$\left(\frac{F_y}{\mu_{y0} \cdot F_z}\right)^2 + \left(\frac{F_x}{\mu_{xp} \cdot F_z}\right)^2 = 1 \tag{2.66}$$

Assuming the slip angle to be small, such that the side force can be expressed as cornering stiffness times slip angle, the relationship (2.66) leads to

$$C_{\alpha,\text{combined}} = C_{\alpha,\text{pure}} \cdot \sqrt{1 - \left(\frac{F_x}{\mu_{xp} \cdot F_z}\right)^2} \tag{2.67}$$

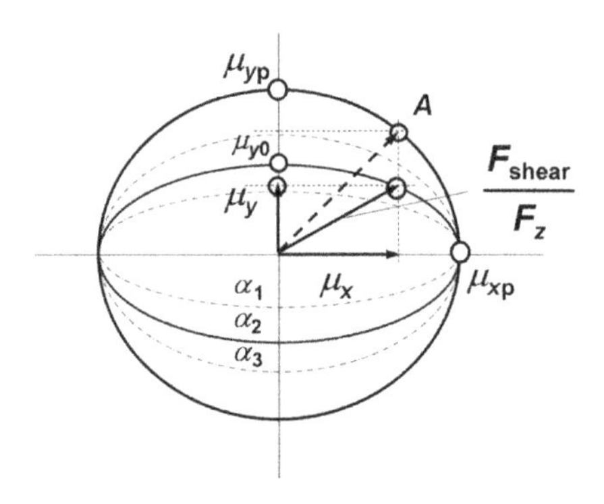

FIGURE 2.29 Elliptic approximation of a tire friction envelope.

with the $C_{\alpha,\text{pure}}$ and $C_{\alpha,\text{combined}}$ indicating the cornering stiffness in case of pure side slip and combined slip, respectively. We refer to Eq. (2.67) as the *elliptic approximation of the cornering stiffness* under combined slip.

Following the approximation of the inner ellipse in Figure 2.29, we can conclude the following:

- The side force F_y is a function of both κ and α. For a fixed slip angle α, this side force has a maximum value at $\kappa = 0$, which reduces with increasing $|\kappa|$ (i.e., in case of either driving or braking). This also means that the peak value of $F_y(\alpha;\ \kappa)$ versus α decreases for increasing $|\kappa|$, as we observed previously.
- Pure longitudinal slip characteristic behavior, as shown in Figure 2.17, shows a local peak value in F_x, followed by a decrease of F_x when κ is further increased. This behavior is not shown in Figure 2.29, which indicates serious limitations in the elliptic approximation.

Let us consider the polar diagram (F_x versus F_y for fixed slip angle), based on the tire parameters included in Appendix 6. The results are shown in Figure 2.30. Negative values for the slip angle correspond to negative F_y-values. Indeed, one observes local maximum values for the longitudinal force $F_x(\kappa;\ \alpha)$ versus κ, with these values decreasing with increasing $|\alpha|$ (Figure 2.31). This behavior is similar to the lateral force for varying κ, as observed previously. The lateral force versus longitudinal slip κ is shown in Figure 2.31 as well, for varying slip angle α. There is a slight lack of symmetry in κ when the side force is very small ($\alpha = 0$). In case of only brake slip,

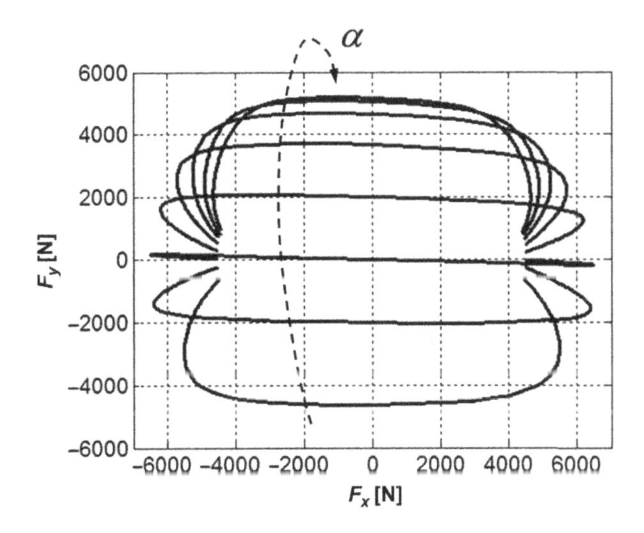

FIGURE 2.30 Polar diagram: F_x vs. F_y for constant slip angle ($\alpha = -6, -2, 0, 2, 4, 6, 8$, and 10 [°]), for $F_z = 6000$ [N].

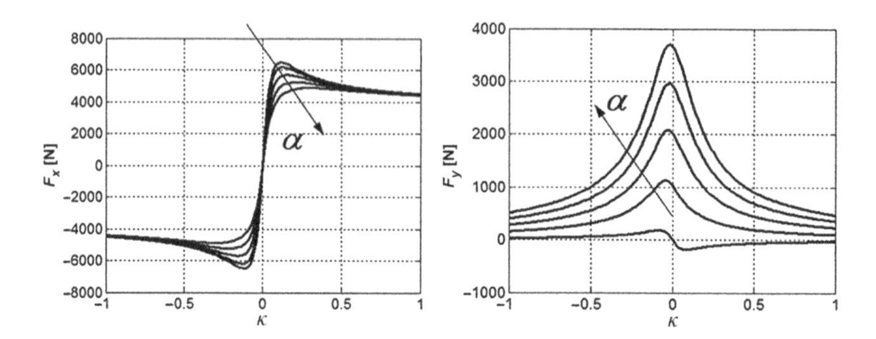

FIGURE 2.31 F_x and F_y vs. longitudinal slip, for varying slip angle, and wheel load 6000 [N].

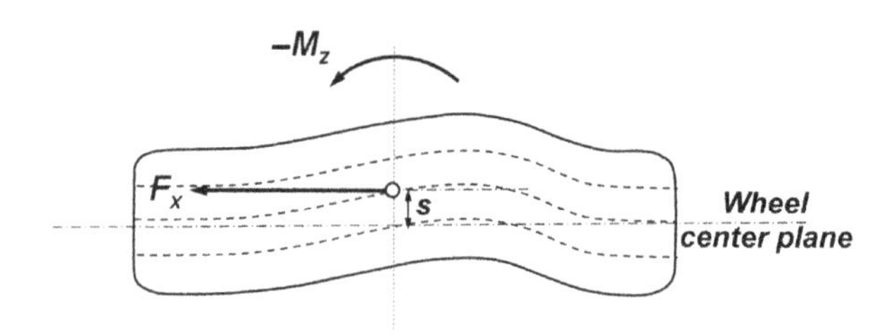

FIGURE 2.32 Drive force induced aligning torque under combined slip conditions.

a small side force arises that is negative for drive slip and positive for brake slip. This is due to irregularities in the tire design.

Next, let us consider the aligning torque under combined slip conditions. When both a lateral and a longitudinal force act on a tire, the longitudinal force acts some distance s away from the tire symmetry plane, as indicated in Figure 2.32.

A positive lateral force leads to this deflection $s > 0$, which in combination with the drive force F_x, results in a contribution to the aligning torque of $-s \cdot F_x$. This must be added to the residual torque minus side force times pneumatic trail, according to expression (2.41), which leads to

$$M_z(\alpha; \kappa) = -t_p \cdot F_y + M_{zr} - s \cdot F_x \tag{2.68}$$

with now all functions at the right-hand side depending on both slips α and κ. This means a negative contribution to the aligning torque for driving and a positive contribution to the aligning torque for braking. The aligning torque is shown versus F_x in Figure 2.33 for the same variation in slip angles as in Figure 2.30. The almost straight line corresponds to $\alpha = 0$. Indeed, a positive longitudinal force leads to a negative torque, with the opposite behavior for braking. For a positive slip angle, the aligning torque becomes more negative for zero longitudinal slip, with some recovery of increasing κ, but not tending

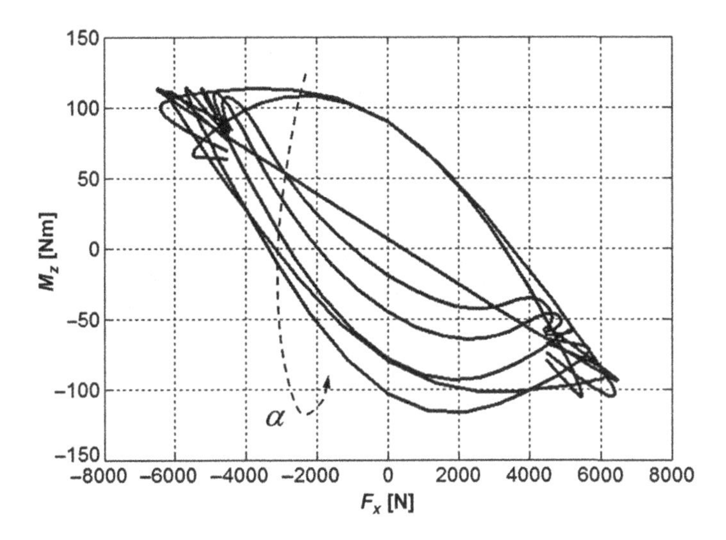

FIGURE 2.33 F_x vs. aligning torque for constant slip angle ($\alpha = -6$, -2, 0, 2, 4, 5, 8, and 10 [°]), for $F_z = 6000$ [N].

to a small positive value as for pure slip (see Figure 2.24). For large κ, the last term in Eq. (2.68) is dominant in the aligning torque.

2.6.2 Modeling Tire Behavior for Combined Slip

Pacejka [32] describes the combined slip characteristics by multiplying the pure slip force with a weighting function $G(\alpha, \kappa)$

$$F_x(\kappa; \alpha) = G_{x\alpha}(\alpha, \kappa) \cdot F_{x,\text{pure}}(\kappa) \tag{2.69}$$

$$F_y(\alpha; \kappa) = G_{y\alpha}(\alpha, \kappa) \cdot F_{y,\text{pure}}(\alpha) + S_{\text{V}y\kappa} \tag{2.70}$$

for vertical shift $S_{\text{V}y\kappa}$, which is a function of both slip values α and κ.

Consider braking in a turn. With small lateral slip, the brake force will be close to the pure slip value $F_{x,\text{pure}}(\kappa)$, which means that the weighting function $G(\alpha, \kappa)$ will be close to 1. If cornering is more extreme and the slip angle increases, the brake force will drop (see Figure 2.31), which means that the weighting function must also drop. Obviously, reduction with α also depends on κ. The reduction will be stronger if $|\kappa|$ is smaller.

This weighting function is shown in Figure 2.34 for varying slip angle α and longitudinal slip κ. Apparently, this function has the qualitative behavior of the pneumatic trail (see Figure 2.28) and can therefore be described well using the cosine function of the Magic Formula description, balanced by a denominator to make the function equal to 1 for $\alpha = 0$

$$G_{x\alpha}(\alpha, \kappa) = \frac{\cos(C_{x\alpha} \cdot \arctan(B_{x\alpha} \cdot \alpha_s - E_{x\alpha} \cdot (B_{x\alpha} \cdot \alpha_s - \arctan(B_{x\alpha} \cdot \alpha_s))))}{\cos(C_{x\alpha} \cdot \arctan(B_{x\alpha} \cdot S_{\text{H}x\alpha} - E_{x\alpha} \cdot (B_{x\alpha} \cdot S_{\text{H}x\alpha} - \arctan(B_{x\alpha} \cdot S_{\text{H}x\alpha}))))}$$

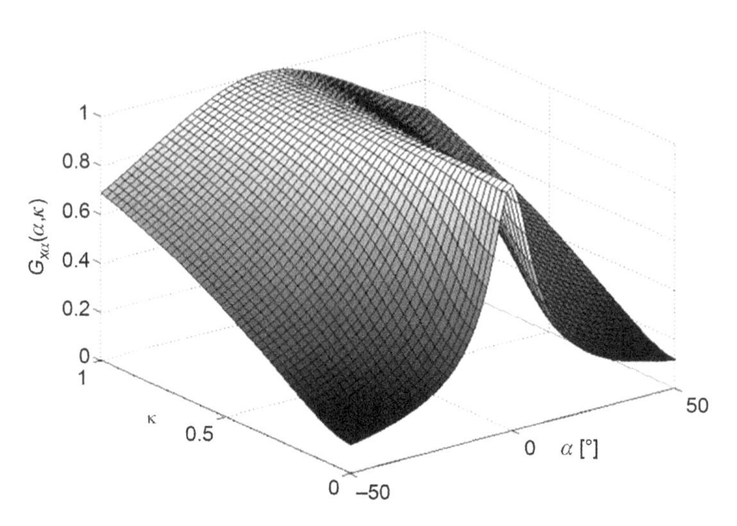

FIGURE 2.34 Weighting function $G_{x\alpha}$, as function of slip values α and κ.

with

$$\alpha_s = \alpha + S_{Hx\alpha}, B_{x\alpha} = (R_{Bx1} + R_{Bx3} \cdot \gamma^2) \cdot \cos(\arctan(R_{Bx2} \cdot \kappa))$$

for horizontal shift $S_{Hx\alpha}$. The other parameters in this weighting function depend on wheel load and we refer to Ref. [32] for more details.

In a similar way, one can describe the weighting function in Eq. (2.70)

$$G_{y\alpha}(\alpha,\kappa) = \frac{\cos(C_{y\kappa} \cdot \arctan(B_{y\kappa} \cdot \kappa_s - E_{y\kappa} \cdot (B_{y\kappa} \cdot \kappa_s - \arctan(B_{y\kappa} \cdot \kappa_s))))}{\cos(C_{y\kappa} \cdot \arctan(B_{y\kappa} \cdot S_{Hy\kappa} - E_{y\kappa} \cdot (B_{y\kappa} \cdot S_{Hy\kappa} - \arctan(B_{y\kappa} \cdot S_{Hy\kappa}))))}$$

with

$$\kappa_s = \kappa + S_{Hy\kappa}, B_{y\kappa} = (R_{By1} + R_{By4} \cdot \gamma^2) \cdot \cos(\arctan(R_{By2} \cdot (\alpha - R_{By3})))$$

for horizontal shift $S_{Hy\kappa}$. The other parameters in this weighting function depend on wheel load, see Ref. [32] again for more details.

The aligning torque follows Eq. (2.68) with certain equivalent slip angles used in the pneumatic trail for combined slip and in the residual torque. We refer to Ref. [32].

2.6.3 Approximations in case of Combined Slip

The *elliptic approximation* (2.67) has been introduced to determine the cornering stiffness when a drive force is known (e.g., to compensate for rolling or aerodynamic and slope resistances). To judge the accuracy of this approximation, we consider it as a function of lateral slip α and longitudinal slip κ through the drive force F_x in case of combined slip. We used the tire

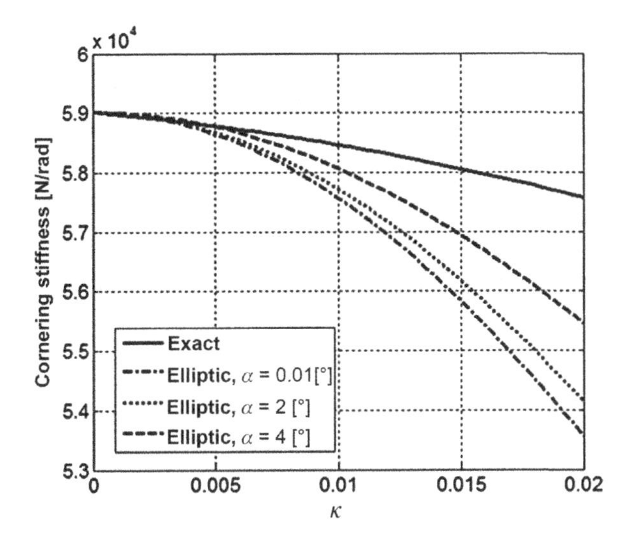

FIGURE 2.35 Elliptic approximation of cornering stiffness.

parameters from Appendix 6. Results are shown in Figure 2.35, where we varied the drive slip from 0 to 0.02. The error in the cornering stiffness amounts about 10% at maximum, especially if lateral slip is low. If lateral slip is larger (it is to be expected that drive slip will be small compared to lateral slip), the error reduces but not that much. The major contribution to accuracy is to have a small drive slip. Consequently, use this approximation for combined slip only if the longitudinal force (drive or brake force) is small.

Another approximation exists that is quite accurate but does not require expensive combined slip measurements. In Section 2.7, a physical modeling approach will be discussed known as the brush model. It is shown that within the restrictions of that physical model, the combined slip shear forces F_x and F_y can be expressed in terms of the pure slip shear forces as follows

$$F_x = \frac{\rho_x}{\rho} \cdot F_{x,\text{pure}}(\rho), F_y = \frac{\rho_y}{\rho} \cdot F_{y,\text{pure}}(\rho) \tag{2.71}$$

for theoretical slip vector (ρ_x, ρ_y), defined in Eq. (2.34) and expressed in practical slip quantities in Eq. (2.37), and with magnitude ρ:

$$\rho = \sqrt{\rho_x^2 + \rho_y^2} \tag{2.72}$$

It has been shown by Pauwelussen and Andress [34] that Eq. (2.71) is also a good approximation of tire contact forces if practical slip quantities are used:

$$F_x = \frac{\kappa}{\sqrt{\kappa^2 + \tan^2\alpha}} \cdot F_{x,\text{pure}}\left(\sqrt{\kappa^2 + \tan^2\alpha}\right) \tag{2.73a}$$

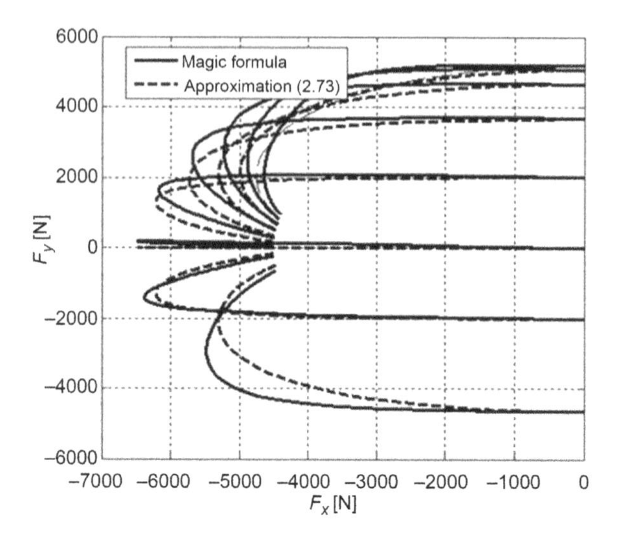

FIGURE 2.36 Tire combined slip characteristics, approximation (2.73), and Magic Formula results.

$$F_y = \frac{\tan\alpha}{\sqrt{\kappa^2 + \tan^2\alpha}} \cdot F_{y,\text{pure}}\left(\sqrt{\kappa^2 + \tan^2\alpha}\right) \tag{2.73b}$$

In Ref. [34], a maximum error of about 8% was observed between Magic Formula estimates and the estimates based on Eqs. (2.73a) and (2.73b), for eight different passenger car tires, for which Magic Formula data were available. In many applications, this is sufficient. A typical example is when one needs to complete verification analyses using tire data for which only plots of pure slip characteristics are known (e.g., derived from a paper), or if only pure slip tests are available. Such plots or test results can easily be transferred (or well estimated) in Magic Formula parameters for pure slip (see Appendix 7). The expressions (2.73a) and (2.73b) will allow approximate combined slip analysis.

We verified this approximation using the same data as used in the previous plot, and for a wheel load of 6000 [N] (see Figure 2.37). This figure corresponds to the polar diagram in Figure 2.30. The behavior for small and large brake slip is quite good. For intermediate brake slip, the brake force is underestimated for the set of Magic Formula data from Appendix 6. Also, observe the symmetry in the approximation based on Eqs. (2.73a) and (2.73b). Changing the sign of the slip angle will only lead to a sign change in F_y without any effect on F_x. The Magic Formula allows a lack of symmetry, as shown in Figure 2.36.

2.7 PHYSICAL TIRE MODELS

In this section, we shall present the theory of steady-state slip with the aid of some simple physical models:

- The brush-type tire model
- The brush-string model

In all cases, it is assumed that the properties of the tire can be described by averaging the local behavior over the tire width, which means that the tire is replaced by a disk of zero width. For both the brush-type and the string models, one may extend the model to account for a finite width. However, this is not discussed here and we refer the reader to Ref. [32].

In the previous sections, we discussed the empirical Magic Formula model, with the mathematical relationships between forces, moments, and slip chosen so that experimental results are reproduced in an accurate way. Such an approach is usually referred to as a similarity approach. In contrast to this, one may try to describe the tire performance in a physical way, i.e., deriving the tire performance characteristics based on an analytical description of belt and treads. Such a description for the belt may be a string model, which is stressed by a tensile force and restricted in lateral or longitudinal deflection by a distributed stiffness acting in the contact area. A model of that kind is known as the *stretched string model*. The stretched string model has been used in various investigations in case of pure slip (mainly out of plane), both for steady-state and transient conditions (see Chapter 3). The latter case is often related to shimmy. The works of von Schlippe [45], Segel [49], Pacejka [31] and [32], and Besselink [2] provide more information on this. Higuchi [17] applied the stretched string model for combined slip.

With focus on only the tread deflection, the tire could be described using a rigid ring with little beams (brushes) that connect radially to the belt with a constant linear stiffness against transversal or longitudinal loading. Such a model is referred to as the *brush model*, originally discussed by Fromm [8].

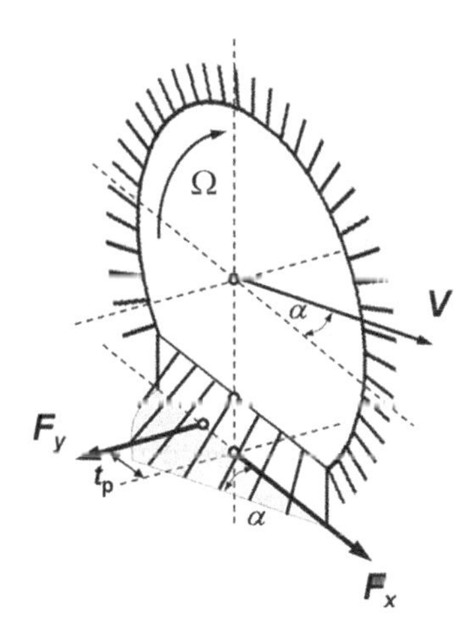

FIGURE 2.37 Schematic layout of the brush tire model.

TABLE 2.4 Status of Various Physical Tire Models

Physical Tire Model	Pure Slip	Combined Slip
Brushes only	Well established	Well established
Bare string	Pacejka [31]	Higuchi [17]
Brush string	Restricted to one sliding region at the rear (Pacejka [31]) Extended to two sliding regions (Pauwelussen [36])	Pauwelussen [36]

Extending the stretched string model with the tread stiffness means that brushes are attached to the flexible belt and therefore, combining belt and tread compliance, leading to the *brush-string model*, which is investigated by Pauwelussen in Ref. [36] for combined slip. When the brushes are absent, the model is referred to as the *bare string model.* Before the work of Pauwelussen [36], the combined brush-string model was not derived and applied for all possible cases of combined slip and for all possible combinations of carcass stiffness, tread stiffness, and string tension. The status for the brush-string model, as well as for its extreme cases of bare string, and brush model are indicated in Table 2.4.

2.7.1 The Brush Model

In this section, we discuss the theory of steady-state combined tire slip with the aid of the simple brush-type tire model. The theory of this section will not consider camber and turning (turn slip) of the wheel (see Ref. [32] for an extensive treatment of the brush model). We refer to Figure 2.36 for a schematic layout of the model. The tire is equipped with small linear beams (brush elements), some of which touch the ground (the contact area) and, as a result, will be deformed as a linear beam. Both a lateral force and a longitudinal force are assumed to act in the contact area. Consequently, the tire is assumed to move sideways with a slip angle α, in combination with a longitudinal slip κ, i.e., we assume the general case of combined slip. Each brush element in the contact area connects the ground (the "tip" of the brush element) with the tire (the "base" of the brush element). A brush element tries to follow the direction of the speed, which means that in case of a nonzero slip angle and as long as the tip is fixed to the road (adhesion), the tips of subsequent brush elements follow a linear pattern. This pattern is obviously defined by this slip angle. Under conditions of pure side slip, the tip moves only laterally with respect to the tire. The longitudinal slip will force the tip to move rearward in the contact area in case of braking and forward in case of driving. When the friction limit at the brush tip is

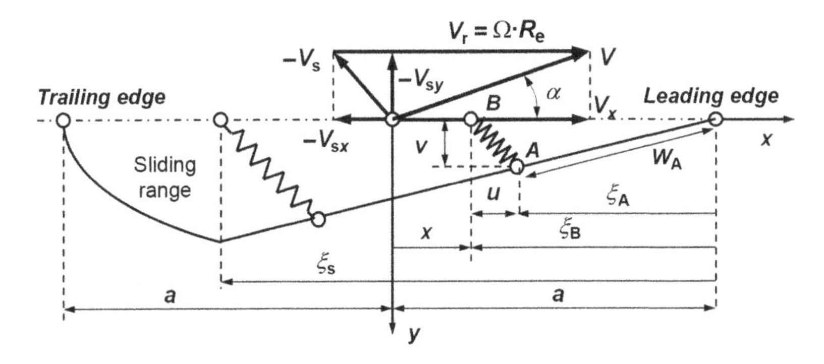

FIGURE 2.38 Top view brush model.

exceeded, the tip begins to slide. These phenomena have been discussed in the preceding sections, and we conclude that two regions are identified: a leading adhesion region where the contact line (connecting the tips of the brush elements) is straight and a sliding region where the shear stress follows Coulomb law

$$\tau = \sqrt{\tau_x^2 + \tau_y^2} = \mu \cdot \sigma_z$$

for shear stresses τ_x, τ_y in the contact area, road friction μ, and normal contact stress σ_z. The tire is moving with speed V, built up from a rolling speed (see Eq. (2.35)) and a slip speed V_s with both a lateral and a longitudinal component.

A top view of the tire under deflection of the tread elements (the brushes) is shown in Figure 2.38. At the leading edge of the contact area, with total length $2a$, the deformation is still zero. The base and the tip of the tread element coincide. With the tire moving with speed V and rolling with rolling speed V_r, the base of the tread is attached to the wheel plane and will move inside the contact area with the rolling speed, say to point B. At the same time, the tip of the tread element will move to point A opposite to the speed V. Figure 2.38 suggests a positive longitudinal (drive) slip, but this is not a restriction and is merely done for the figure only. The discussions given next do not depend on the sign of slip κ.

Assuming a time interval Δt, this means that the displacement w_A in the actual contact area along the deformed treads can be written as

$$w_A = V \cdot \Delta t$$

The new positions ξ_A (tip) and ζ_B (base) are found from

$$\xi_A = V \cdot \cos(\alpha) \cdot \Delta t, \quad \xi_B = V_r \cdot \Delta t$$

from which expressions for the deflections u and v (cf. Figure 2.38) can be derived

$$u = (V_r - V \cdot \cos\alpha) \cdot \Delta t$$

$$v = V \cdot \sin\alpha \cdot \Delta t$$

This means that the displacements can be expressed in terms of either the position in the deformed tread situation ξ_A or in the undeformed tread coordinate ξ_B as follows:

$$\begin{pmatrix} u \\ v \end{pmatrix} = \begin{pmatrix} \dfrac{V_r - V_x}{V_x} \\ \tan\alpha \end{pmatrix} \cdot \xi_A = \begin{pmatrix} \dfrac{V_r - V_x}{V_r} \\ \dfrac{V_x}{V_r} \cdot \tan\alpha \end{pmatrix} \cdot \xi_B \tag{2.74}$$

The vector of coefficients corresponds to either practical slip or theoretical slip, as previously defined. The expression (2.74) is of the general form

$$\text{Displacement} = \text{slip} \times \text{position}$$

where slip is defined on the basis of either the position ξ_A with respect to the deformed tire, or the position ξ_B with respect to the undeformed tire. Consequently, practical slip quantities are related to the deformed tire quantities, whereas the theoretical slip quantities are derived based on undeformed tire quantities.

The contact area is taken as a square with length $2a$ and width $2b$. It is common to assume a parabolic pressure distribution $\sigma_z(x)$, that is taken uniform over the contact width $2b$

$$\sigma_z(x) = \sigma_{z0} \cdot \left[1 - \left(\frac{x}{a}\right)^2\right] \tag{2.75}$$

with σ_{z0} following from the condition that

$$F_z = \iint_{\text{contact area}} \sigma_z(x) \cdot \mathrm{d}x\, \mathrm{d}y$$

and thus

$$\sigma_{z0} = \frac{3 \cdot F_z}{8ab}$$

Please note that the parabolic contact pressure distribution (2.75) is not a real restriction and one may easily use other distributions.

We shall now derive expressions for the total displacement in the contact area, where a distinction is made between adhesion and sliding. In the *adhesion region*, it follows that

$$e = \sqrt{u^2 + v^2} = \rho \cdot \xi_B = \frac{\xi_B}{1+\kappa} \cdot \sqrt{\kappa^2 + \tan^2\alpha} \tag{2.76}$$

with total theoretical slip ρ. In the *sliding region*, assuming Coulomb friction with friction coefficient μ, the shear stress $\tau(x,y)$ is bounded by $\mu \cdot \sigma(x)$. The displacement e is therefore bounded as well, and it follows from the stiffness of the tread, denoted as k:

$$e = e_{\max} = \frac{\tau(x,y)}{k} = \frac{\mu \cdot \sigma_z(x)}{k} = \frac{3 \cdot \mu \cdot F_z}{8 \cdot a^3 \cdot b \cdot k} \cdot (a^2 - x^2)$$

Note that k will not be the same in x and y directions. Considering this analysis for pure slip in case of either lateral or longitudinal direction, one can distinguish between different stiffnesses k_x and k_y. For combined slip, we restrict our analysis to equal tread stiffnesses $k = k_x = k_y$ (isotropic model).

We introduce the *tire parameter* θ by

$$\theta = \frac{4}{3} \cdot \frac{a^2 \cdot b \cdot k}{\mu \cdot F_z} \tag{2.77}$$

resulting in

$$e_{\max} = \frac{\xi_B \cdot (2 \cdot a - \xi_B)}{2 \cdot a \cdot \theta}$$

The breakaway point ξ_s (indicated in Figure 2.38), at which adhesion becomes sliding, is found by taking $e_{\max}$ equal to the expression (2.76) yielding:

$$\xi_s = 2 \cdot a \cdot (1 - \theta \cdot \rho)$$

Consequently, for $\rho = 0$, the breakaway point is given by $\xi_s = 2a$ and the full contact area is in the state of adhesion. With increasing ρ, the breakaway point ξ_s moves to a value $\xi_s = 0$, attained at $\rho = 1/\theta$. In other words, the parameter $\theta > 1$ is the reciprocal total slip, for which the full contact area is sliding. When the total theoretical slip exceeds the magnitude $1/\theta$, the tire remains in a state of complete sliding. In case of pure slip, this situation is reached for either

$$\alpha_m = |\alpha| = \arctan(1/\theta)$$

or

$$\kappa_{\mathrm{m}} = \frac{1}{\theta - 1}, \text{ in case of driving } (\kappa > 0)$$

$$\kappa_{\mathrm{m}} = -\frac{1}{\theta + 1}, \text{ in case of braking } (\kappa < 0)$$

In the *adhesion region*, the shear stresses follow from the deflections (2.76):

$$\underline{\tau}_{\mathrm{adh}} = k \cdot \underline{e} = k \cdot \xi_{\mathrm{B}} \cdot \underline{\rho} \equiv k \cdot \xi_{\mathrm{B}} \cdot \begin{pmatrix} \rho_x \\ \rho_y \end{pmatrix}$$

In the *sliding region*, the shear stresses follow Coulomb rule, assuming that the shear stress vector has the same orientation as the theoretical slip vector,

$$\underline{\tau}_{\mathrm{sliding}} = \mu \cdot \sigma_z(x) \cdot \frac{\underline{\rho}}{\rho} = \frac{k \cdot e_{\max}}{\rho} \cdot \begin{pmatrix} \rho_x \\ \rho_y \end{pmatrix} = \frac{k}{\rho} \cdot \frac{\xi_{\mathrm{B}} \cdot (2 \cdot a - \xi_{\mathrm{B}})}{2 \cdot a \cdot \theta} \cdot \begin{pmatrix} \rho_x \\ \rho_y \end{pmatrix}$$

Clearly, for pure slip, the shear stress vector and the slip vector will have the same orientation. In case of combined slip, these orientations will not be identical, but the difference is small.

With these expressions, we are now able to derive expressions for the contact forces F_x and F_y, as well as the aligning torque M_z by integrating (the moment of) the shear stresses over the contact area. For the shear force vector, one arrives at

$$\underline{F}_{\mathrm{shear}} \left[\int_0^{\xi_s} \underline{q}_{\mathrm{adh}}(\xi_{\mathrm{B}}) \mathrm{d}\xi_{\mathrm{B}} + \int_{\xi_s}^{2 \cdot a} \underline{q}_{\mathrm{sliding}}(\xi_B) \mathrm{d}\xi_{\mathrm{B}} \right]$$

and the force components (lateral and longitudinal force) are obtained from

$$\underline{F}_{\mathrm{shear}} = \frac{\underline{\rho}}{\rho} \cdot F \equiv \frac{\underline{\rho}}{\rho} \cdot \sqrt{F_x^2 + F_y^2} \tag{2.78}$$

with F found to be given by

$$F = \mu \cdot F_z \cdot [3 \cdot \theta \cdot \rho - 3 \cdot (\theta \cdot \rho)^2 + (\theta \cdot \rho)^3] \;\; \rho < 1/\theta \tag{2.79a}$$

$$= \mu \cdot F_z; \quad \rho \geq 1/\theta \tag{2.79b}$$

where $\underline{q}_{\mathrm{adh}}$ and $\underline{q}_{\mathrm{sliding}}$ denote the integrated shear stress over the tire width

$$\underline{q} = \begin{pmatrix} q_x \\ q_y \end{pmatrix} = \int_{\text{tire width}} \underline{\tau}.\mathrm{d}y \tag{2.80}$$

Note that the slip stiffness C_ρ (with ρ referring to either lateral slip or longitudinal slip) is found by linearization of Eqs. (2.79a) and (2.79b) near $\rho = 0$:

$$C_\rho = 4 \cdot a^2 \cdot b \cdot k$$

Selecting different tread stiffnesses in lateral and longitudinal directions (no isotropic model), one arrives at different values for the cornering and longitudinal slip stiffness, as expected.

The longitudinal and lateral force are shown in Figure 2.39 for varying brake slip and selected values of the slip angle, where we have chosen:

$k = 2 \times 10^7$ [N/m^3]
$b = 0.1$ [m]
$\mu = 1.0$

We select an unloaded tire radius $R = 0.32$ [m], and assume the loaded tire radius R_1 behavior to conform to Figure 2.5. The half-contact length can be approximated by:

$$a = \sqrt{R^2 - R_1^2}$$

or, in terms of tire radial deflection d

$$a = \sqrt{2 \cdot R \cdot d - d^2} \approx \sqrt{2 \cdot R \cdot d} = \sqrt{2 \cdot R \cdot F_z / C_{Fz}} \tag{2.81}$$

for tire stiffness C_{Fz}. This expression will overestimate the contact length. The tire in Figure 2.5 has a stiffness of $C_{Fz} = 2.10^5$ [N/m]. Assuming $F_z = 4000$ [N], one finds $a = 0.11$ [m].

Observe that Figures 2.31 and 2.39 are very similar, where the results in Figure 2.31 are based on the empirical Magic Formula model and the results in Figure 2.39 are based on the simple brush model. However, two

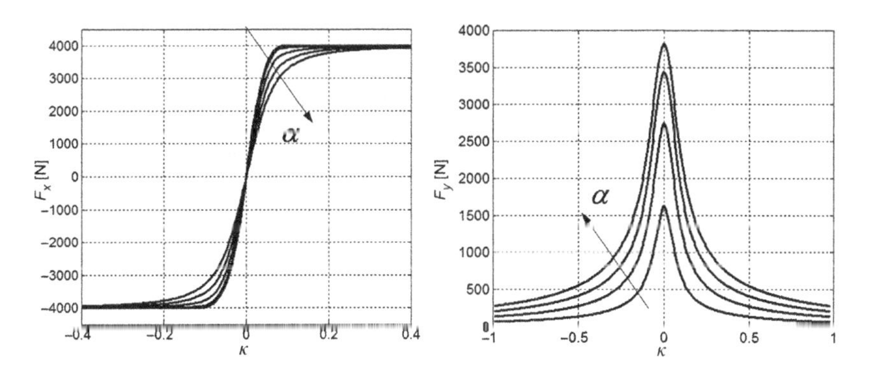

FIGURE 2.39 F_x and F_y versus longitudinal slip, for varying slip angle, and wheel load 4000 [N], based on the physical brush model cf. Eqs. (2.79a) and (2.79b).

differences can be observed. The longitudinal force according to the brush model saturates without showing the typical local peak in the longitudinal force characteristic, followed by the decay for larger slip κ. Consequently, the saturation level of the brake or drive force is reached much sooner and does not change beyond this slip value. Tire irregularities are not accounted for in the lateral force for small slip angle.

In Figures 2.40 and 2.41, we show the top view of brush deflections in case of pure lateral speed (Figure 2.40) and combined slip with a fixed slip angle and increasing brake slip (Figure 2.41). Slip is increasing from the image at the top to the image at the bottom of these figures. In Figure 2.40, the extension of the sliding region is clearly shown. At first, the major part of the brush deflection develops in a linear way from the leading edge of the contact area. With slip angle increasing, the intersection point between sliding and adhesion region moves to the front, which means that the adhesion region is reduced and the sliding region is extended.

In Figure 2.41, the lateral deflections of the brushes reduce in size when the brake slip is increased. There is a major deflection in-plane because of this brake slip, which cannot be seen in a top-down projected view. At the same time, one observes a change in orientation of the brush elements from purely lateral to a rearward deflection. Tread elements, entering the contact area, are stretched because of the brake contact force, as explained in Section 2.4.1. The phenomenon shown in Figure 2.41 is consistent with the general physical considerations in Section 2.4.1. Finally, when increasing the brake slip,

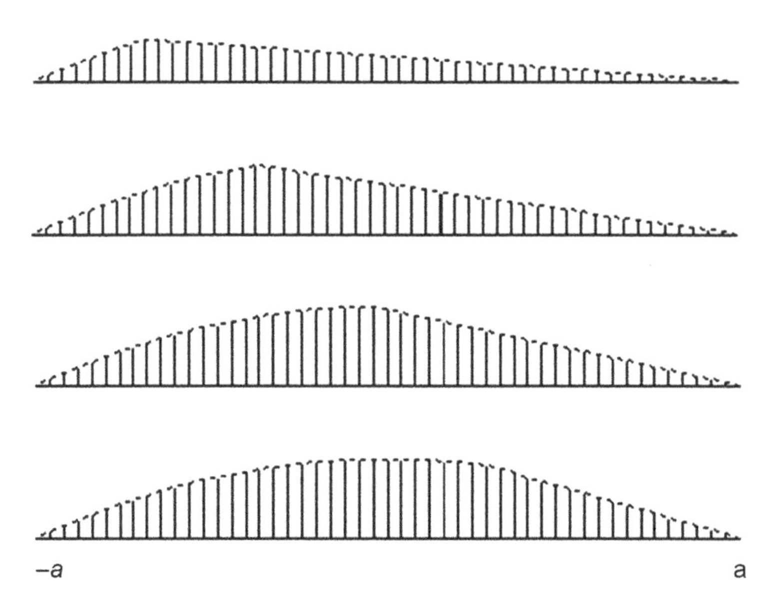

FIGURE 2.40 Brush deflections along the contact area for pure side slip with increasing slip angle.

the total slip and the sliding are increased, as shown in Figure 2.41. A side view for pure longitudinal slip (braking) is shown in Figure 2.42. Vertical brushes remain vertical at the leading edge, with the base moving rearward along the contact area, and this deflection vanishing at the trailing edge of the contact area.

In the same way as for the shear forces, one arrives at a closed form expression for the aligning torque M_z:

$$M_z = -\int_0^{2\cdot a} q_y(\xi_B)\cdot(a-\xi_B)\mathrm{d}\xi_B =$$

$$= \frac{-\rho_y}{\rho}\cdot\mu\cdot F_z\cdot a\cdot\left[\theta\rho - 3\cdot(\theta\rho)^2 + 3\cdot(\theta\rho)^3 - (\theta\rho)^4\right];\quad \rho < 1/\theta \tag{2.82}$$

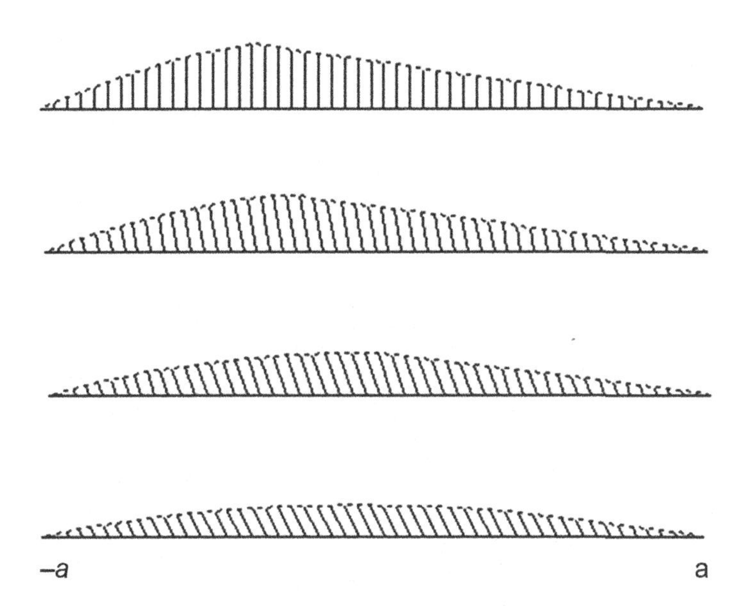

FIGURE 2.41 Brush deflections along the contact area for combined slip with increasing brake slip and fixed slip angle.

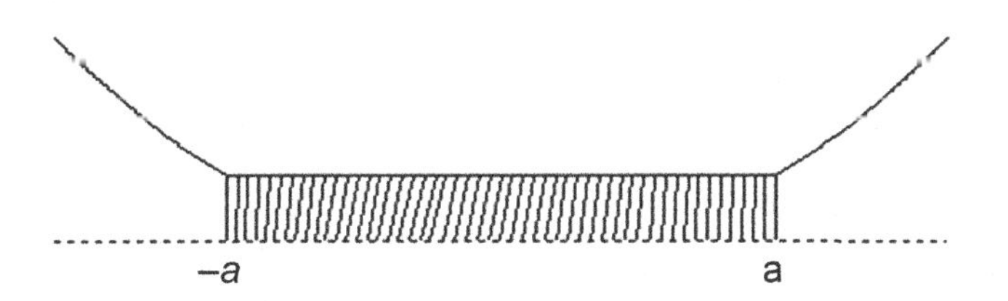

FIGURE 2.42 Brush deflections along the contact area for pure brake slip.

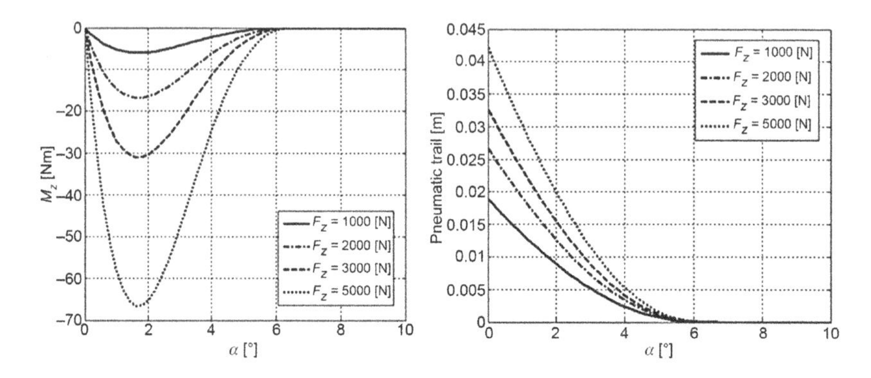

FIGURE 2.43 Aligning torque and pneumatic trail versus slip angle in case of pure slip, for different wheel loads, based on the physical brush model, cf. Eqs. (2.82), (2.83a) and (2.83b).

In case $\rho \geq 1/\theta$, M_z will vanish. Note that this can either be a result of increasing slip angle α or increasing brake slip or drive slip $|\kappa|$. The pneumatic trail follows from the ratio of F_y and $-M_z$:

$$t_p = \frac{1}{3} \cdot a \cdot \frac{1 - 3 \cdot \theta \cdot \rho + 3 \cdot (\theta \cdot \rho)^2 - (\theta \cdot \rho)^3}{1 - \theta \cdot \rho + \frac{1}{3} \cdot (\theta \cdot \rho)^2}; \quad \rho < 1/\theta \tag{2.83a}$$

$$= 0; \quad \rho \geq 1/\theta \tag{2.83b}$$

Aligning torque and pneumatic trail versus slip angle are shown in Figure 2.43 for the same values of the road friction and model parameters as chosen previously. The wheel load has been varied from 1000 to 5000 [N].

The aligning torque reaches a peak at $\alpha = \arctan(1/(4 \cdot \theta))$, after which it reduces in absolute size to zero at $\alpha = \arctan(1/\theta)$. The aligning torque does not change sign with increasing slip angle in contrast to results based on the Magic Formula description (see Figures 2.24 and 2.28). The pneumatic trail is a monotonous function in α, starting with a nonzero slope at $\alpha = 0$. It tends to zero, which value is reached at $\alpha = \arctan(1/\theta)$. Its value at vanishing slip angle

$$t_p \rightarrow \frac{a}{3}; \quad \alpha \downarrow 0$$

is smaller than normally encountered (around $a/2$), see also Ref. [32].

We close this section with some remarks concerning the effect of brake/drive force on the aligning torque and the approximations for combined slip contact force according to Eq. (2.71).

Remarks

1. The polar plot of F_x versus F_y, as depicted in Figure 2.30 for the empirical Magic Formula, can also be derived for the physical brush model. Because the force characteristics based on the brush model saturate

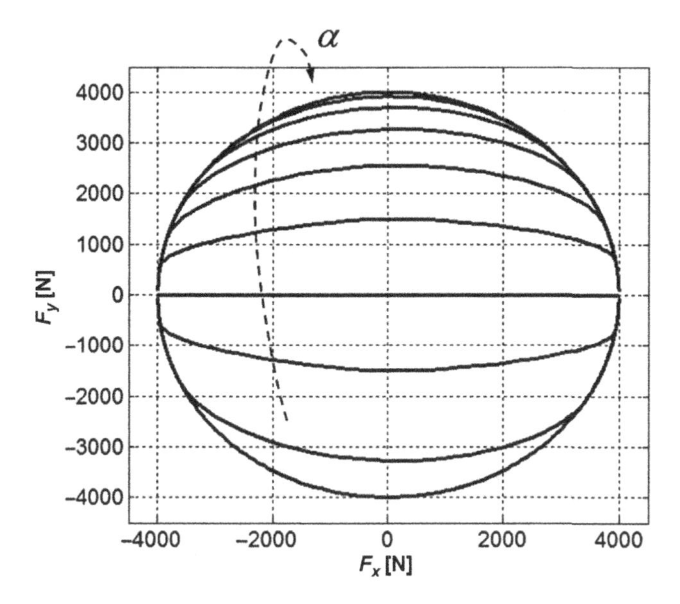

FIGURE 2.44 Polar diagram of F_x versus F_y for constant slip angle for $F_z = 4000$ [N] and $\mu = 1.0$ based on the brush model.

without decay for large slip, this polar plot will be similar to the one in Figure 2.29. We determined this polar plot for the brush model for $F_z = 4000$ [N] (see Figure 2.44). When we plot the aligning torque versus F_x, expression (2.82) leads to a plot that is symmetric in F_x, unlike Figure 2.33. This can be corrected by adding simple carcass flexibility to the brush model. This means that the entire carcass is pinned to the projected center of the wheel through springs acting in lateral and longitudinal direction with different stiffness values. Just as in the discussion on Figure 2.33, the resulting carcass deflections will then contribute to the aligning torque, leading to the loss of symmetry, as indicated in Figure 2.33. We refer to Ref. [32] for further details.

2. The combined equations (2.79a) and (2.79b) show that the explicit expression of contact shear force versus slip does not change if we move from pure slip to combined slip. The only difference is that ρ equals α or $-\kappa/(1+\kappa)$ in case of pure lateral or longitudinal slip, or the total theoretical slip according to Eq. (2.76). That means that expression (2.78) can be interpreted as

$$\begin{pmatrix} F_x(\rho_x, \rho_y) \\ F_y(\rho_x, \rho_y) \end{pmatrix} = \begin{pmatrix} \rho_x \\ \rho_y \end{pmatrix} \cdot \frac{F_{\text{pure}}(\rho)}{\rho}$$

Thus far, we discussed an isotropic brush model. The preceding expression has been the inspiration for the approximation (2.71), which gave accurate results that were sufficient in many cases, especially if qualitative analysis (trends, sensitivity) is the objective.

2.7.2 The Brush-String Model

In this section, we shall discuss the brush-string model. In this model, the brushes are attached to a flexible belt, in contrast to the preceding section where brushes were attached to a rigid ring. The brush-string model combines belt and tread compliance. Two deflections are distinguished in longitudinal and lateral directions, which we shall denote as u and v, respectively. These deflections refer to the tread deflection and the belt deflection, indicated with subscripts t and b, respectively:

$$u = u_{\mathrm{t}} + u_{\mathrm{b}}, v = v_{\mathrm{t}} + v_{\mathrm{b}} \tag{2.84}$$

Only the variation of deflections with the longitudinal coordinate x is studied here, i.e., all deflections and forces are assumed the averaged value over the width of the contact area. Conditions such as turn-slip of very large camber angles, for which this assumption is not correct, are not considered.

A schematic outline is shown in Figure 2.45. The belt is modeled as an infinite string under a tension force and is connected to the tire symmetry plane through longitudinal c_{cx} and lateral carcass stiffnesses c_{cy} per unit length. In the contact area, brush elements are attached to the belt. It is assumed that both longitudinal and lateral slip are present, with resulting shear forces F_x and F_y and aligning torque M_z. The leading edge of the adhesion part of the contact area (usually equal to the leading edge of the entire contact area) is displaced with belt deflections u_1 and v_1 in longitudinal and lateral directions, respectively. Because of the presence of carcass stiffnesses per unit length c_{cx} and c_{cy}, any part of the belt with length dx experiences resistance forces in x and y direction, equal to $c_{cx} \cdot u \cdot \mathrm{d}x$ and $c_{cy} \cdot v \cdot \mathrm{d}x$, respectively. With the definition of q_x and q_y according to Eq. (2.80) (i.e., being the longitudinal and lateral external forces on the tire per unit length) and considering the equilibrium of a part of the belt with length dx

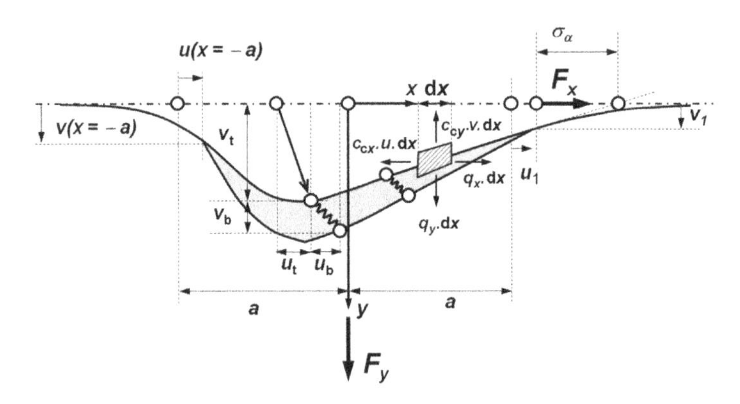

FIGURE 2.45 Schematic outline of the brush-string model.

(see Figure 2.45), one is able to derive the following differential equations for the belt defections u_b and v_b

$$S_1 \cdot \frac{d^2 u_b}{dx^2} - c_{cx} \cdot u_b = -q_x; \quad -a < x < a \tag{2.85a}$$

$$S_2 \cdot \frac{d^2 v_b}{dx^2} - c_{cy} \cdot v_b = -q_y; \quad -a < x < a \tag{2.85b}$$

where S_1 is the longitudinal elastic resistance of the tread band (Young's modulus E times cross sectional area) and S_2 is the effective tension force in the belt. We refer to Pacejka [31] and [32] and Higuchi [17] for more details about the derivation of Eqs. (2.85a) and (2.85b). Clearly, the Eqs. (2.85a) and (2.85b) are not restricted to the contact area, but hold on the entire belt, where the noncontact part is "rolled out to infinity." Considering Eq. (2.85b) for $q_y = 0$, and assuming the belt deflection to be finite for $x > a$, one finds from integration for $x > a$ that

$$v_b(x) = C \cdot e^{-x \cdot \sqrt{c_{cy}/S_2}}; \quad x > a$$

and therefore, at $x = a$

$$\frac{dv_b}{dx}(x = a) = -v_1 \cdot \sqrt{\frac{c_{cy}}{S_2}} = -v_b(x = a) \cdot \sqrt{\frac{c_{cy}}{S_2}} \tag{2.86}$$

Consider Figure 2.45, in which we indicate the distance σ_α between the leading edge of the contact area and the intersection of the line through the straight part of the contact zone (adhesion) with the wheel center plane. This distance is referred to as the relaxation length for the belt deflection in the lateral direction. Further, it is related to the distance a tire needs to travel before a significant percentage of steady-state shear force is reached following a sudden change in the slip angle. With the line through the straight part of the contact zone tangent to the deflection profile, one should have

$$\frac{dv_b}{dx}(x = a) = -\frac{v_1}{\sigma_\alpha} \tag{2.87}$$

A similar reasoning holds for longitudinal deflection, in which we introduce the relaxation length σ_κ for the belt deflection. As a result, we can rewrite Eqs. (2.85a) and (2.85b) as follows:

$$\sigma_\kappa^2 \cdot \frac{d^2 u_b}{dx^2} - u_b = -\frac{q_x}{c_{cx}}; \quad -a < x < a \tag{2.88a}$$

$$\sigma_\alpha^2 \cdot \frac{d^2 v_b}{dx^2} - v_b = -\frac{q_y}{c_{cy}}; \quad -a < x < a \tag{2.88b}$$

with boundary conditions:

$$\frac{\mathrm{d}u_\mathrm{b}(\pm a)}{\mathrm{d}x} = \mp \frac{u_\mathrm{b}(\pm a)}{\sigma_\kappa}, \quad \frac{\mathrm{d}v_\mathrm{b}(\pm a)}{\mathrm{d}x} = \mp \frac{v_\mathrm{b}(\pm a)}{\sigma_\alpha} \tag{2.89}$$

With the treads again modeled as linear massless springs, the tread deflections are proportional to the shear forces q_x and q_y, with tread stiffnesses denoted by k_x and k_y:

$$q_x = k_x \cdot u_\mathrm{t}, \quad q_y = k_y \cdot v_\mathrm{t}; \quad -a < x < a \tag{2.90}$$

Equations (2.88a), (2.88b) and (2.90) describe the relationship between belt and tread deflections and the local shear stress in the contact area. In the preceding section (which covered the brush model), we distinguished between adhesion and sliding in the contact area. Depending on the level of slip, one of the following situations will occur:

- For large slip, the full tire is sliding and no adhesion occurs within the contact patch.
- For other combinations of slip, the tire is partly sliding and partly in adhesion.

In contrast to the brush model, it has been shown by Pacejka (under pure slip conditions [31]) and extended by Higuchi [17] that the bare string model may show two sliding regions: a small leading region (a_1,a) and a trailing sliding region ($-a$, a_2), surrounding an adhesion region (a_2, a_1). Therefore, for the brush-string model, one might expect both the possibility of one and two sliding regions, depending on the model parameters (specifically the stiffnesses of carcass and treads). The breakaway points between adhesion and sliding parts of the contact area follow from the condition of continuity of shear forces. The shear forces are in general not differentiable at these points. Consequently, an explicit expression for a single breakaway point ξ_s, as derived for the brush model, does not hold for the brush-string model.

Let us consider the sliding part of the contact area in more detail, where the normal tire force per unit length q_z is defined by

$$q_z(x) = \int_{\text{tire width}} \sigma_z \cdot \mathrm{d}y$$

We shall assume the same parabolic pressure distribution as introduced in Eq. (2.75), which means that

$$q_z(x) = q_{z0} \cdot \left[1 - \left(\frac{x}{a}\right)^2\right]; \quad q_{z0} = \frac{3 \cdot F_z}{4 \cdot a} \tag{2.91}$$

This is not a real restriction, as Higuchi [17] showed by also considering higher-order behavior. Again assuming Coulomb friction with road friction μ, the shear forces in the sliding region are linearly related to q_z:

$$q_{\text{shear}} = \sqrt{q_x^2 + q_y^2} = \mu \cdot q_z \tag{2.92}$$

Consequently, there exists an angle φ (that may depend on x) that describes the orientation of the shear force components, as well as of the sliding speed components, such that under conditions of sliding, the shear forces can be parameterized by

$$q_x(x) = \mu \cdot \cos(\phi) \cdot q_z(x), \quad q_y(x) = \mu \cdot \sin(\phi) \cdot q_z(x) \tag{2.93}$$

In the adhesion region (a_2, a_1), the tire is attached to the road, which means that the sliding speed vanishes. We have seen in the preceding section that, for the brush model, the deflection behaves linearly along the contact area with the theoretical slip acting as the proportionality factor, cf. Eq. (2.74). Neglecting the tire yaw velocity, and assuming small slip angles and restricting to steady-state behavior, the sliding speed components V_{gx} and V_{gy} within the contact area can be expressed as (see Ref. [32])

$$V_{gx} = -V(1+\kappa) \cdot \left(\rho_x + \frac{\mathrm{d}u}{\mathrm{d}x}\right), \quad V_{gy} = -V(1+\kappa) \cdot \left(\rho_y + \frac{\mathrm{d}v}{\mathrm{d}x}\right) \tag{2.94}$$

for tire forward speed V, and theoretical slip values ρ_x and ρ_y, defined by Eq. (2.37). These equations express the difference of the effective tire sliding speed and the deflection speed in the contact area, building up the local sliding speed. Expressions (2.94) are referred to as the fundamental differential equations of a rolling and slipping body, and hold in general, regardless of the tire model used, with u and v the total deflections in the contact area depending on the position (x, y) within the contact area. Hence, in the adhesion region with vanishing sliding speeds V_{gx} and V_{gy}, one obtains

$$\frac{\mathrm{d}u}{\mathrm{d}x} = -\rho_x, \quad \frac{\mathrm{d}v}{\mathrm{d}x} = -\rho_y; \quad a_2 < x < a_1 \le a \tag{2.95}$$

which agrees with Eq. (2.74). In combination with Eq. (2.90), the following expressions for the belt deflections can be derived in case of adhesion:

$$u_b(x) = -\rho_x(x - a_1) + u_1 - \frac{q_x}{k_x}; \quad a_2 < x < a_1 \le a \tag{2.96a}$$

$$v_b(x) = -\rho_y(x - a_1) + v_1 - \frac{q_y}{k_y}; \quad a_2 < x < a_1 \le a \tag{2.96b}$$

Substituting these expressions in the belt Eqs. (2.88a) and (2.88b), one finds the following equations for q_x and q_y:

$$\sigma_{t\kappa}^2 \cdot \frac{d^2 q_x}{dx^2} - q_x = -\varepsilon_x^2 \cdot k_x \cdot [u_1 - \rho_x \cdot (x - a_1)]; \quad a_2 < x < a_1 \leq a$$

$$\sigma_{t\alpha}^2 \cdot \frac{d^2 q_y}{dx^2} - q_y = -\varepsilon_y^2 \cdot k_y \cdot \left[v_1 - \rho_y.(x - a_1)\right]; a_2 < x < a_1 \leq a$$

with

$$\varepsilon_x^2 = \frac{c_{cx}}{c_{cx} + k_x}, \quad \varepsilon_y^2 = \frac{c_{cy}}{c_{cy} + k_y}$$

$$\sigma_{t\kappa} = \varepsilon_x \cdot \sigma_\kappa, \quad \sigma_{t\alpha} = \varepsilon_y \cdot \sigma_\alpha$$

The general solutions of these equations are given by

$$q_x(x) = A_x \cdot e^{x/\sigma_{t\kappa}} + B_x \cdot e^{-x/\sigma_{t\kappa}} + \varepsilon_x^2 \cdot k_x \cdot [u_1 - \rho_x \cdot (x - a_1)]; a_2 < x < a_1 \leq a$$

$$q_y(x) = A_y \cdot e^{x/\sigma_{t\alpha}} + B_y \cdot e^{-x/\sigma_{t\alpha}} + \varepsilon_y^2 \cdot k_y \cdot [v_1 - \rho_y \cdot (x - a_1)]; a_2 < x < a_1 \leq a$$

with integration constants A_x, B_x, A_y, B_y that, in combination with Eqs. (2.96a) and (2.96b), result in explicit expressions for the belt deflection in case of adhesion. In case of sliding, Eq. (2.93) holds and the solution of Eqs. (2.88a) and (2.88b) can be written as

$$u_b(x) = C_x \cdot e^{x/\sigma_\kappa} + D_x \cdot e^{-x/\sigma_\kappa} + \frac{Q \cdot \cos(\phi)}{c_{cx} \cdot a^2} \cdot (a^2 - 2 \cdot \sigma_\kappa^2 - x^2) \quad (2.97a)$$

$$v_b(x) = C_y \cdot e^{x/\sigma_\alpha} + D_y \cdot e^{-x/\sigma_\alpha} + \frac{Q \cdot \sin(\phi)}{c_{cy} \cdot a^2} \cdot (a^2 - 2 \cdot \sigma_\alpha^2 - x^2) \quad (2.97b)$$

with integration constants C_x, D_x, C_y, D_y. In case of two sliding regions, one must distinguish between φ_1 and φ_2 for the leading and trailing sliding region. In that case, four more integration constants must be introduced (and eliminated from boundary conditions at $x = -a$, a_2, a_1, and a). In conclusion, assuming one sliding region (i.e., $a_1 = a$), we derived explicit expressions for the belt deflections and shear forces in the adhesion and sliding parts of the contact area, involving unknowns A_x, B_x, A_y, B_y, C_x, D_x, C_y, D_y, u_1, v_1, φ, and a_2. These unknowns are found from the mixed boundary conditions Eq. (2.89), vanishing shear force at $x = a$, continuity of the shear forces and the belt deflections at $x = a_2$, and differentiability of the belt deflections at $x = a_2$. In case of $a_1 \neq a$, similar continuity and differentiability conditions hold at $x = a_1$, and we demand differentiability of the shear forces at $x = a_1$.

TABLE 2.5 Parameters for the Brush-String Model Calculations

Parameter	Large σ	Small σ	Moderate σ
a [m]	0.05	0.05	0.05
c_{cx} [N/m^2]	2.0×10^5	2.0×10^5	2.0×10^5
c_{cy} [N/m^2]	1.0×10^5	1.0×10^5	1.0×10^5
F_z [N]	300×9.81	300×9.81	300×9.81
μ	1	1	1
σ_κ [m]	0.3	0.01	0.15
σ_α [m]	0.5	0.02	0.25

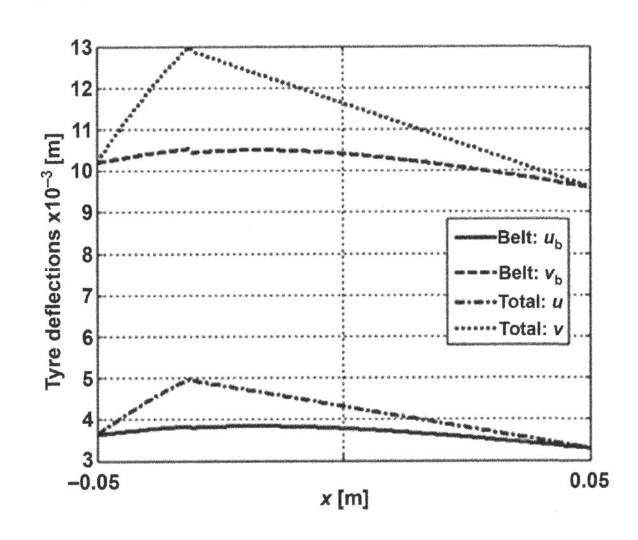

FIGURE 2.46 Belt and total tire deflections for large σ, $(\kappa,\alpha) = (0.02, 0.04)$, and $k_x = k_y = 1.0 \times 10^7$ [N/m^2].

We first examine the output for two sets of parameters for small and large relaxation lengths, respectively, in accordance with Refs. [17] and [36]. We included one more dataset for moderate relaxation lengths. The data are listed in Table 2.5. The sliding speed is proportional to the tire speed V, see Eq. (2.94). The tire speed V is chosen as 10 [km/h].

The first results are obtained for the large values of σ_κ and σ_α for tread stiffnesses k_x, $k_y = 1.0 \times 10^7$ [N/m^2], and for slip values $\kappa = 0.02$ and $\alpha = 0.04$.

Figures 2.46–2.48 show the belt and total tire deflections, shear forces, and sliding speeds. Observe the smooth transition in the belt deflections and the sharp transition in the total tire deflections at the breakaway point near $x = -0.031$, due to the attached brushes. Tread deflections vanish at the edges of the contact area, but the belt deflections do not. Clearly, with

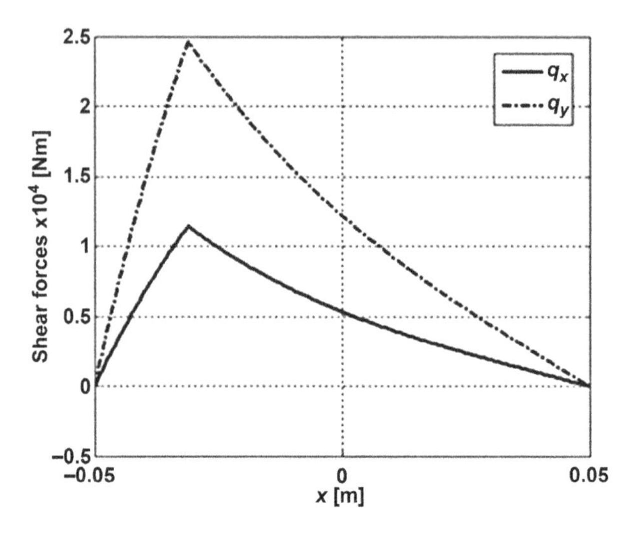

FIGURE 2.47 Shear forces for large σ, $(\kappa,\alpha)=(0.02,\ 0.04)$ and $k_x=k_y=1.0\times 10^7$ [N/m^2].

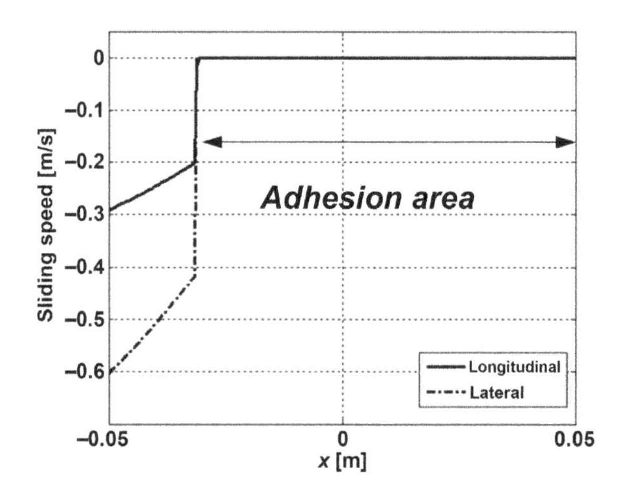

FIGURE 2.48 Sliding speeds for large σ, $(\kappa,\alpha)=(0.02,\ 0.04)$ and $k_x=k_y=1.0\times 10^7$ [N/m^2].

$|\alpha|>|\kappa|$, the lateral belt deflections and local shear force are expected to exceed their longitudinal counterparts. In the adhesion part, the total tire deflection is linear in the x position and the sliding speeds are zero. In the sliding part, the shear forces follow the normal contact stress cf. Eq. (2.93), with the value for φ found as 65 [°].

Figures 2.49–2.51 show the belt and total tire deflections, shear forces, and sliding speeds for small values of the relaxation lengths σ_κ and σ_α for the same tread stiffness and practical slip values.

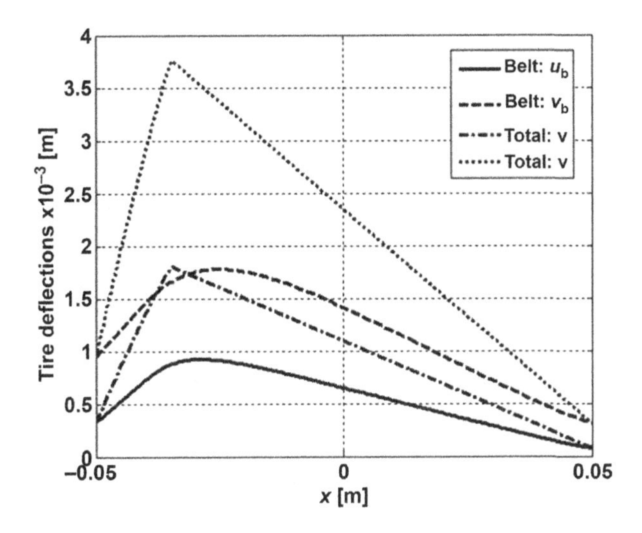

FIGURE 2.49 Belt and total tire deflections for small σ, $(\kappa,\alpha) = (0.02, 0.04)$ and $k_x = k_y = 1.0 \times 10^7$ [N/m^2].

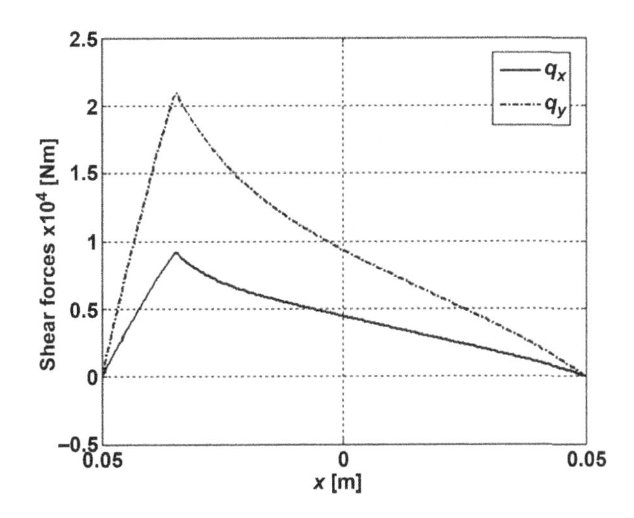

FIGURE 2.50 Shear forces for small σ, $(\kappa,\alpha) = (0.02, 0.04)$ and $k_x = k_y = 1.0 \times 10^7$ [N/m^2].

The sliding part has slightly decreased in size compared to the large σ-case. The tire deflections are reduced and the relative contribution of the belt deflections in the total deflections has been reduced as well.

Consequently, the tire deflections only occur near the contact area, whereas for large σ, the deflections extend over a large part of the tire circumference. In mathematical terms, this is obvious because the exponential part of the solution for the belt Eqs. (2.88a) and (2.88b) is reduced with smaller relaxation lengths. In physical terms, smaller values of relaxation lengths

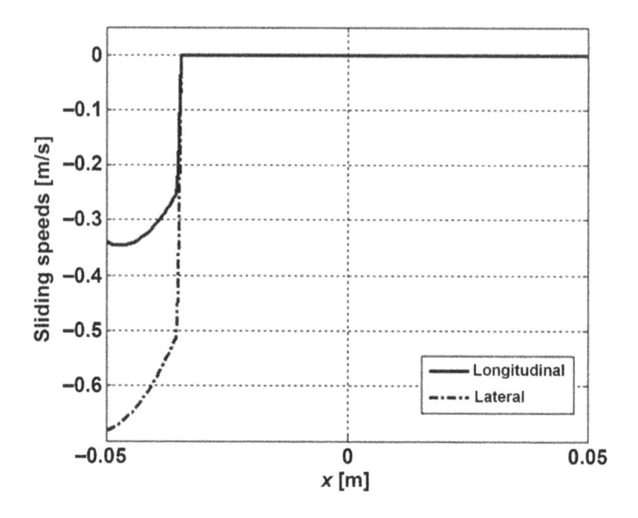

FIGURE 2.51 Sliding speeds for small σ, $(\kappa,\alpha) = (0.02,\ 0.04)$ and $k_x = k_y = 1.0 \times 10^7$ [N/m^2].

correspond to larger values of the carcass stiffnesses c_{cx} and c_{cy}, and therefore, smaller belt deflections. The sliding speeds do not change much, except for exhibiting a more curved behavior.

Next, we will vary the tread stiffnesses k_x and k_y. Note that the shear forces in the adhesion area depend on the parameters ε_x and ε_y, varying between 0 (for large tread stiffness) and 1 (for small tread stiffness). That means that, for large tread stiffnesses, the effective relaxation lengths in the expressions for the shear forces in case of adhesion are quite small, and the third term becomes proportional to c_{cx} and c_{cy} for q_x and q_y, respectively. Therefore, one should expect the solution to be dominated by the belt. For small tread stiffnesses, this third term becomes proportional to this tread stiffness and a more brush-type behavior is expected.

The lateral belt deflection v and the lateral shear force q_y are shown in Figures 2.52 and 2.53 for large relaxation lengths σ_κ and σ_α and for different values of tread stiffness $k_x = k_y$, as listed in Table 2.6. The corresponding values for small relaxation lengths are depicted in Figures 2.54 and 2.55.

Table 2.6 also includes the edge values of the adhesion part of the contact area, which indicate a front sliding region for small relaxation values and high tread stiffness (as clearly shown in Figure 2.55). Indeed, with increasing tread stiffness, the tire behavior becomes more belt-like and the results correspond well with those obtained by Higuchi [17], whereas small tread stiffness typically leads to the sharp transition between adhesion and sliding, corresponding to a brush-type response. Comparing Figures 2.50–2.55, one may conclude that softer treads lead to an enlarged adhesion region, and therefore, an increased cornering and braking potential

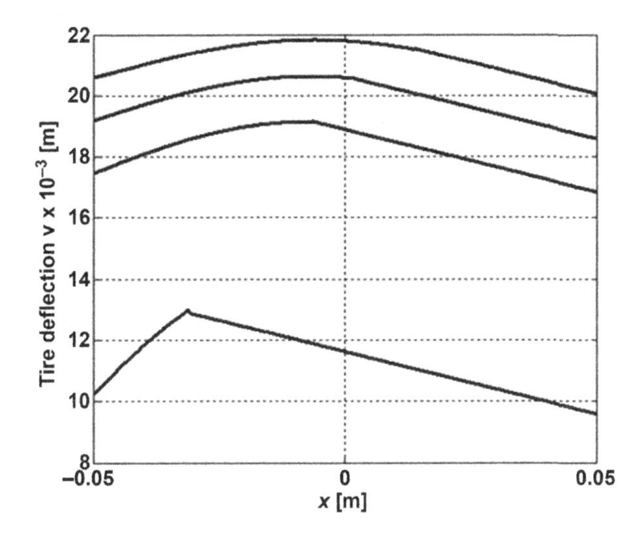

FIGURE 2.52 Total tire deflection v for large σ, $(\kappa,\alpha) = (0.02, 0.04)$ and $k_x = k_y$ according to Table 2.6.

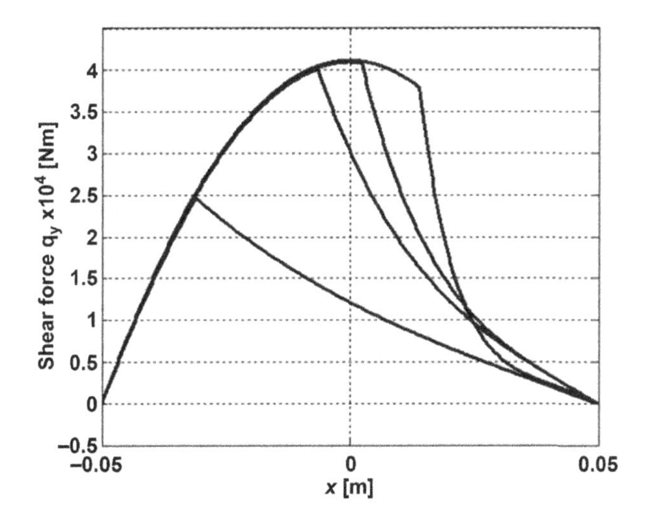

FIGURE 2.53 Shear force q_y for large σ, $(\kappa,\alpha) = (0.02, 0.04)$ and $k_x = k_y$ according to Table 2.6.

of the tire (e.g., winter tires versus summer tires). In the limit of belt-type tire behavior, a discontinuous behavior arises in the local shear forces, which was also observed by Higuchi [17]. A larger relaxation value results in a smaller adhesion region and a sharper decrease of the local shear force in the adhesion region to increase tread stiffness.

We close this section with an investigation of the local tire deflections and shear forces for varying slip. We selected a fixed value of $\alpha = 0.025$ and have varied κ from 0.02 to 0.16. The tire parameters were chosen according

TABLE 2.6 Varying Tread Stiffness

$k_x = k_y$ [N/m^2]	Large σ		Small σ	
	a_2 [m]	a_1 [m]	a_2 [m]	a_1 [m]
5.0×10^6			−0.0402	0.05
1.0×10^7	−0.3100	0.05	−0.0346	0.05
5.0×10^7	−0.0064	0.05	−0.0222	0.05
1.0×10^8	0.0024	0.05	−0.0187	0.05
5.0×10^8	0.0140	0.05	−0.0143	0.0490
1.0×10^9			−0.0134	0.0483

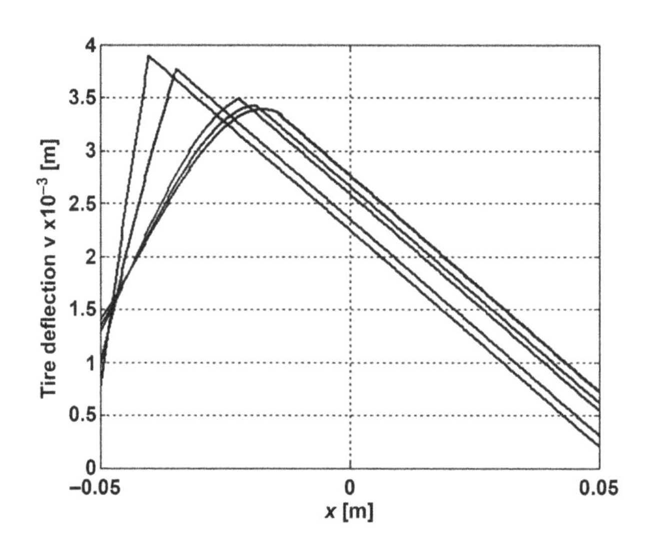

FIGURE 2.54 Total tire deflection (v) for small σ, $(\kappa,\alpha) = (0.02, 0.04)$, and $k_x = k_y$, according to Table 2.6.

to the last column of Table 2.5 (moderate σ) and $k_x = k_y = 1.0 \times 10^7$ [N/m^2]. Results are given in terms of total tire deflections u and v (Figure 2.56), local shear forces (Figure 2.57), and sliding speeds (Figure 2.58).

Starting with the tire deflections, one observes an increase of u and a decrease of v with increasing κ, as expected. The slope of u versus x in the adhesion region increases with κ as well, whereas the slope of the lateral tire defection v versus x decreases. The local shear force q_x increases with κ, where we note that the local stress orientation angle φ in the rear sliding region will decrease, which has an effect on q_x and q_y. It is clear from Figure 2.55 that the total longitudinal tire shear force F_x (the local force q_x integrated over the contact area) will saturate, just as we found for the brush model. The lateral tire shear force F_y decreases at the same time, eventually vanishing at very large κ.

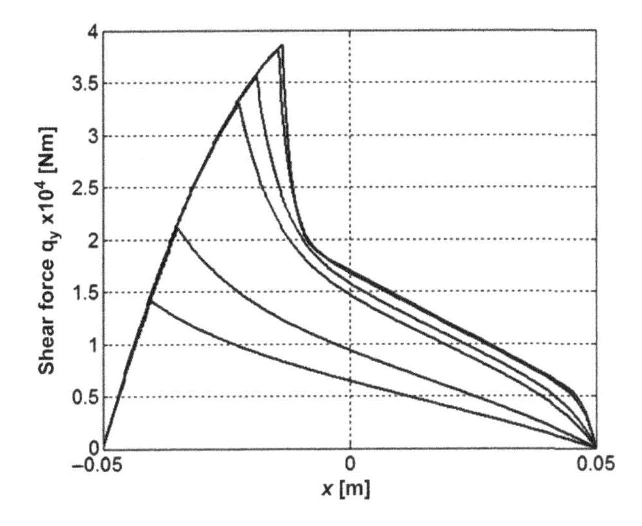

FIGURE 2.55 Shear force q_y for small σ, $(\kappa,\alpha) = (0.02, 0.04)$, and $k_x = k_y$, according to Table 2.6.

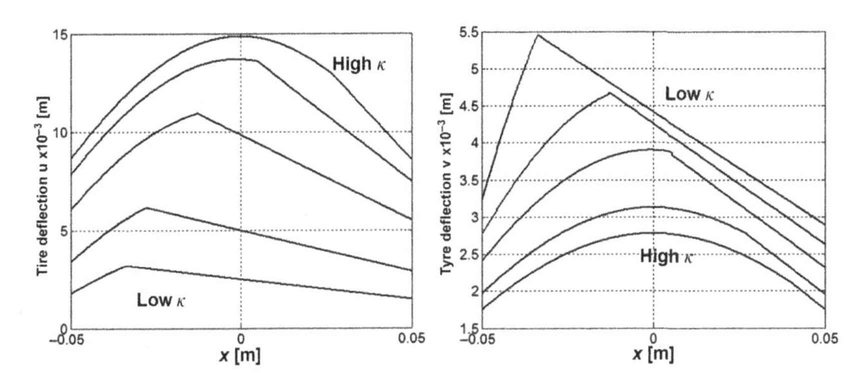

FIGURE 2.56 Total tire deflections u and v for $\alpha = 0.025$ and varying slip (κ) from 0.02 to 0.16 (for moderate relaxation lengths, $k_x = k_y = 1.0 \times 10^7$ [N/m^2]).

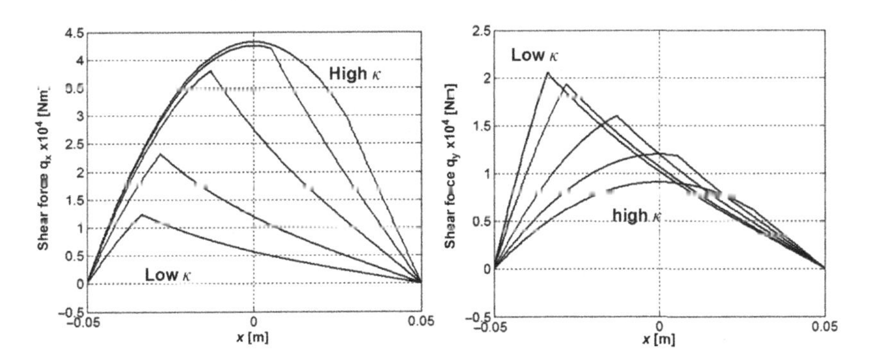

FIGURE 2.57 Local shear forces q_x and q_y for $\alpha = 0.025$ and varying slip (κ) from 0.02 to 0.16 (for moderate relaxation lengths, $k_x = k_y = 1.0 \times 10^7$ [N/m^2]).

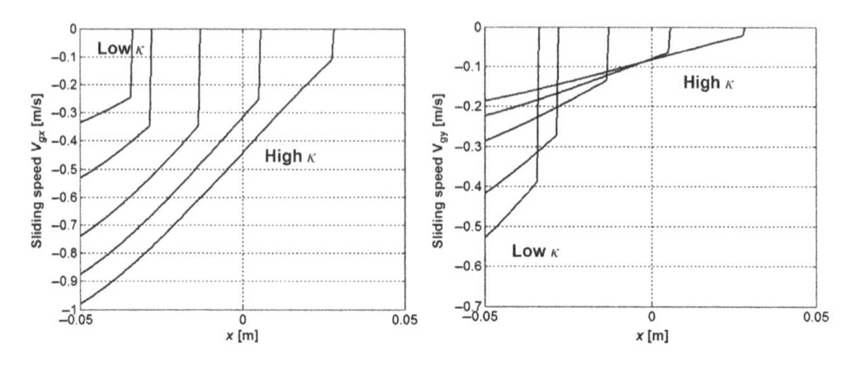

FIGURE 2.58 Sliding speeds V_{gx} and V_{gy} for $\alpha = 0.025$ and varying slip (κ) from 0.02 to 0.16 (for moderate relaxation lengths, $k_x = k_y = 1.0 \times 10^7$ [N/m^2]).

Figure 2.58 shows an increase of the sliding region, as well as an increase of the sliding speeds in this region in longitudinal direction. At the same time, the lateral sliding speeds are reduced. In the total sliding limit, the tire will follow the resulting shear force. According to Figure 2.58, the lateral contribution in this sliding speed is expected to correspond to an almost constant local lateral sliding speed along the contact area. In longitudinal direction, the local longitudinal rubber sliding is expected to vary linearly along this contact area.

Remark

Consider pure lateral slip conditions and no treads (i.e., only belt behavior). Assuming there is no sliding at the leading edge of the contact area, the initial slope of the deflection near $x = a$ satisfies $dv_b/dx\ (x = a) = -\rho_y$, according to Eq. (2.96b). That means that $v_1\ (x = a) = \sigma_\alpha \cdot \rho_y$, according to Eq. (2.87).

With a lateral force F_y acting on the tire, a lateral defection arises that, as shown in Figures 2.46 and 2.52, doesn't change much along the contact area in case of large σ_y. Let us define the lateral spring stiffness of the tire as

$$C_{Fy,\xi} \equiv \frac{F_y}{v(x = \xi)},$$

which will also not vary much along the contact area. Taking $\xi = 0$, we can approximate the ratio of cornering stiffness and slip stiffness as follows:

$$\frac{C_\alpha}{C_{Fy,0}} \approx \frac{F_y/\rho_y}{F_y/\rho_y.\sigma_\alpha} = \sigma_\alpha \tag{2.98}$$

We conclude that the relaxation length σ_α can be interpreted as the ratio of the tire cornering stiffness and the tire spring stiffness. We will use this in Chapter 3, where we treat tire transient behavior.

Chapter | Three

Nonsteady-State Tire Behavior

In this chapter, the nonsteady-state behavior of a tire is discussed. For fast maneuvering of the tire, the behavior cannot be assumed to be steady state. If the slip angle changes instantaneously (a step input), the effective slip angle at the contact between tire and ground changes from a zero value to the input value within a nonzero time, due to the compliance of the tire carcass, as the carcass is moving away from the wheel plane. This effective slip angle corresponds to shear stress, which means that the tire shear force needs time to build up as well. A similar phenomenon is observed for a tire under a sudden brake torque, where the local brake slip in the contact area, and therefore the contact force between tire and road, needs time to build up because of the rotational compliance of the tire. This delay effect is known as *transient tire behavior*, which is discussed in Section 3.1.

In addition to these transient effects, a tire can be considered as a belt with finite mass, which is being suspended to the rim by the sidewalls. As a result, the belt may move separately from the rim, with the sidewall stiffnesses (radial, tangential, camber, etc.) and damping values determining the eigenfrequencies of this motion and the sensitivity of the belt behavior for these eigenfrequencies. These inertia effects have to be accounted for in case of relatively short and sharp road undulations with a single step as an obvious example, when control algorithms interfere with the tire dynamic behavior such as in case of Active Brake Control and Electronic Stability Control, and in case of intermediate frequency ride and comfort studies when the longitudinal contact force has to be considered. This dynamic tire behavior is treated in Section 3.2.

3.1 TIRE TRANSIENT BEHAVIOR

3.1.1 The Tire Transient Model

A tire responding to a sudden change of the slip angle α is schematically shown in Figure 3.1. The carcass is modeled as a simple spring with stiffness

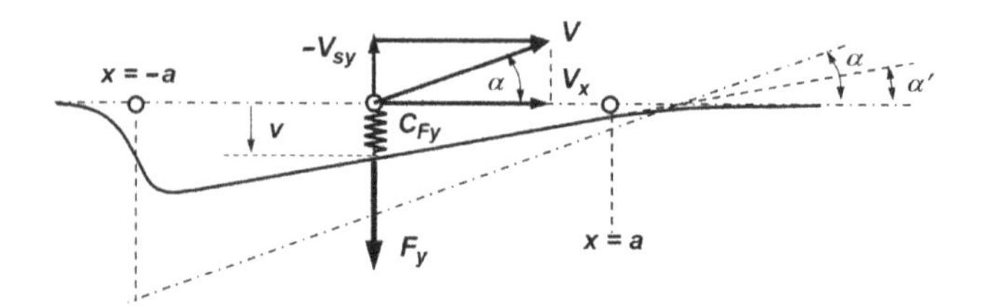

FIGURE 3.1 Tire responding to sudden change in slip angle.

C_{Fy}, linking the tire at the center of the contact area to the undeformed wheel plane. The slip angle at the wheel center (the axle) is denoted by α, and the effective slip angle at the contact area is denoted by α'.

The deflection of the tire at ground level is denoted by v, pointing in negative y direction. The lateral speed of the tire at this point is given by:

$$\dot{y} = V_{sy} + \dot{v} \tag{3.1}$$

Dividing this expression by constant forward speed (V) and assuming small slip, one may derive

$$\frac{F_y}{C_\alpha} = \alpha' = -\frac{\dot{y}}{V} = -\frac{V_{sy} + \dot{v}}{V} = \alpha - \frac{\dot{F}_y}{V \cdot C_{Fy}}$$

for cornering stiffness C_α. Replacing F_y with $C_\alpha \cdot \alpha$ and multiplying the equation by V, this relationship can be rewritten as:

$$\frac{C_\alpha}{C_{Fy}} \dot{\alpha}' + V \cdot \alpha' = V \cdot \alpha \tag{3.2}$$

We conclude that the effective slip angle α' under transient conditions follows from a first-order lag equation with lag (relaxation) time

$$\tau_{lag} = \frac{1}{V} \cdot \frac{C_\alpha}{C_{Fy}} \equiv \frac{\sigma_\alpha}{V} \tag{3.3}$$

with σ_α being the corresponding relaxation distance, which is also referred to as the lateral relaxation length. Compare this expression with Eq. (2.98).

In case of large slip α, we replace Eq. (3.3) with

$$\sigma_\alpha(\alpha') = \frac{(\partial F_y / \partial \alpha)(\alpha')}{C_{Fy}} \tag{3.4}$$

meaning that Eq. (3.2) now becomes nonlinear in α'

$$\sigma_\alpha(\alpha') \cdot \dot{\alpha}' + V \cdot \alpha' = V \cdot \alpha \tag{3.5}$$

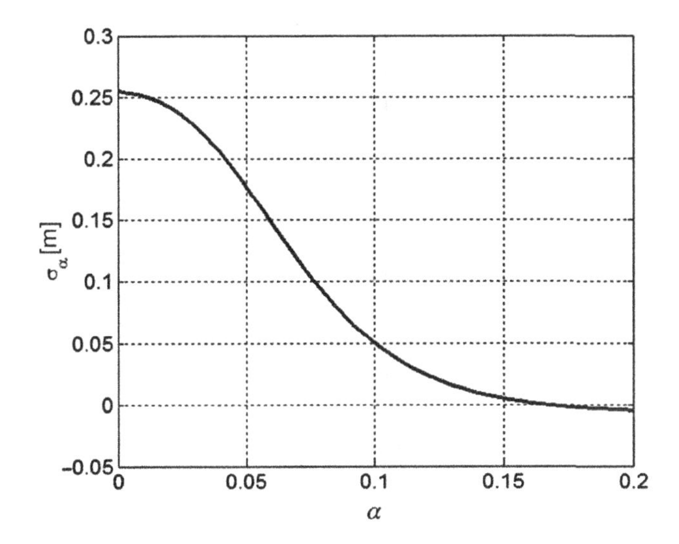

FIGURE 3.2 Function σ_α versus slip angle based on F_y cf. Figure 2.21 and for $C_{Fy} = 2 \times 10^5$ [N/m].

We determined the relaxation function σ_α from the cornering force, as depicted in Figure 2.21 for $F_z = 4000$ [N]. The lateral spring stiffness has been chosen as 2×10^5 [N/m^2] and the result is shown in Figure 3.2. As long as the slip angle (α) does not exceed 0.03, one might take σ_α as constant. Beyond that value, one should take the dependency on α into account. Due to the lower value of σ_α, the tire will respond quicker to sharp changes in slip angle when it is closer to extreme slip conditions (Figure 3.2).

Transient tire behavior is considered in practice as follows. In case of a rapidly changing slip angle α at the wheel axle, Eq. (3.2) is solved to obtain α'. The slip angle α is directly related to the vehicle states (yaw rate and lateral speed), cf. Eq. (2.40). The tire model (e.g., the empirical Magic Formula model) is then used to determine F_y from α', which is then used in the vehicle equilibrium equations (such as Eqs. (2.38) and (2.39)). In case of large lateral slip, one must include a feedback loop based on the relationship (3.4). We have illustrated this process in Figure 3.3.

A similar relationship can be derived for longitudinal slip response in the contact area. Consider a wheel under varying longitudinal slip κ and a resulting longitudinal force F_x between tire and road, as shown in Figure 3.4. Due to the sidewall compliance, the rotation angles of tire and wheel axle will not be the same. As a result, the slip as applied at the axle (denoted as κ) will differ from the slip based on the rotation of the tire (denoted as κ').

We indicated the difference in angle with θ. If we write the circumferential stiffness of the tire as $C_{Fx} \cdot R_l^2$, with loaded radius R_l, then the longitudinal force F_x is proportional to $R_l \cdot \theta$ (which is the longitudinal displacement

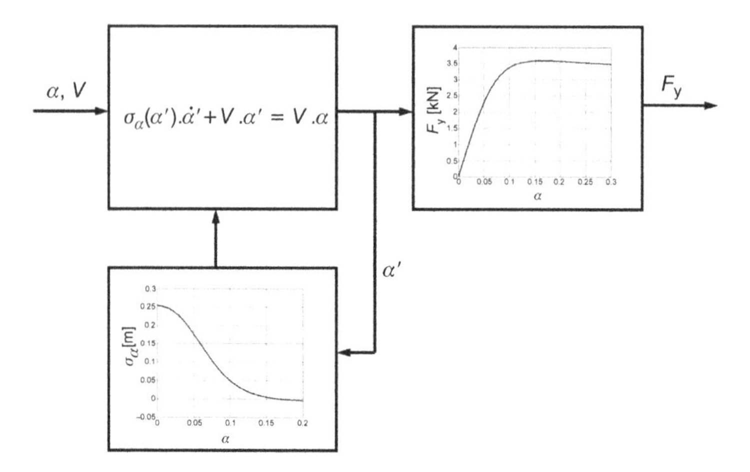

FIGURE 3.3 From slip angle to lateral force under transient conditions, schematic layout.

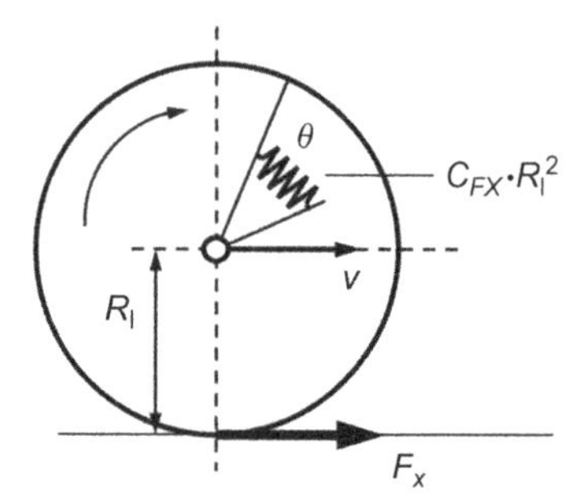

FIGURE 3.4 Tire, responding to sudden change in longitudinal slip.

of the tire in the contact area, with respect to the axle position). Under linear tire conditions, the longitudinal force is also proportional to the ratio of slip speed and forward speed. One must account for nonstationary rotation angle θ, and therefore, a different rotational speed at the contact between tire and road compared to the rotation speed at the axle. As a result, we can write for F_x

$$F_x = C_{Fx} \cdot R_l \cdot \theta \approx C_{Fx} \cdot R_e \cdot \theta \tag{3.6}$$

with C_{Fx} as the proportionality factor and R_e as the effective tire radius under free rolling. We can also write

$$F_x = -C_\kappa \cdot \frac{R_e.\dot{\theta} + V_{sx}}{V} = -C_\kappa \cdot \frac{R_e}{V} \cdot \dot{\theta} + C_\kappa \cdot \kappa \tag{3.7}$$

with C_κ as the longitudinal slip stiffness and V_{sx} as the slip speed (taken at the wheel axle). Combining expressions (3.6) and (3.7), and writing (assuming small slip),

$$\kappa' = \frac{F_x}{C_\kappa}$$

one finds

$$\frac{C_\kappa}{C_{Fx}} \dot{\kappa}' + V \cdot \kappa' = V \cdot \kappa \tag{3.8}$$

which is the counterpart of expression (3.2) in the longitudinal direction. This means we can substitute the relaxation length σ_κ in Eq. (3.8), which is defined similar to Eq. (3.3):

$$\sigma_\kappa = \frac{C_\kappa}{C_{Fx}} \tag{3.9}$$

yielding

$$\sigma_\kappa \cdot \dot{\kappa}' + V \cdot \kappa' = V \cdot \kappa \tag{3.10}$$

For large slip, the relaxation length should be taken as a function of κ', similar to the situation for large lateral slip. The longitudinal spring stiffness C_{Fx} is, in general, larger than the lateral spring stiffness. On the other hand, the longitudinal slip stiffness exceeds the lateral slip stiffness as well. For the tire data used in the preceding sections, the relaxation length σ_κ is slightly smaller than its lateral counterpart and falls off more rapidly with increasing slip.

Remarks

1. The practical longitudinal slip is defined in expression (2.19) as the slip speed divided by the forward speed. That means that small speeds, or even zero speed, will lead to numerical problems (dividing by zero) in case of steady-state slip conditions. This problem can be handled by assuming transient tire behavior and applying Eq. (3.10), with the right-hand side being the slip speed at the axle. This equation does not suffer from a situation with small speed. A similar remark holds for lateral slip with small speed.
2. As previously described, there are two mechanisms resisting tire deflection, (1) the tire spring resistance and (2) the slip stiffness. The first mechanism acts, by definition, as a spring. The second mechanism, slip stiffness, means that the longitudinal tire force is proportional to the rotational deflection speed (see Eq. (3.7)). Consequently, this mechanism can be assumed to act

as a damper. In conclusion, the tire–road contact can be considered as a spring and damper in series. This observation was previously made by Pacejka and we refer reader to Ref. [32] for further details.

3.1.2 Applications of the Tire Transient Model

In this section, we shall address applications of the tire transient model in either lateral or longitudinal direction. These cases are:

1. Shimmy of a trailing wheel
2. A single wheel vehicle under repetitive braking

SHIMMY OF A TRAILING WHEEL

This problem has been investigated extensively by Besselink and Pacejka, see [2], [31], and [32]. A schematic layout of a trailing wheel system is shown in Figure 3.5, with caster length e and a constant speed V, with the rotational motion of the swivel axis restricted by a torsional spring with stiffness c and a torsional damper with damping value k. Under certain conditions, such a system may show extreme oscillations, which is known as shimmy. Shimmy can lead to significant wear on tire and wheel alignment, and can be very dangerous if it occurs, for example, during landing of an airplane. In comparison to the research previously mentioned, we neglect the aligning torque and describe the lateral motion of the wheel using the lateral single-point transient model (expression (3.2)). The yaw inertia J_{w} of the trailing arm system around the kingpin will, in general, include a contribution that is proportional to e^2; however, we will treat e and J_{w} independently.

The equation of equilibrium for the trailing wheel system can be written as follows:

$$J_w \cdot \ddot{\psi} + k \cdot \dot{\psi} + c \cdot \psi = -e \cdot F_y$$

We assume a small swivel angle and therefore, small slip. Consequently, we have a linear relationship between F_y and the slip angle α' at ground level. Hence,

$$J_{\mathrm{w}} \cdot \ddot{\psi} + k \cdot \dot{\psi} + c \cdot \psi = -e \cdot C_\alpha \cdot \alpha' \tag{3.11}$$

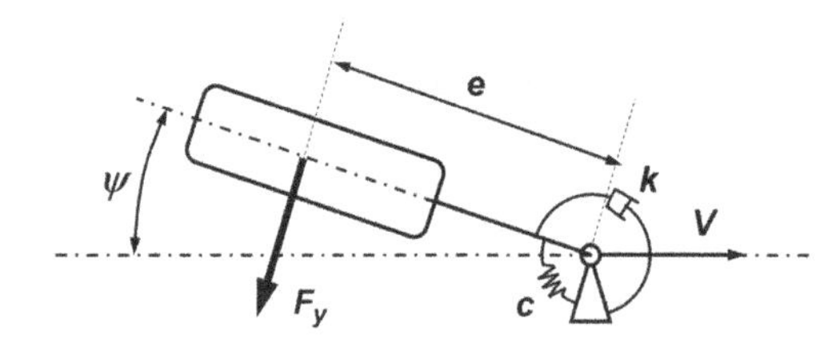

FIGURE 3.5 A trailing wheel system.

for cornering stiffness C_α. The slip angle α at axle level equals the ratio of local lateral speed and forward speed, with the local lateral speed following from rotation and forward speed

$$\alpha = \frac{e \cdot \dot{\psi} + V \cdot \sin \psi}{V} \approx \frac{e \cdot \dot{\psi}}{V} + \psi \tag{3.12}$$

As a result, and using Eq. (3.2), one arrives at the following equation:

$$\sigma_\alpha \dot{\alpha}' + V \cdot \alpha' = e \cdot \dot{\psi} + V \cdot \psi \tag{3.13}$$

with relaxation length σ_α. Equations (3.11) and (3.12) form a homogenous system, the characteristic equation of which can be written as:

$$\begin{vmatrix} J_\mathrm{w} \cdot \lambda^2 + k.\lambda + c & e \cdot C_\alpha \\ e \cdot \lambda + V & -(\sigma_\alpha \cdot \lambda + V) \end{vmatrix} = 0 \tag{3.14}$$

or

$$\begin{aligned} &J_\mathrm{w} \cdot \sigma_\alpha \cdot \lambda^3 + (J_\mathrm{w} \cdot V + k \cdot \sigma_\alpha) \cdot \lambda^2 + (k \cdot V + c \cdot \sigma_\alpha + e^2 \cdot C_\alpha) \cdot \lambda \\ &+ c \cdot V + e \cdot V \cdot C_\alpha = 0 \end{aligned} \tag{3.15}$$

for eigenvalue λ, of which the real part should be negative to guarantee stability, i.e., to avoid shimmy. The conditions for stability are based on the Routh–Hurwitz criterion, stating that stability is preserved in case of a third-order characteristic equation:

$$p_0 \cdot \lambda^3 + p_1 \cdot \lambda^2 + p_2 \cdot \lambda + p_3 = 0$$

if, and only if, the following conditions hold

Condition (i):
All coefficients are positive: $p_i > 0$ for $i = 0, 1, 2, 3$

Condition (ii):
The so-called Hurwitz determinants should be positive, which means for a third-order equation that

$$p_1 \cdot p_2 - p_0 \cdot p_3 > 0$$

The nontrivial conditions, resulting from Eq. (3.15), are:

(i) $c \cdot V + e \cdot V \cdot C_\alpha > 0$
(ii) $(J_\mathrm{w} \cdot V + k \cdot \sigma_\alpha) \cdot (k \cdot V + c \cdot \sigma_\alpha + e^2 \cdot C_\alpha) - J_\mathrm{w} \cdot V \cdot \sigma_\alpha \cdot (c + e \cdot C_\alpha) > 0$

The first condition indicates that, in order to avoid instability, the caster length should exceed $-c/C_\alpha$. In case of no yaw stiffness, e should be positive. A negative caster length is allowed for nonzero c if either c is large enough or if the cornering stiffness is not too large. Instability will then be of the nonoscillatory divergent type.

The more interesting case is when condition (ii) is violated. Let us choose the case of $c = 0$ and a very small speed V. Then, condition (ii) is reduced to

$$\text{(ii)}_{c=0,\ V\downarrow 0} \quad e^2 > 0$$

which is satisfied as long as $e \neq 0$, meaning that stability is secured. A second special case is when $c = 0$ and k is small, for which condition (ii) becomes

$$\text{(ii)}_{c=0,\ k\downarrow 0} \quad e^2 - \sigma_\alpha . e > 0$$

Hence, if $k \downarrow 0$, then e should exceed σ_α. Clearly, a small relaxation length will improve stability. It is easily shown that $e > \sigma_\alpha$ is a sufficient condition for stability for arbitrary values of c, k, and V. The condition (ii) is fulfilled if the following condition holds

$$J_W \cdot V \cdot (c \cdot \sigma_\alpha + e^2 \cdot C_\alpha) - J_W \cdot V \cdot \sigma_\alpha \cdot (c + e \cdot C_\alpha) > 0$$

which is identical to the condition for $c = 0$ and $k = 0$. Therefore, we will restrict ourselves to values of the caster length e less than σ_α.

Next, let us consider the general case for the parameters as listed in Table 3.1.

Results are shown in Figure 3.6 for damping value (k) varying from 0 to 60 [Nms] and for two values of c as listed in Table 3.1. One observes areas of instability between the lines $e = 0$ and $e = \sigma_\alpha = 0.25$. Increasing the damping reduces this area and therefore improves the stability (avoiding shimmy). This type of instability is oscillatory. The plots indicate a stable behavior for large speeds and positive caster lengths, where the minimal speed depends on

TABLE 3.1 Parameters for Shimmy Analysis

Parameter	Value Range
J_W [kgm^2]	5.0
c [Nm]	0.0, 500.0
k [Nms]	0 (10) 60
σ_α [m]	0.25
C_α [N]	5×10^4
V [m/s]	0...30
e [m]	0...σ_α

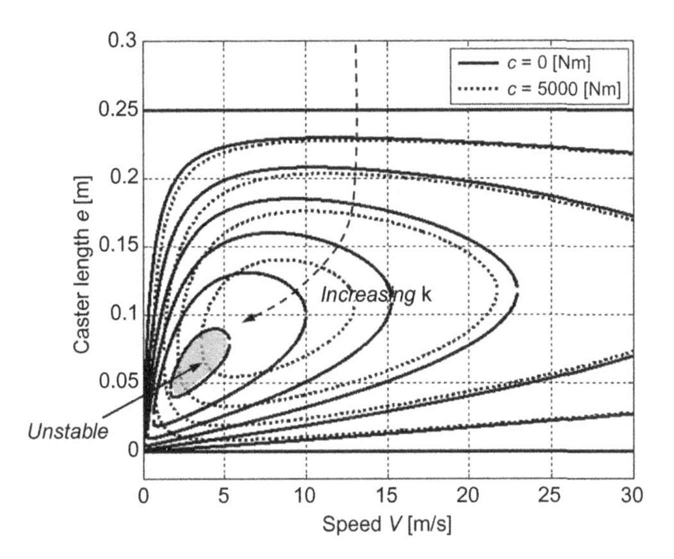

FIGURE 3.6 Areas of (oscillatory) loss of stability for the trailing wheel system, for varying rotational damping value (k), and for two values of rotational stiffness (c).

the damping value k. A zero speed corresponds to stability, as long as $k>0$. If we increase the stiffness c, the instability areas reduce in size, which means that the system will become more stable. This is especially true for large damping.

In case of oscillatory loss of stability, the solution of the characteristic equation will be purely imaginary, i.e., it will be equal in size to the radial frequency ω. We write $\lambda = i \cdot \omega_x / V$, which means that we consider the path frequency ω_x (oscillations along the traveled distance). Substitution in the characteristic equation provides

$$\omega_x^2 = \frac{p_3}{p_1} = \frac{e \cdot C_\alpha}{V \cdot (J_\mathrm{w} \cdot V + k \cdot \sigma_\alpha)} \tag{3.16}$$

That means that a larger caster length will correspond to a larger path frequency. If we account for dependency of J_w on the caster length e, the same conclusion holds for not too large values of e. Starting with a fixed caster length, the path frequency will decrease with increasing speed.

SINGLE WHEEL VEHICLE UNDER REPETITIVE BRAKING

Consider a single wheel system, as shown in Figure 3.7. A mass m is attached to a wheel, moving with speed V and rotational speed Ω. When a brake moment M_b is applied, a slip κ' arises and a corresponding contact force F_x. It is assumed that the slip is small enough to consider the relationship between F_x and κ' as linear.

We use a repetitive brake torque, which means that M_b will be active for 2 s, followed by an inactive period of the same duration, after which the brakes are activated again and this pattern continues. As a result, the

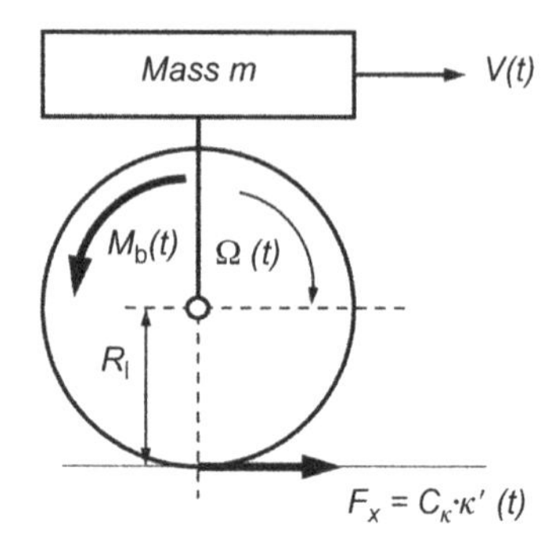

FIGURE 3.7 Single wheel vehicle.

single wheel system will slow down in separate steps, with sharp transitions between the active and inactive parts of the braking process. This problem is described by three equations:

i. *equilibrium of vehicle in forward direction*

$$m \cdot \dot{V} = C_\kappa \cdot \kappa' \tag{3.17}$$

ii. *equilibrium of wheel in rotational direction*

$$J_{\text{wheel}} \cdot \dot{\Omega} = -R_l \cdot C_\kappa \cdot \kappa' - M_b \approx -R_e \cdot C_\kappa \cdot \kappa' - M_b \tag{3.18}$$

iii. *transient tire behavior*

$$\sigma_\kappa \cdot \dot{\kappa}' + V \cdot \kappa' = V \cdot \kappa = \Omega \cdot R_e - V \tag{3.19}$$

with three unknowns: speed V, wheel rotational speed Ω, and longitudinal slip at ground level κ'. We have solved this system for the repetitive brake torque M_b with the resulting speed V shown in Figure 3.8. The speed V drops to almost zero in four steps, with the transitions indicated by the numbers ①,…,④. It is expected that oscillations will occur in Ω and κ' at these transitions between $M_b \neq 0$ and $M_b = 0$, with the frequency and damping depending on the speed V and the model parameters. Due to the large vehicle mass, the speed V will change much slower than Ω and κ', and oscillations will not be present in the speed time history. The parameter values for this analysis are listed in Table 3.2, where J_{wheel} represents the wheel rotational moment of inertia. The other parameters were defined previously.

To investigate the response in Ω and κ', assume V to be constant and combine Eqs. (3.18) and (3.19), which leads to the following second-order equation:

$$J_{\text{wheel}} \cdot \sigma_\kappa \cdot \ddot{\Omega} + J_{\text{wheel}} \cdot V \cdot \dot{\Omega} + R_e^2 \cdot C_\kappa \cdot \Omega = R_e \cdot C_\kappa \cdot V - V \cdot M_b - \sigma_\kappa \cdot \dot{M}_b$$

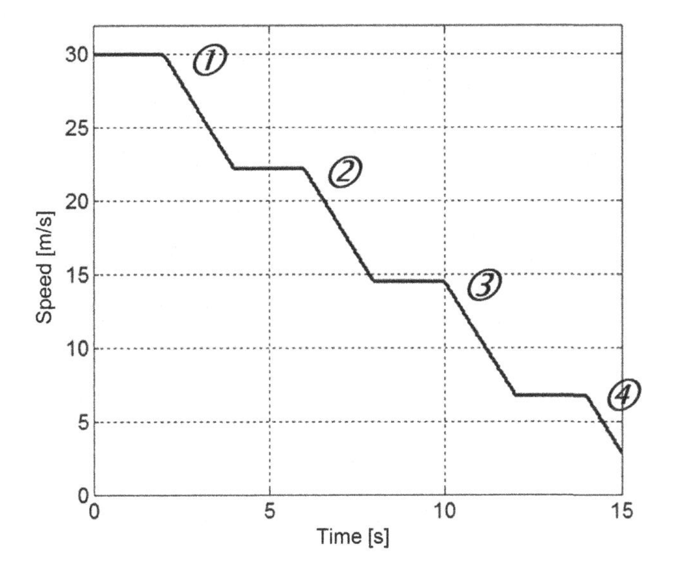

FIGURE 3.8 Speed versus time for repetitive braking.

TABLE 3.2 Parameters for Single Wheel Vehicle Analysis

Parameter	Value
J_{wheel} [kgm^2]	1.225
σ_κ [m]	0.40
C_κ [N]	8×10^4
m [kg]	400
R_e [m]	0.35

The undamped natural frequency ω and damping ratio ζ (see Appendix 2) are found to be

$$\omega = \sqrt{\frac{R_e^2.C_\kappa}{J_{wheel}.\sigma_\kappa}}, \quad \zeta = \frac{V}{2.R_e}.\sqrt{\frac{J_{wheel}}{\sigma_\kappa.C_\kappa}} \sim \frac{V}{\sqrt{\sigma_\kappa}}\sqrt{\frac{m_{wheel}}{m}}$$

where J_{wheel} is proportional to the wheel mass m_{wheel} times the square of the tire radius, and the longitudinal slip stiffness is nearly proportional to the vertical load, i.e., to the vehicle mass m.

Note that the natural frequency does not depend on the vehicle speed V, but the damping ratio does. Lowering the speed (there are different speed ranges during the braking procedure) will decrease the damping and therefore, increase the oscillations in the rotational speed. This typically occurs during ABS (Anti-lock

Braking System) braking. At the end of the braking process, when the vehicle speed is low, the wheel rotational speed variations show the largest amplitudes. In addition, the lag between the actual slip κ' at the tire–road contact and the slip κ at the axle will increase, moving from transition ① to transition ④.

Results are shown in Figures 3.9 and 3.10. From Figure 3.9, which shows the slip values for κ and κ' at transition ②, one observes a much larger oscillation

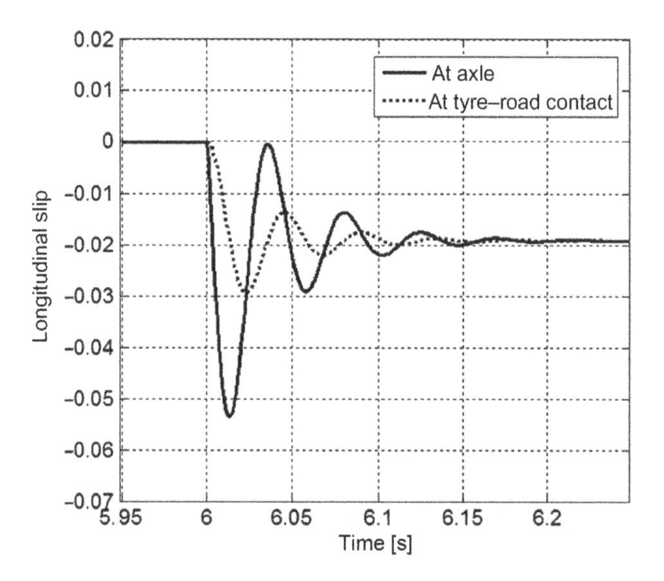

FIGURE 3.9 Longitudinal slip κ (axle) and κ' (tire-road) for transition 2 for $V = 22.2$ [m/s].

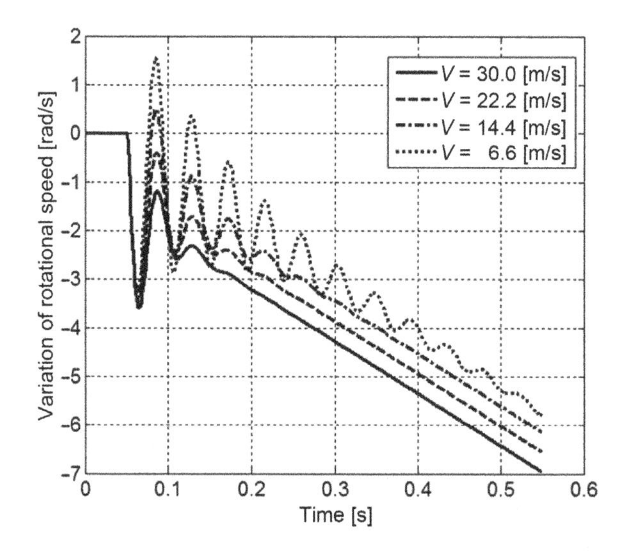

FIGURE 3.10 Rotational speed response at the four transitions between $M_b = 0$ and $M_b \neq 0$.

amplitude for the slip at the axle (due to the oscillations in the rotational wheel speed), and a significant lag of κ' with respect to κ. The wheel speeds for the four transitions from $M_b = 0$ to $M_b \neq 0$ are shown in Figure 3.10, where the time is taken from 0.05 [s] before this change in M_b. It is clear from the results that the damping is significantly reduced for lower speed V.

3.2 DYNAMIC TIRE RESPONSE TO ROAD DISTURBANCES

3.2.1 Introduction to the Rigid Ring Tire Model

Thus far, we have discussed steady-state and first-order dynamic behavior of tires. That means a tire responds immediately to external impacts such as changes in slip angle, brake torque, smoothly varying road undulations, or the response of the tire–wheel system is delayed with the lag time depending on the tire relaxation length. Typical cases for which this (first order) transient tire model can be applied are braking or power off in a turn, cornering and/or braking on a bumpy road, and shimmy phenomena.

However, a tire is more than a first-order system. For instance, the belt may move with respect to the rim in different in-plane and out-of-plane directions, where the tire wall stiffnesses and damping values (radial, tangential, camber, yaw, etc.), in combination with the belt mass, determine this behavior. These dynamic effects occur in case of relatively short and sharp road disturbances, where one may be interested in comfort, or in the impact of road disturbances on fatigue in the vehicle chassis structure. Another situation, when belt dynamics are important, is when control algorithms are applied such as in case of Active Brake Control and Electronic Stability Control. Belt behavior may be misinterpreted by the controller, leading to non-optimal control actions.

Historically, one refers to transient behavior if only the first order relaxation effect of the tire is considered, in spite of the fact that this is part of the dynamic behavior of the tire too. One distinguishes between the following application ranges for tire behavior:

1. *Steady-state behavior*: the tire contact forces respond immediately to changes in slip.
2. *Transient behavior*: first-order relaxation (lag) of slip at the tire–road contact (and contact forces) to changes in slip at axle level.
3. *Dynamic behavior*: belt dynamics are accounted for, in lateral, longitudinal, vertical, and combined slip conditions.

Clearly, the eigenfrequencies of the belt behavior against the rim play an important role when belt dynamics must be accounted for. These eigenfrequencies were studied by Gong [14] using a flexible ring tire model and were discussed for in-plane applications by Zegelaar [59]. A distinction can be made between two types of vibration modes for a free tire (i.e., not standing on the road):

1. Rigid ring modes: the belt moves around the rim as a rigid ring.
2. Flexible ring modes: the belt deflects around the circumference.

The qualitative behavior of rigid ring and flexible ring modes for a free tire are shown in Figure 3.11 (see also Figure 2.13).

The difference between a free tire and a standing tire is the static deflection in vertical and longitudinal (also lateral and yaw) directions. Pacejka [32] introduced the residual stiffnesses, which are the static tire stiffnesses, to ensure that these deflections were correctly described. A tire can be considered a combination of the free tire behavior (depicted in Figure 3.11) and these residual stiffnesses. Therefore, the modes illustrated in Figure 3.11 also arise for a standing tire, with the deformation shape being quite similar, but with more and different representations. For example, the second free mode ($n = 1$) leads to both a longitudinal and a vertical mode for the standing tire, with different natural frequencies.

These natural frequencies for the standing (passenger) tire begin in the range of 35–50 Hz for the circumferential mode ($n = 0$). The longitudinal and vertical ($n = 1$) modes are around 70–100 Hz, and the higher modes have natural frequencies far above 100 Hz (see Refs. [46] and [59]). This leads to an important conclusion:

If the relevant excitation of the tire is sufficiently below 80 Hz, the tire response can be modeled using a rigid ring approach, if the static deflections are accounted for by residual stiffnesses.

Using this conclusion, the tire model can be rather elegant and simple, and maintain good performance up to frequency levels, being relevant for the analysis of road disturbances, active chassis control, fatigue loading to the chassis components, etc. This idea has resulted in the so-called industrial SWIFT project (SWIFT: Short Wavelength Intermediate Frequency Tyre Model), being carried out in joint cooperation between de Delft University of Technology and the Dutch TNO contract research organization, with a major support by many automotive companies and suppliers, and with its results included in the theses by Zegelaar [59] and Maurice [23], and of course in [32]. This work has been continued by Schmeitz [46], finally yielding a complete well-established treatment of dynamic tire behavior for intermediate frequencies, and for longitudinal, lateral, and combined conditions.

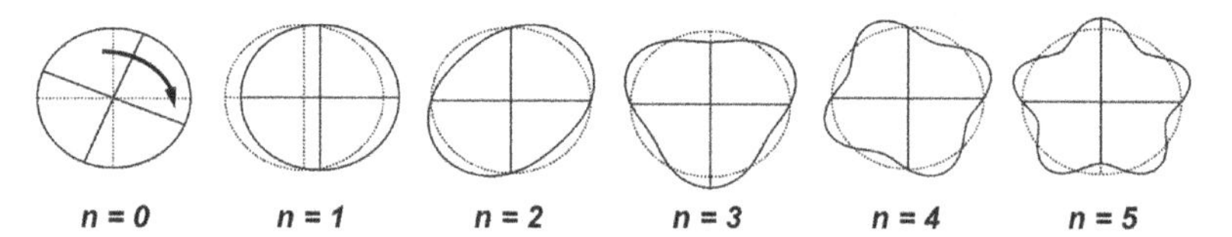

FIGURE 3.11 Vibration modes of free tire.

A complete discussion on dynamic tire modeling for all possible combinations of slip is too lengthy to be included here. However, one should have an understanding of this subject to understand the various software tools and the need to apply them. For this reason, this section addresses the dynamic tire model for in-plane applications (i.e., only longitudinal slip) in case of specific road disturbances. The extension to out-of-plane and combined slip applications uses similar model elements and ideas as the in-plane case, which means that a solid basis is derived through the in-plane discussion.

Using a rigid ring model for in-plane applications requires us to build our tire model as a combination of a rigid belt, attached to residual stiffnesses in vertical and longitudinal (circumferential) directions. The ring itself is attached to the wheel rim with radial and circumferential springs and dampers. In the contact area, a slip model must be used, which will, in general, correspond to the first-order transient model (see Section 3.2). Zegelaar schematically described the complete in-plane model, as shown in Figure 3.12. Note that the transient model can be considered as a spring (carcass spring stiffness) and a damper (slip stiffness) in a series, as discussed in the preceding section. The model is a single-point model, with slip properties and forces transferred between road and tire through this single contact point.

A single-point model has many advantages over models with multiple contact points, but cannot be realistically used, as shown in Figure 3.13 in which we depict a rigid ring on an arbitrary road surface (lower left picture).

In this case, the speed is assumed to be low. The enveloping model (shown in the upper left image) is expected for a tire rolling over such a surface, meaning that the tire adapts to the road surface, which further leads to a certain profile for the axle position. The tire envelops the *actual road surface*, acting like a filter, and transfers the road surface to the pattern for the wheel axle, which is referred to as the *effective road surface* (i.e., the axle position profile).

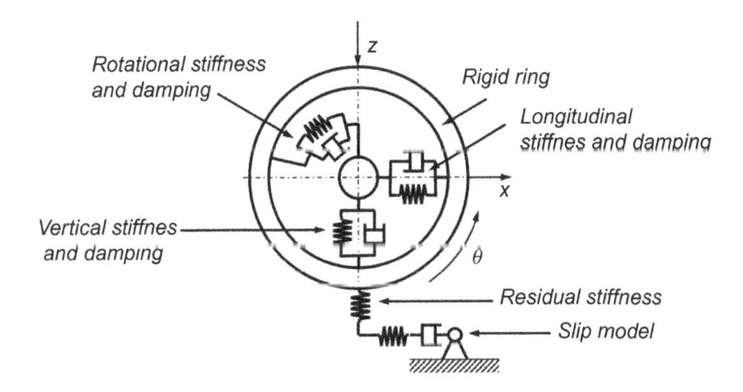

FIGURE 3.12 Schematic layout of dynamic tire model, consisting of a rigid ring belt description, a residual stiffness, and a contact slip model (based on Ref. [59]).

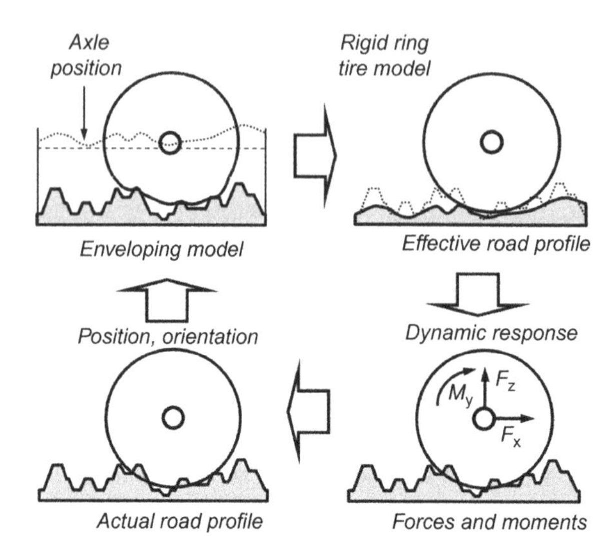

FIGURE 3.13 Assessment of dynamic tire performance using a tire enveloping model.

However, suppose we determined this effective road surface. Then, one might expect the following to hold true:

> *A rigid ring single-point tire model (combined with residual elastic behavior and a slip model) following the effective road surface gives the same response, with sufficient accuracy up to 80 Hz, compared to the real flexible tire following the actual road surface.*

This is the main idea behind the rigid ring dynamic tire model (schematically depicted in the upper right image in Figure 3.13). The dynamic analysis based on the effective road surface will lead to the dynamic forces and moments, which act on the tire while passing road disturbances. The resulting change in position and tire load can then be used to update the effective road surface from the actual road surface, and so forth. Hence, this approach requires the derivation of the quasi-static enveloping behavior of the tire, which will be the subject of the next section.

3.2.2 Enveloping Properties of Tires to Road Disturbances

The enveloping properties of a tire have been investigated with different models [23], [46], and [59]. The simplest approach by far is the single-point contact follower approach, as it only considers the vertical variation of the road profile. The classic theory of comfort and ride using combinations of quarter vehicle models is based on this approach. Because this approach only accounts for the vertical spring, and neglects the tire geometry, it cannot describe any enveloping properties. One possible improvement to this approach is obtained if the radial variation of the road profile, with respect to the wheel

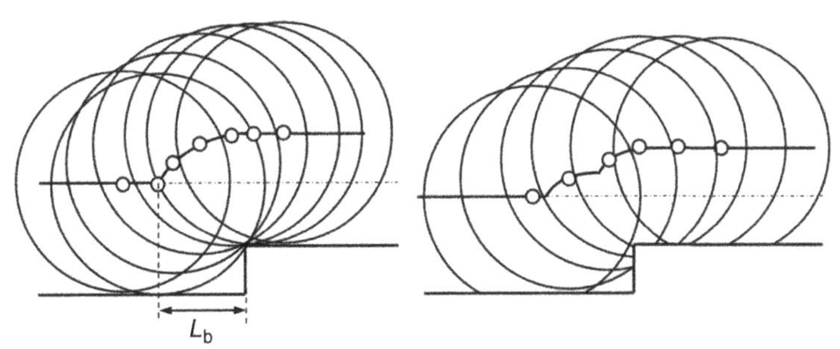

FIGURE 3.14 A rigid and flexible wheel, rolling over a step.

center, is considered instead of the vertical variation. This would result in both vertical and longitudinal contact forces, but would still lack accuracy. Another possible improvement would be to describe the tire as a flexible system; for example, using a suspended flexible ring model, a finite element model, or a model with flexibly interconnected radial springs (radial–interradial spring model). However, these models are not able to determine the tire axle behavior with either sufficient accuracy or within the required calculation time, making this approach unsuitable for handling analysis.

Let us consider a tire rolling over a step obstacle, as indicated in Figure 3.14. Suppose we could describe the effective road surface for such an obstacle, then one might expect that, when an arbitrary road surface is approximated by subsequent (upward and downward) steps, the effective road surface could be derived from a superposition of step-based effective road surfaces.

The following situations are considered in Figure 3.14:

- A rigid wheel rolling over the step obstacle.
- A flexible wheel rolling over the step obstacle.

For the rigid tire, the wheel center is lifted at a distance L_b before the step (being at position X_{step}), when the wheel first hits the step. The shape of the axle position Z_{axle} above road level versus position X with X, varying from $X_{step}-L_b$ to X_{step}, is found to be

$$Z_{axle} = \sqrt{R^2 - (X_{step} - X)^2} + h_{step} - R \tag{3.20}$$

for wheel radius R, and step height h_{step}. The length L_b follows from:

$$L_b = \sqrt{R^2 - (R - h_{step})^2} \tag{3.21}$$

For the flexible wheel, however, the axle position appears to change according to a shifted superposition of two sinus-type curves. Zegelaar's research provides references on this observation and its experimental validation. This observation was then used to develop a pragmatic approach to derive the effective road surface. This approach, shown in Figure 3.15, consisted of the following steps:

1. Use the step input to derive a quarter sine wave type basic road function. This basic function is defined for an X interval of length L_b. It is found that this function is shifted over a distance L_f in the direction of the step obstacle.
2. Use a two-point follower model with length L_s, moving along this basic function.
3. The midpoint of this two-point follower describes the height w_e of the effective road input.
4. The orientation of the two-point follower describes the effective road slope β_e.

The two-point follower should be considered as the contact of the tire to the road. Indeed, the length L_s of the two-point follower (the "skateboard" in Figure 3.15) appears to be on the order of 80% of the tire contact length. This also means that L_s depends on the wheel load. It was shown by Zegelaar [59] that the shift L_f and the "length" L_b of the basic function only depend on the step height (and on the tire) and not on the tire load. Consequently, the approach by Zegelaar allowed a distinction between the road-dependent parameters and the tire load-dependent parameters. Note that the dynamic tire response will also include variation in tire load, and therefore, the need to update the skateboard length with every new time step. On the other hand, the values of L_f and L_b can be determined beforehand, regardless of the actual tire behavior.

In summary, we introduced three types of road profiles:

1. *The actual road profile*: the real road profile.
2. *The effective road profile*: the position of the wheel axle, to be used as the input for the dynamic analysis using the single-point rigid ring tire model.
3. *The basic road function*: a road profile, defined previously for a step obstacle, to be used as an intermediate step from an actual road to an effective road.

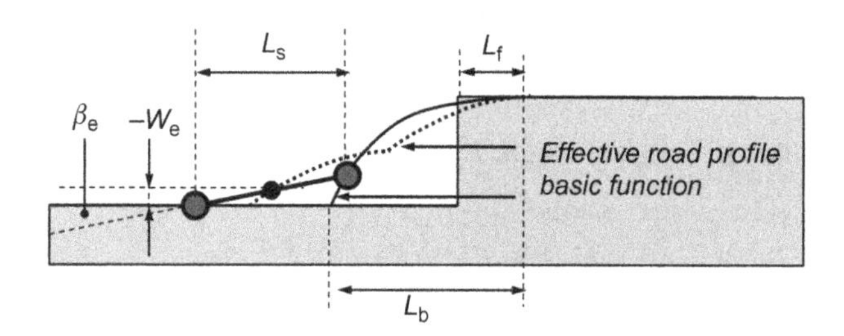

FIGURE 3.15 Derivation of the effective road surface based on a basic road function and a two-point follower.

The approach by Zegelaar appeared to lead to a number of considerations, which made this not an ideal approach for an effective dynamic tire software tool:

- The shift L_f was difficult to interpret in terms of physical phenomena, and an explicit expression for the basic road function that would end at the step obstacle is preferred.
- A sine function for the description of the basic road function appeared not to be a very accurate approach, also with the shift L_f in mind.
- The superposition of steps to build up an arbitrary road shape appeared to be cumbersome, especially in case of a downward slope.

Schmeitz [46] suggested a very elegant alternative for this sinus-type basic function by introducing elliptical cams.

He found that the basic function can be well described by introducing an elliptical cam with half axis lengths a_e and b_e (see Figure 3.16), with this elliptical cam following the road profile. This leads to a basic function (X_b, Z_b) serving as the center of the cam. This approach can be justified by the observation that the tire shape near the contact area is elliptically shaped in a similar way (see also Figure 3.15). The elliptic cam is described by shape parameters a_e, b_e, and c_e as follows:

$$\left(\frac{x}{a_e}\right)^{c_e} + \left(\frac{y}{b_e}\right)^{c_e} = 1 \tag{3.22}$$

the expression (3.21) for L_b now changes into

$$L_b = a_e \cdot \left(1 - \left(1 - \frac{h_{step}}{b_e}\right)^{c_e}\right)^{\frac{1}{c_e}} \tag{3.23}$$

Expression (3.20) changes into

$$Z_{axle} = b_e \cdot \left(1 - \left(\frac{X_{step} - X}{a_e}\right)^{c_e}\right)^{1/c_e} + h_{step} - b_e \tag{3.24}$$

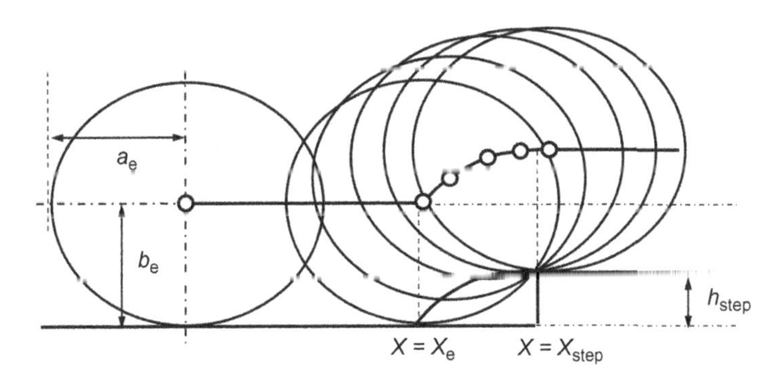

FIGURE 3.16 Deriving a basis road curve using elliptical cams.

A two-point follower can still be used but now, two elliptic cams are "rolling" as a tandem over the step obstacle, with the lowest points connected by this two-point follower. Figure 3.17 shows the "enveloping" tire to illustrate the relationship of the tandem with the tire.

Referring to these cams with index f for front and index r for rear, and denoting the global heights of the front and rear ellipse centers by Z_f and Z_r, we find for the effective road height w_e and effective forward slope β_e that

$$-w_e(X) = \frac{Z_f(X_f) + Z_r(X_r)}{2} - b_e \tag{3.25}$$

$$\beta_e(X) = \frac{Z_f(X_f) - Z_r(X_r)}{L_s} \tag{3.26}$$

where X_f and X_r denote the positions of the ellipse centers along the road, i.e., where X is exactly between X_f and X_r, at distance $L_s/2$.

Two important conclusions were drawn in Ref. [46], based on experimental research:

- *The tandem (two-point follower) base length L_s depends solely on the contact length of the tire (i.e., on the vertical load). Usually, L_s equals about 80% of the contact length.*
- *The shape of the elliptical cam (i.e., the parameters a_e, b_e, and c_e) is not affected by the vertical load or the step height.*

Hence, the only parameter changing during simulation is L_s. Schmeitz uses a more accurate estimation for half the contact length a, dependent on F_z, compared to Eq. (2.81):

$$a = q_2 \cdot F_z + q_1 \cdot \sqrt{F_z} \tag{3.27}$$

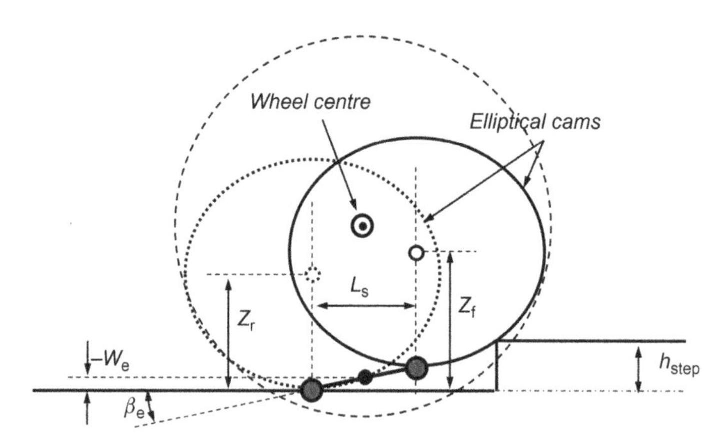

FIGURE 3.17 Tandem of elliptical cams, replacing a flexible tire to derive the effective road input.

where q_1 and q_2 are determined from matching Eq. (3.27) with experimental results. The cam parameters vary by tire, within average (according to Ref. [46]):

$a_e \approx 1.05-1.20$ times the free tire radius R
$b_e \approx 1.00-1.10$ times the free tire radius R, with $a_e > b_e$
$c_e \approx 1.6-1.8$

We used the elliptic cam approach to determine the effective road input w_e and road slope β_e for a number of specific road shapes:

i. a step-up
ii. a sinusoidal bump
iii. a pothole

for different wheel loads $F_z = 2000$, 4000, and 6000 [N]. The additional parameters were chosen cf. Ref. [46], with the order of magnitude listed in Table 3.3. Results are shown in Figures 3.18–3.20.

Using the elliptic cam approach to determine the effective road profile (i.e., the tire enveloping properties) appears to be effective for different road shapes. This has been confirmed in Ref. [46] by comparison of the analytical results with experiments.

Figure 3.18 shows the typical superposition of two functions that resemble quarter sine functions that, together, build up the effective road height corresponding to the wheel axle height. Increasing the load means a larger tandem length L_s, which means that the effective road is initiated earlier and lasts longer. The slope is typically shaped like a shifted sine function that is subtracted from another one. This slope defines the orientation of the steady-state wheel force while passing this step. A short obstacle, such as the sinusoidal bump, is enveloped quite nicely with a large wheel load.

The 15 [mm] obstacle pushes into the tire for about 50%, whereas the lowest wheel load only shows an indentation of the obstacle into the tire for

TABLE 3.3 Parameters to Determine Enveloping Tire Behavior

Parameter	Value
R [m]	0.31
a_e [m]	0.32
b_e [m]	0.32
c_e	1.82
$L_s/(2 \cdot a)$	0.88
q_1 [m/N$^{1/2}$]	6.6×10^{-4}
q_2 [m/N]	4.5×10^{-6}

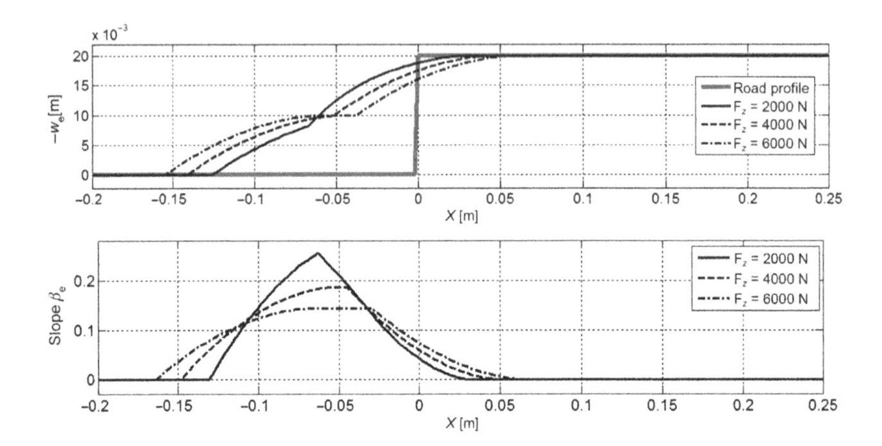

FIGURE 3.18 Effective road input (height, slope) for a step-up obstacle.

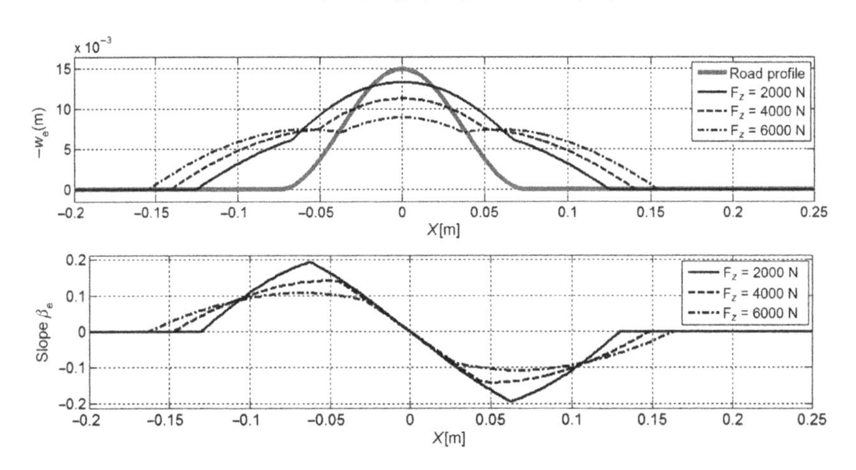

FIGURE 3.19 Effective road input (height, slope) for a sinusoidal bump obstacle.

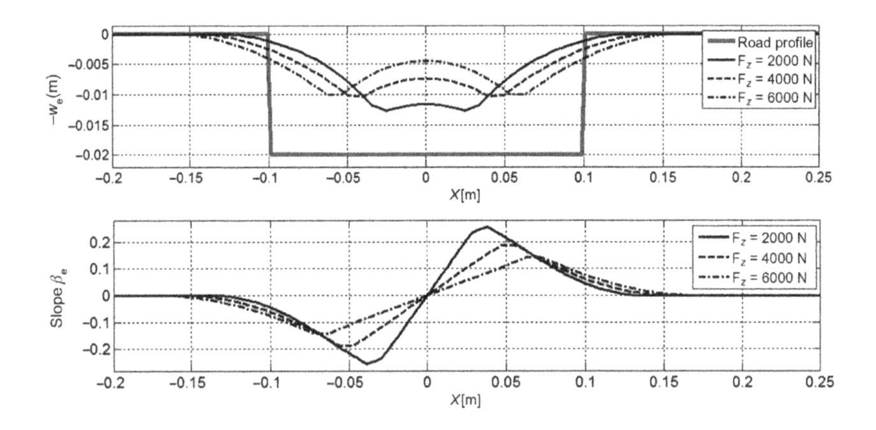

FIGURE 3.20 Effective road input (height, slope) for a pothole.

only about 2.5 [mm]. The effective road slope changes sign while passing this obstacle. As illustrated, the pothole is the opposite of the step-up obstacle because the pothole is too short for the tire to hit the bottom. If the length of the pothole had been extended, it would yield a similar shape as in Figure 3.18, but from inverted as high to low. The larger the wheel load, the more constant the axle height while passing this pothole.

3.2.3 Dynamic Response to Road Disturbances

In this section, we discuss the dynamic response of the rigid ring model (shown in Figure 3.12) following the effective road surface derived using the cam-based two-point follower, which has been described in the preceding section. We argued that this response corresponds to the dynamic response of a real tire operating on an uneven road. This has been validated extensively in the research [23, 46, 59].

The center of the rigid ring-shaped belt is denoted as (x_b, z_b) and the location of the rim is denoted as (x_a, z_a). Consider a vehicle moving at a constant speed V. The position of the rim will move vertically and longitudinally with respect to the vehicle due to elastokinematic wheel suspension deflections. With the belt suspended to the rim, the center of the belt (x_b, z_b) will move with respect to the rim. In addition to these displacements, both rim and belt will show rotational deflections θ_a and θ_b, respectively, with respect to the angular wheel position φ on a flat road. Denoting the rotation speed on a flat road with Ω, we obtain (see also Eq. (2.1))

$$V = \Omega \cdot R_\mathrm{e} = \dot{\varphi} \cdot R_\mathrm{e}$$

for an effective rolling radius R_e under free rolling. That means that the rotational speeds of rim and belt are described by:

$$\begin{aligned} &\text{rim} \quad : \Omega - \dot{\theta}_a \\ &\text{belt} \quad : \Omega - \dot{\theta}_b \end{aligned}$$

Note that Ω acts in the negative y direction, whereas the rotational speeds of rim and belt are defined in the positive y direction. With both the speed of the belt center along the effective road profile and the rotational speed of the belt varying in time, the longitudinal slip in the contact area will also vary. The belt motions and the contact slip are initiated by normal and tangential contact forces between belt and road, denoted as $F_{c\mathrm{T}}$ and $F_{c\mathrm{N}}$, respectively.

The rigid ring belt equations in case of a tire moving over an uneven surface with effective road height w_e and effective road slope β_e are second-order equations and are described as follows:

$$\begin{aligned} &m_\mathrm{b} \cdot \ddot{x}_b + k_{\mathrm{b}x} \cdot (\dot{x}_b - \dot{x}_a) + c_{\mathrm{b}x} \cdot (x_b - x_a) \\ &+ k_{\mathrm{b}z} \cdot (\Omega - \dot{\theta}_a) \cdot (z_b - z_a) = \cos(\beta_\mathrm{e}) \cdot F_{c\mathrm{T}} + \sin(\beta_\mathrm{e}) \cdot F_{c\mathrm{N}} \end{aligned} \tag{3.28}$$

$$m_{\mathrm{b}} \cdot \ddot{z}_b + k_{\mathrm{b}z} \cdot (\dot{z}_b - \dot{z}_a) + c_{\mathrm{b}z} \cdot (z_b - z_a) \\ - k_{\mathrm{b}x} \cdot (\Omega - \dot{\theta}_a) \cdot (x_b - x_a) = -\sin(\beta_{\mathrm{e}}) \cdot F_{c\mathrm{T}} + \cos(\beta_{\mathrm{e}}) \cdot F_{c\mathrm{N}} \tag{3.29}$$

$$J_{\mathrm{b}y} \cdot \ddot{\theta}_b + k_{\mathrm{b}\theta} \cdot (\dot{\theta}_b - \dot{\theta}_a) + c_{\mathrm{b}\theta} \cdot (\theta_b - \theta_a) = R_{\mathrm{e}} \cdot F_{c\mathrm{T}} + M_{cy} \tag{3.30}$$

for belt mass m_{b}, belt polar moment of inertia $J_{\mathrm{b}y}$, sidewall damping values $k_{\mathrm{b}x}$, $k_{\mathrm{b}z}$, and $k_{\mathrm{b}\theta}$, and sidewall stiffness values $c_{\mathrm{b}x}$, $c_{\mathrm{b}z}$, and $c_{\mathrm{b}\theta}$. In the right-hand side of Eq. (3.30), we replaced the loaded tire radius with the effective tire radius R_{e}, as discussed by Zegelaar [59]. The second term M_{cy} in the right-hand side corresponds to rolling resistance, which will be neglected here. The local forces in the contact area are shown in Figure 3.21. Note that β_{e} denotes the effective road slope, as discussed in Section 3.2.2.

The last terms in the left-hand sides of Eqs. (3.28) and (3.29) are because moving the belt with respect to the rim leads to a net velocity of elements of the belt, perpendicular to the direction of the belt displacement. Consider Figure 3.22, where the belt is displaced with respect to the rim over a distance $z_a - z_b$. One can show that an arbitrary point of the belt at angle γ at some time has a velocity component in the x direction of

$$V_x(\gamma) = (\Omega - \dot{\theta}_a) \cdot (R \cdot \sin\gamma + z_a - z_b)$$

where quadratic and higher-order terms in the deflection have been omitted. Integration over the belt circumference results in a net velocity between belt and rim in x direction and therefore, a net damping force, which explains the

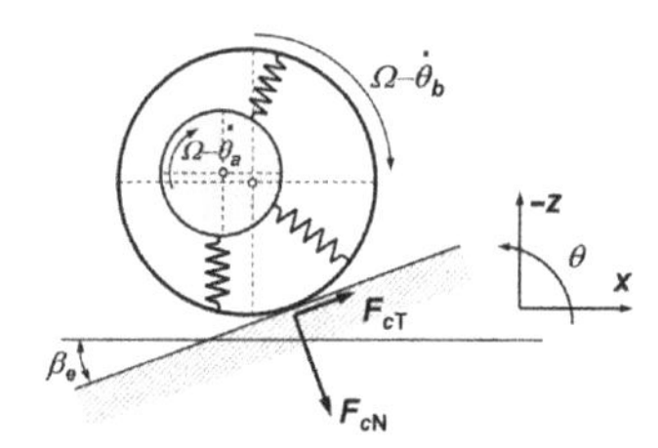

FIGURE 3.21 Local contact forces and rotational speeds.

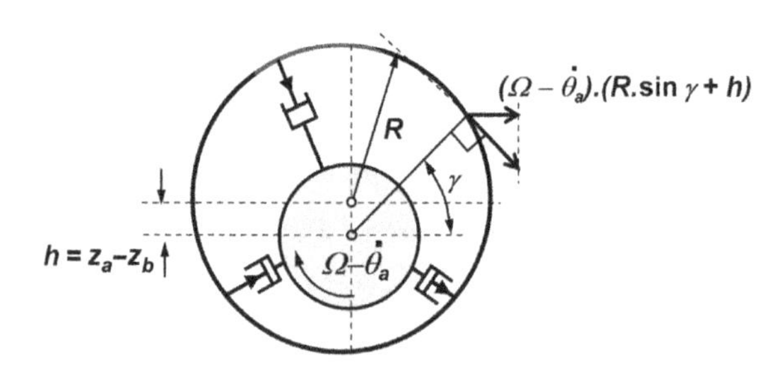

FIGURE 3.22 Vertical belt deflection leads to a horizontal net deflection speed and an additional damping force.

last term in the right-hand side of Eq. (3.28). A similar explanation can be found for the additional vertical damping force in Eq. (3.29).

The inertia and damping coefficients in Eqs. (3.28)–(3.30) are assumed constant. The stiffness coefficients (sidewall stiffness) will, in general, be dependent on the tire rotational velocity. This is described in Ref. [59] by the following expression for c_{bx}

$$c_{bx} = c_{bx0} \cdot \left(1 - q_{bVx} \cdot \sqrt{|\Omega - \dot{\theta}_a| \cdot \|\underline{x}_b - \underline{x}_a\|}\right)$$

for constants c_{bx0} and q_{bVx} and where $\| \cdot \|$ is used to indicate the total displacement of the tire ring, with respect to the rim. The second constant is typically on the order of 0.2–0.3 [$s^{1/2}/m^{1/2}$]. Similar relationships are used for c_{bz} and $c_{b\theta}$, with the same order of magnitude for q_{bVz} and $q_{bV\theta}$, respectively.

The tangential contact force F_{cT} depends on the longitudinal slip at the contact area, and the normal force F_{cN}

$$F_{cT} = F_{cT}(\kappa', F_{cN}) \tag{3.31}$$

with F_{cT}, described by the Magic Formula, (see Eq. (2.22)) with the model parameters given in Appendix 6. As we have seen in Section 3.1.1, the slip satisfies the transient Eq. (3.10), which is expressed here as follows:

$$\sigma_\kappa \cdot \dot{\kappa}' + \left|V_{r,\text{belt}}\right| \cdot \kappa' = -V_{s,\text{belt}} \tag{3.32}$$

with belt rolling speed $V_{r,\text{belt}}$ and slip velocity $V_{s,\text{belt}}$ in the contact area. The rolling speed equals the effective tire radius times the belt rotational speed, i.e.,

$$V_{r,\text{belt}} = R_e \cdot (\Omega - \dot{\theta}_b) \tag{3.33}$$

Please note that the relaxation length σ_κ varies with slip (see Figure 3.2). In this section, we assume σ_κ to be constant. If the tangential residual stiffness is neglected, then the relaxation length at free rolling is equal to half the contact length a (see Eq. (3.27)). The slip velocity equals the difference between the belt velocity parallel to the road surface and the belt rolling speed, i.e.,

$$V_{s,\text{belt}} = \dot{x}_b \cdot \cos \beta_c - \dot{z}_b \cdot \sin \beta_c - V_{r,\text{belt}} \tag{3.34}$$

The normal deflection of the tire has two components: the deflection between the rim and the rigid ring that describes the belt due to sidewall stiffnesses, and the residual deflection between belt and contact patch (see also Figure 3.12). The first component follows from the preceding differential equations. The residual deflection, denoted by ρ_{zr}, is related to the normal tire force in a nonlinear way, which Zegelaar introduced as a third-order relationship

$$F_{cN} = q_{Fzr3} \cdot \rho_{zr}^3 + q_{Fzr2} \cdot \rho_{zr}^2 + q_{Fzr1} \cdot \rho_{zr} \tag{3.35}$$

assuming $\rho_{zr} > 0$. Negative deflection means there is no contact and therefore, $F_{cN} = 0$. The coefficients in expression (3.35) depend on the rotational speed of the rim and the belt stiffness c_{bz}. We refer reader to Refs. [32], [46], and [59] for further details. We determined these coefficients using the tire data, as presented by Schmeitz (see Section 3.7 in Ref. [46]), for different forward speeds $V = \Omega \cdot R_e$. The results are listed in Table 3.4.

The residual deflection consists of different contributions. First, it includes the difference between the effective road height w_e and the height of the belt z_b. A second contribution comes from the fact that the tire radius increases with increasing rotational velocity Ω, also known as the centrifugal phenomenon. Finally, it is stated in Refs. [32] and [59] that the vertical force appears to decrease with the horizontal displacement of the contact patch. This displacement is given by:

$$\rho_x = (x_b - x_a) + R_0 \cdot (\theta_b - \theta_a) \tag{3.36}$$

where R_0 is the unloaded tire radius. The completed vertical tire deflection can be written as:

$$\begin{aligned} \rho_{zr} &= (z_b - w_e) + \Delta R - q_{Fcx} \cdot \rho_x^2 \\ &\equiv (z_b - w_e) + q_{V1} \cdot (\Omega - \dot{\theta}_a)^2 - q_{Fcx} \cdot \rho_x^2 \end{aligned} \tag{3.37}$$

where w_e is the effective road height for coefficients q_{V1} (on the order of 10^{-7} [m/s^2] for a passenger car) and q_{Fcx} (on the order of 3 [1/m]).

The increase of tire radius with increasing Ω also affects the effective tire radius R_e. As we have seen in Section 2.1, this radius R_e also depends on the normal tire load. This value could be described using expression (2.3), however, we follow Refs. [32] and [59] and obtain

$$R_e = q_{re3} \cdot \sqrt{F_{cN}^3} + q_{re2} \cdot F_{cN} + q_{re1} \cdot \sqrt{F_{cN}} + q_{re0} + \Delta R \tag{3.38}$$

TABLE 3.4 Coefficients in Normal Force Versus Residual Deflection Relationship

Wheel Velocity V [km/h]	q_{Fzr1} [N/m]	q_{Fzr2} [N/m^2]	q_{Fzr3} [N/m^3]
30	1.957×10^5	8.372×10^5	7.595×10^5
50	1.992×10^5	8.553×10^5	7.911×10^5
70	2.027×10^5	8.735×10^5	8.236×10^5

for coefficients q_{re0}, q_{re1}, q_{re2}, and q_{re3}. Schmeitz [46] determined the values of these parameters for a passenger car. The order of magnitude for these parameters is given in Table 3.5.

We have plotted F_{cN} versus R_e in Figure 3.23. Comparing this plot with Figure 2.5, they both show a similar qualitative behavior.

We have now introduced four differential equations, three of which describe the belt deformations and one that describes the longitudinal slip in the contact area. We still must describe the rotational rim performance, i.e., give an equation from which θ_a can be determined. This equation reads (see Refs. [46] and [59])

$$J_{ay} \cdot \ddot{\theta}_a + k_{b\theta} \cdot (\dot{\theta}_a - \dot{\theta}_b) + c_{b\theta} \cdot (\theta_a - \theta_b) = M_{ay} \quad (3.39)$$

for moment of inertia J_{ay}, including all rotating axle and brake system parts and forced torque M_{ay}. Zegelaar [59] shows the performance of the tire under repetitive brake torque oscillations, as we have shown in Section 3.1.2 using the transient model. Such torque oscillations will not be considered here, as we focus on driving over various road shapes. Therefore, in this section, we will take $M_{ay} = 0$. The extension to nonzero torque is straightforward and we refer reader to Refs. [32] and [59]. The horizontal and vertical motions of the rim depend on the vehicle suspension characteristics. Schmeitz [46] investigated the dynamic

TABLE 3.5 Coefficients in Normal Force Versus Residual Deflection Relationship

q_{re0} [m]	q_{re1} [m/N$^{1/2}$]	q_{re2} [m/N]	q_{re3} [m/N$^{3/2}$]
0.31	-3.0×10^{-4}	2.5×10^{-6}	-1.0×10^{-8}

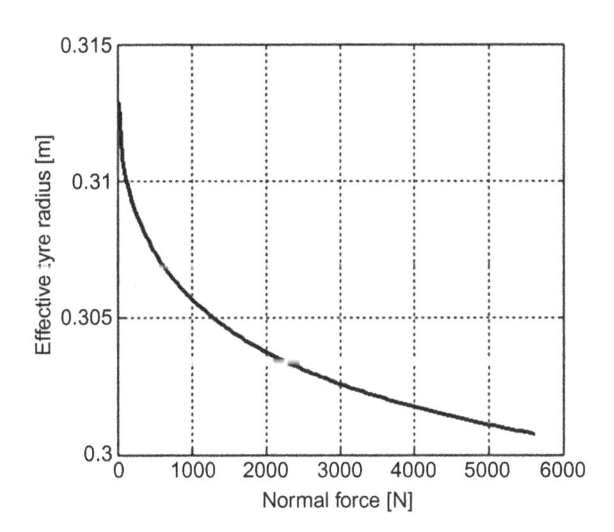

FIGURE 3.23 Effective tire radius depending on the normal contact force (tire load).

tire model as a component of a vehicle system and found a strong resemblance between the rigid ring model and experimental results, up to large frequencies (90–100 Hz). He investigated the effect of enveloping and found that significant deviations from experimental results start to arise from about 20 to 30 Hz if the enveloping properties of the tire are neglected. He also considered a linearized model (important for parameter estimation and eigenfrequency assessment), and found that this model provides good results up to about 60–70 Hz.

In this section, we follow Zegelaar's approach, where the rim is fixed at a certain height and tire is moving over the road at a fixed speed.

This case corresponds to experiments being carried out in the laboratory on a 2.5 m diameter drum that is equipped with specific road obstacles (such as those discussed in Section 3.2.2). Therefore, when we take the tire as moving in the x direction instead of along the road, the longitudinal rim speed equals V and the vertical motion is zero:

$$\dot{x}_a = V; \quad \dot{z}_a = 0$$

According to Ref. [59], the belt stiffness and damping coefficients can be estimated as follows. We assume the following equalities:

$$c_{\mathrm{b}} \equiv c_{\mathrm{b}x} = c_{\mathrm{b}z}; \quad k_{\mathrm{b}} = k_{\mathrm{b}x} = k_{\mathrm{b}z}$$

In his discussion on the flexible ring model, Zegelaar distinguishes between radial sidewall stiffness c_{r} and tangential sidewall stiffness c_φ, that latter of which is defined as per unit length of the flexible belt with dimension N/m^2. The translational sidewall stiffness c_{b} is the combined effect of radial deflection and tangential deflection. The rotational sidewall stiffness $c_{\mathrm{b}\theta}$ follows from integration of the contribution of the tangential deflection of each element of the belt to the total moment over belt circumference. This leads to (see also Ref. [59])

$$c_{\mathrm{b}} = \pi \cdot R \cdot (c_{\mathrm{r}} + c_\varphi); \quad c_{\mathrm{b}\theta} = 2 \cdot \pi \cdot c_\varphi R^2 \tag{3.40}$$

The radial and tangential sidewall stiffnesses can be estimated from the undeformed carcass geometry; here we follow Ref. [59]. It is thereby assumed that the deflections of the carcass are small and that the inner pressure p_{i} remains constant in case of tire deflection. In Figure 3.24, the sidewall height between the rim and the tread area is denoted by h_{s}, the half angle describing the sidewall shape is denoted by φ_{s}, and the sidewall thickness is denoted by t_{s}. According to Ref. [59], the radial and tangential stiffnesses depend on the angle φ_{s}, the inner tire pressure p_{i}, and the shear modulus G_{r} of the sidewall rubber. From Ref. [59], the following expressions for these stiffnesses can be derived:

$$c_{\mathrm{r}} = p_{\mathrm{i}} \cdot \frac{\cos\varphi_{\mathrm{s}} + \varphi_{\mathrm{s}} \cdot \sin\varphi_{\mathrm{s}}}{\sin\varphi_{\mathrm{s}} - \varphi_{\mathrm{s}} \cdot \cos\varphi_{\mathrm{s}}}; \quad c_\varphi = \frac{G_{\mathrm{r}} \cdot t_{\mathrm{s}}}{h_{\mathrm{s}}} \cdot \frac{\varphi_{\mathrm{s}}}{\sin\varphi_{\mathrm{s}}} + \frac{p_{\mathrm{i}}}{\tan\varphi_{\mathrm{s}}} \tag{3.41}$$

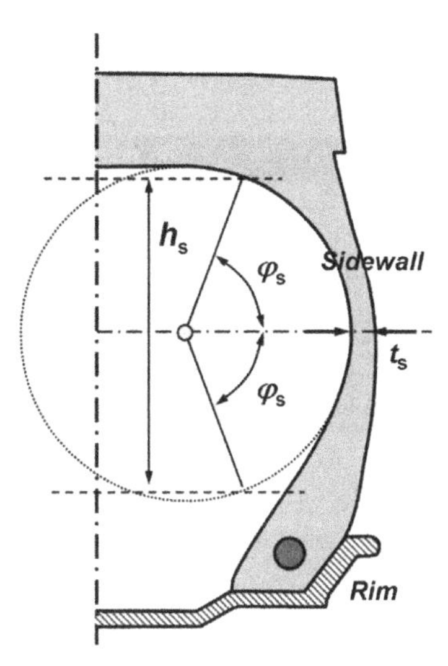

FIGURE 3.24 Carcass geometry layout.

Using a combination of Eqs. (3.40) and (3.41) shows us that the belt stiffnesses reduce with increasing angle φ_s (especially stiffness c_b), as well as with the sidewall height h_s. This suggests a significantly higher belt stiffnesses for a low-profile tire. The stiffnesses clearly increase with an increase in the inner pressure. One might think of truck tires, with higher inner pressure compared to passenger car tires, but also with a larger sidewall height. Zegelaar derived an order of magnitude for the preceding parameter values as follows: $G \approx 1.6 \times 10^6$ [N/m^2], $\varphi_s \approx 60-65$ [°], and $t_s \approx 0.1$ [m], which are for a passenger car tire with an inner pressure of 2.2 bar in $c_b \approx 1.1 \times 10^6$ [N/m] and $c_{b\theta} \approx 4.8 \times 10^4$ [Nm/rad]. Schmeitz [46] gives parameter values that are approximately 50 larger. This difference might be explained by a different choice of tires. Zegelaar refers to Gong's research [14], which was published in 1993, whereas Schmeitz's research was defended in 2004. Nevertheless, the expressions (3.40) and (3.41) might be helpful in estimating belt parameters for dynamic tire response studies. The belt parameters may also be estimated based on the belt eigenfrequencies. As indicated in Ref. [32] and the relevant PhD theses [46, 59], the vertical belt eigenfrequency is usually around 70–80 Hz for a passenger car. The in-phase rotational mode (rim and belt moving in the same direction) has an eigenfrequency between 30 and 40 Hz.

Let us now consider the dynamic belt performance for a fixed axle height, where we choose the road disturbances as introduced in Section 3.2.2, i.e.,

- step-up with height 0.2 [m]
- sinusoidal bump with height 0.15 [m]
- pothole with depth 0.2 [m].

TABLE 3.6 Parameters Used in the Dynamic Tire Response Analysis for Uneven Roads

Parameter	Value	Parameter	Value
V [km/h]	36	c_{bx0}, c_{bz0} [N/m]	1.3×10^6
R [m]	0.31	$c_{b\theta 0}$ [N/m]	7.5×10^4
m_b [kg]	7.5	k_{bx}, k_{bz} [Ns/m]	250
m_a [kg]	18	$k_{b\theta}$ [Nms/rad]	7.5×10^4
m_v [kg]	400	q_{bVx} [s$^{1/2}$/m$^{1/2}$]	0.25
J_{by} [kgm^2]	0.64	q_{bVz} [s$^{1/2}$/m$^{1/2}$]	0.25
J_{ay} [kgm^2]	0.15	$q_{bV\theta}$ [s$^{1/2}$/m$^{1/2}$]	0.25
q_{V1} [ms^2]	8.0×10^{-8}	q_{Fcx} [1/m]	3.0

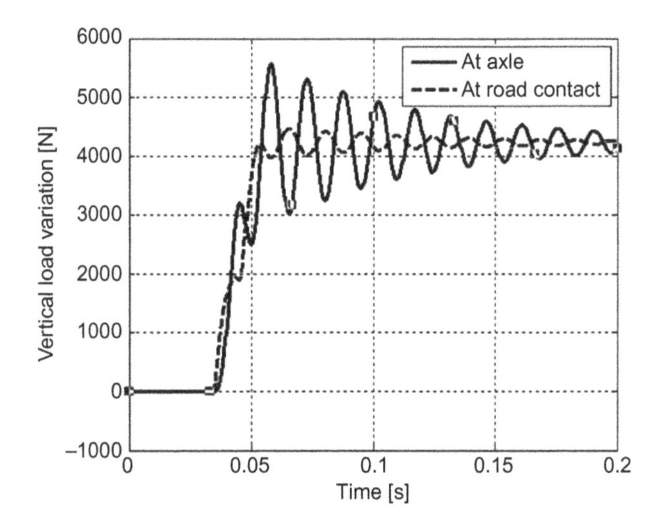

FIGURE 3.25 Vertical force variations for a step-up obstacle.

We have chosen the following parameters for our analysis (Table 3.6).

The mass m_v is the contribution of the vehicle mass to the single wheel. Other parameters are given in Tables 3.3–3.5. For each type of obstacle, we give three plots:

- The variation of the vertical force versus time, with distinction between the axle force and the contact force between road and tire.
- The variation of the longitudinal force versus time, also with distinction between axle force and contact force between road and tire.
- The variation of wheel speed in rad/s for belt and rim.

The force variation for the step-up obstacle is shown in Figures 3.25 and 3.26. Observe the large oscillations in both the vertical and the longitudinal

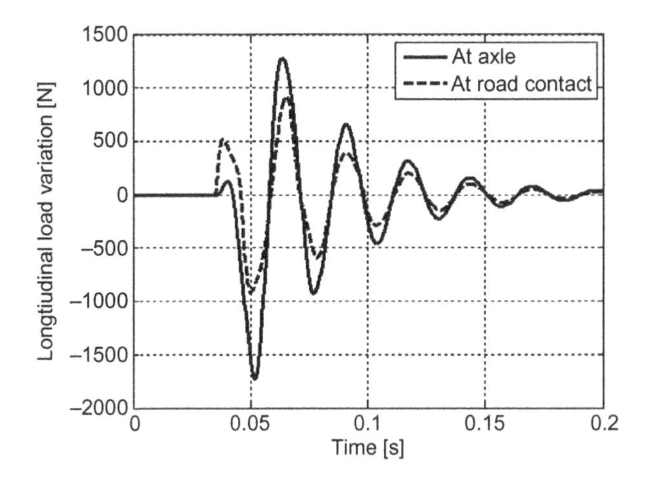

FIGURE 3.26 Longitudinal force variations for a step-up obstacle.

axle force variation, where the vertical force shows an eigenfrequency of about 70 Hz and the longitudinal force shows a lower eigenfrequency of about 35 Hz. These values are of the same order of magnitude as those derived for the vertical mode and the in-phase rotational mode by Zegelaar [59] and Schmeitz [46].

The vertical eigenfrequency is slightly lower (Schmeitz finds 79 Hz), but we have chosen the belt stiffness slightly lower compared to Ref. [46]. Compare this also with the in-plane transient tire behavior, shown in Figures 3.9 and 3.10, which has the oscillation frequency near 32 Hz. The force variation is larger at the axle than in the contact area. The force variation in the contact area may be of interest for road wear assessment. Note that we constrained the axle height and therefore, restricted the vertical axle motion, which is not a realistic condition to examine road wear due to dynamic wheel load variation. Nevertheless, one observes a significant slip force in the contact area, as well as oscillations in the vertical contact force. With an unrestrained axle, one must add a larger force oscillation with a lower frequency due to wheel hob. The wheel speed variation is shown in Figure 3.27. There appears to be a minor difference between the belt and the rim rotation. A maximum wheel speed of 1.5 [rad/s] corresponds to about a 1% longitudinal slip.

The force variation for the sinusoidal bump obstacle is shown in Figures 3.28 and 3.29. Similar observations can be compared with the step-up obstacle. The vertical contact force appears to follow the shape of the effective road input, as expected. The first peaks in the vertical force are approximately of the same magnitude. Increasing the speed will lead to a larger vertical axle load. Because the sine bump obstacle is higher than the step-up obstacle, a larger variation in the longitudinal force is obtained compared to Figure 3.26. This phenomenon corresponds with the findings in

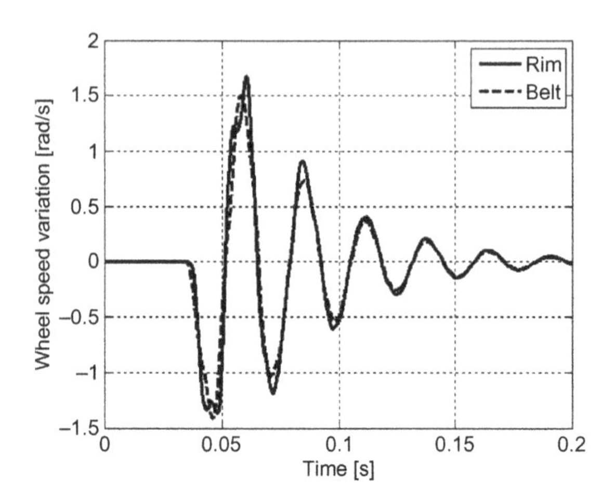

FIGURE 3.27 Wheel speed variation for a step-up obstacle.

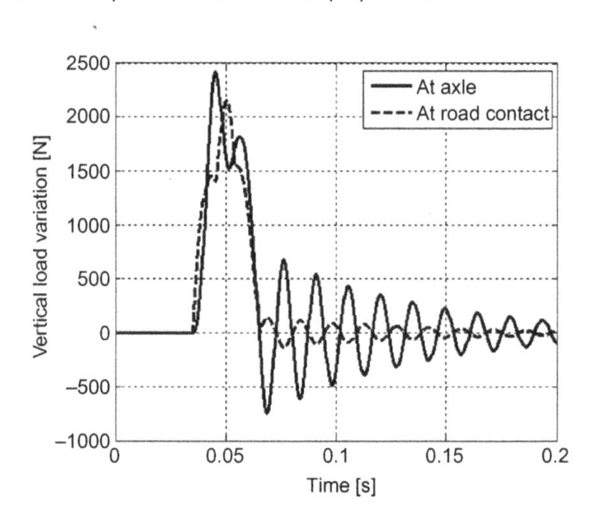

FIGURE 3.28 Vertical force variations for a sine bump obstacle.

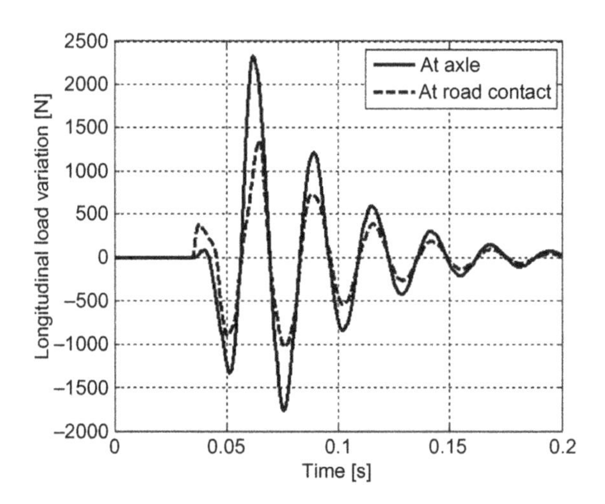

FIGURE 3.29 Longitudinal force variations for a sine bump obstacle.

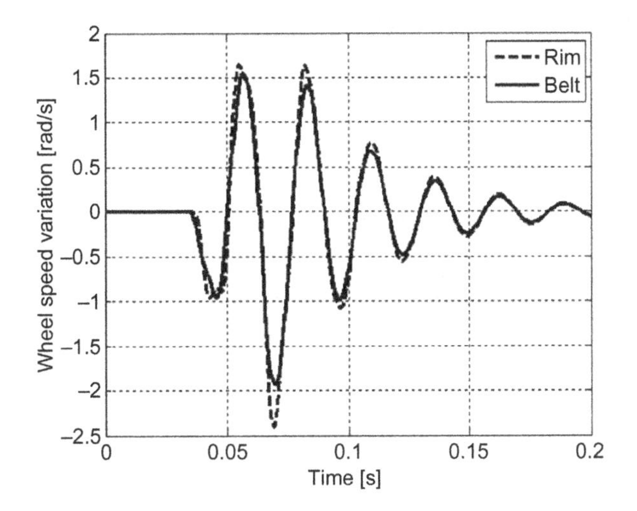

FIGURE 3.30 Wheel speed variation for a sine bump obstacle.

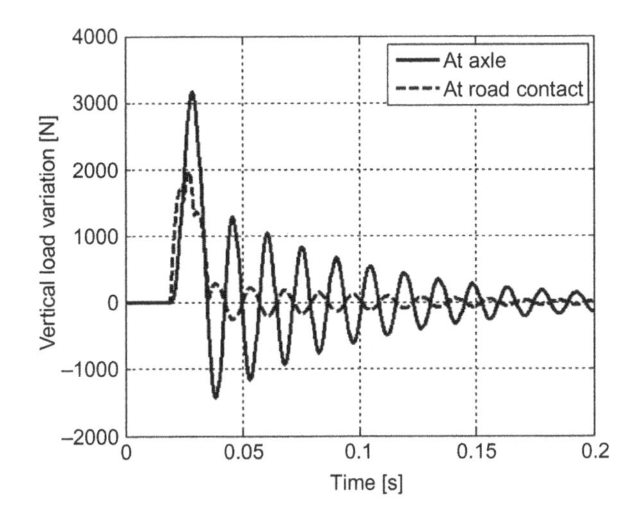

FIGURE 3.31 Vertical force variations for a sine bump obstacle (66 [km/h]).

Refs. [46] and [59]. Zegelaar discusses the reasons for the first small peak in Figure 3.29, where one expected a negative value. He explains this result with the rotational velocity variations of the wheel (Figure 3.30, showing the wheel speed variation) being more important than the effective road slope when the tire first makes contact with the bump. We repeated the simulation for a higher speed, $V = 66$ [km/h], and the resulting force variations are shown in Figures 3.31 and 3.32.

Increasing the velocity leads to more damping in the longitudinal force variation, which is similar to the case of transient behavior

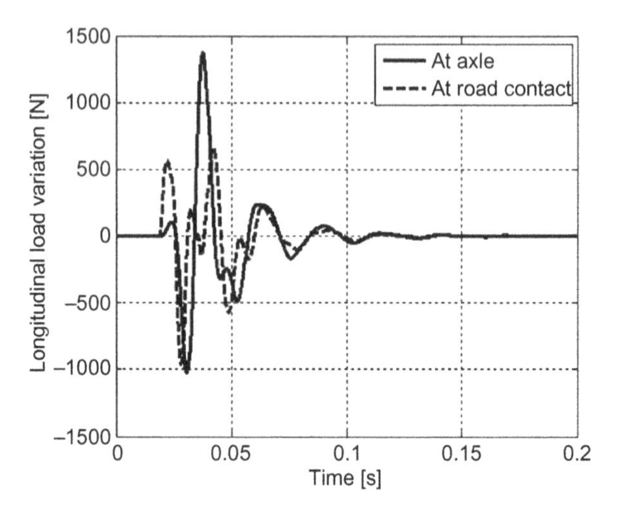

FIGURE 3.32 Longitudinal force variations for a sine bump obstacle (66 [km/h]).

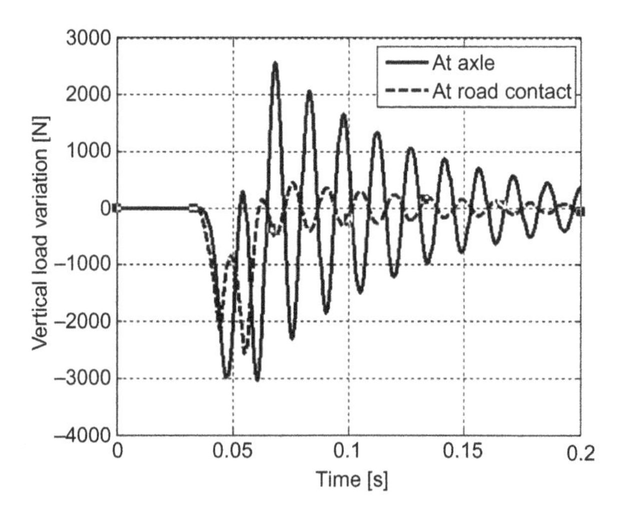

FIGURE 3.33 Vertical force variations for a pothole.

(see Section 3.1.2). As mentioned previously, the first peak in vertical axle force has been significantly increased with this higher velocity. The oscillation amplitudes, after the sine bump has passed, are also increased.

The force variation for the pothole (for 36 [km/h]) is shown in Figures 3.33 and 3.34.

In this case, the initial vertical force variation shows an amplitude of 2000 [N]. Note that, in this case, the effective road slope β_e varies much faster than in the other two cases, as one may conclude from Figures 3.18–3.20. The longitudinal force variation is more or less the opposite behavior

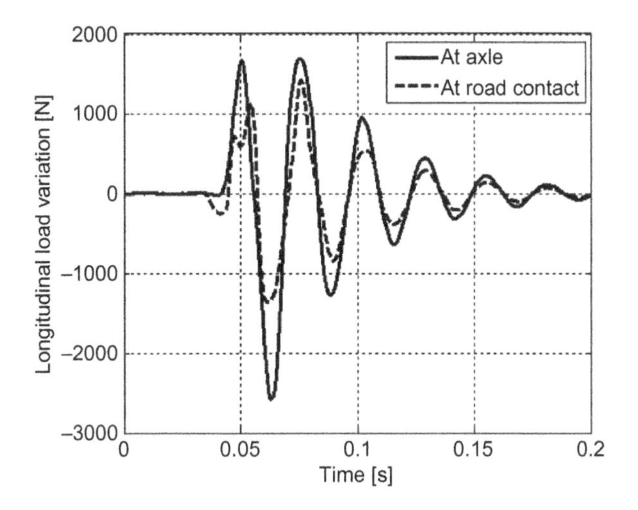

FIGURE 3.34 Longitudinal force variations for a pothole.

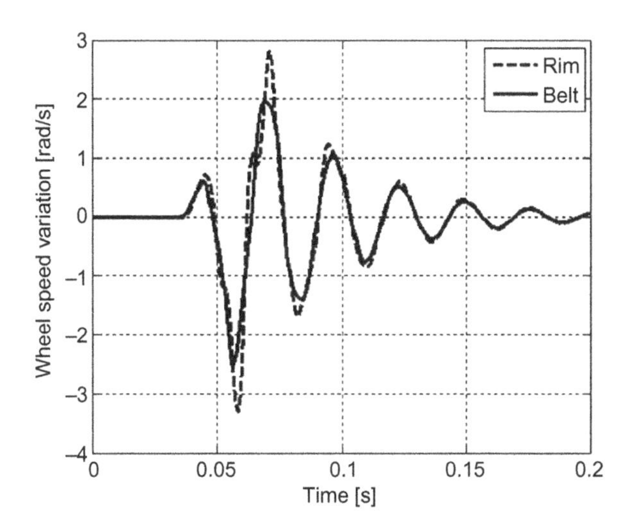

FIGURE 3.35 Wheel speed variation for a pothole.

compared to the sine bump, with a small change of the slip force in negative direction followed by a large positive change, when first entering the pothole. The wheel speed is shown in Figure 3.35. The maximum values correspond to slip and exceed 1.5%. In combination with braking on a slippery road, this value is large enough to have an effect on the ABS control.

Chapter | Four

Kinematic Steering

In Chapters 4 and 5, attention is paid to vehicle behavior. This chapter discusses when tire forces are neglected, which is usually referred to as Ackermann steering. Chapter 5 covers handling and stability behavior.

4.1 AXIS SYSTEMS AND NOTATIONS

To study the response of a vehicle in order to control inputs or disturbances, it is necessary to specify one or more coordinate systems to measure the position of the vehicle. The SAE method (SAE: Society of Automotive Engineer) [61] will be followed here. There are two main systems to measure the position of the vehicle. The first is an earth-fixed system, denoted by *XYZ*. The second system is a vehicle-fixed system *xyz* (lower case), as indicated in Figure 4.1. The origin of the *xyz* system is usually taken at the vehicle's center of mass. The orientation of the vehicle's axis system *xyz* with respect to *XYZ* is given by a sequence of three angular motions:

φ: roll rotation angle about the vehicle's x-axis
θ: pitch rotation angle about the vehicle's y-axis
ψ: yaw rotation angle about the vehicle's z-axis

A reference situation is considered in the case of earth-fixed and vehicle-fixed coordinate systems coinciding. The x-axis is taken from the vehicle's central plane, which is pointing forward and horizontal in the reference situation. The y-axis points to the driver's right-hand side and is horizontal in the reference situation. The z-axis points downward. The *velocity* of the vehicle is taken as the velocity of the center of mass, as measured in the *XYZ* system. Its components in the local *xyz* system are referred to as:

u: longitudinal velocity along the x-axis
v: side velocity along the y-axis
w: normal velocity along the z-axis

Essentials of Vehicle Dynamics.

Clearly, these velocity components, in general, will not be parallel to the ground plane. Therefore, we define the following velocity components:

forward velocity: horizontal velocity component $\perp y$-axis
lateral velocity: horizontal velocity component $\perp x$-axis

The *angular velocities* relative to the local *xyz* system are denoted as:

p: roll angular speed
q: pitch angular speed
r: yaw angular speed (yaw rate)

A top-down view of a car following a path with speed V is shown in Figure 4.2. Three angles are distinguished here that describe the projected

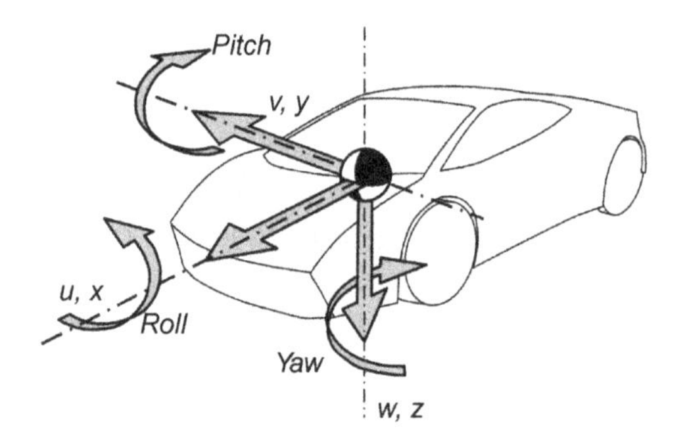

FIGURE 4.1 Vehicle local axis *xyz*.

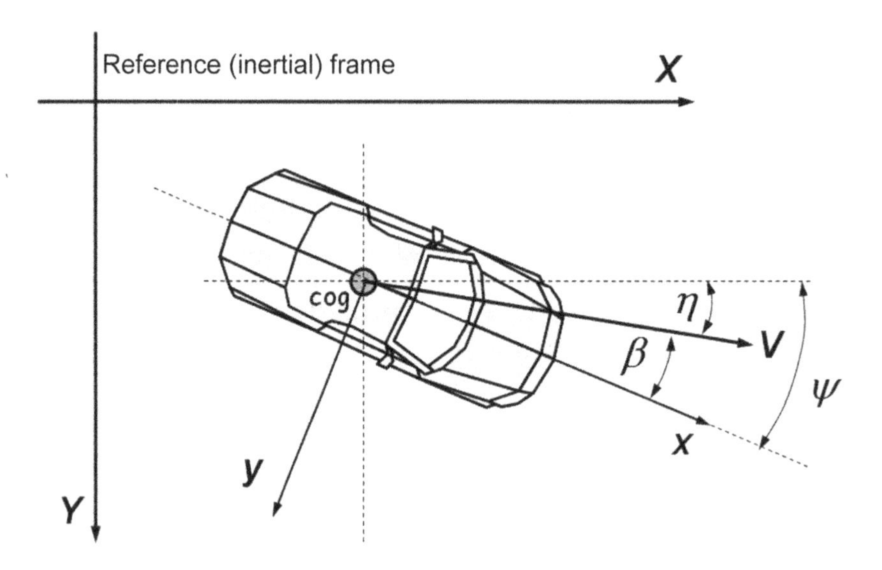

FIGURE 4.2 Projected view.

orientation of the vehicle, with respect to the earth-fixed coordinate system:

- *The heading angle* ψ between the projected x-axis and the global X-axis.
- *The side slip angle* β between the forward velocity (path tangent) and the projected x-axis (taken positive clockwise).
- *The course angle* η between the forward velocity and the global X-axis.

Consequently,

$$\eta = \psi + \beta = \psi(t=0) + \int_0^t r(\tau)\cdot d\tau + \beta \tag{4.1}$$

4.2 ACKERMANN STEERING

In general, steering of a vehicle at a finite velocity leads to side forces at the wheels, which will counteract the lateral force, acting on the vehicle. These side forces correspond to tire slip angles, according to certain lateral tire characteristics. In this chapter, we discuss a situation where the vehicle velocity is very small, such that no significant lateral forces act on the vehicle. Consequently, no side forces are required at the wheels and, ideally, maneuvering of the vehicle can be done by pure rolling of the wheels if we neglect tire turn slip. Any steering mechanism that satisfies these conditions is referred to as Ackermann steering (Figure 4.3). The pole, or center, of the vehicle rotation is on the line connecting the rear wheels.

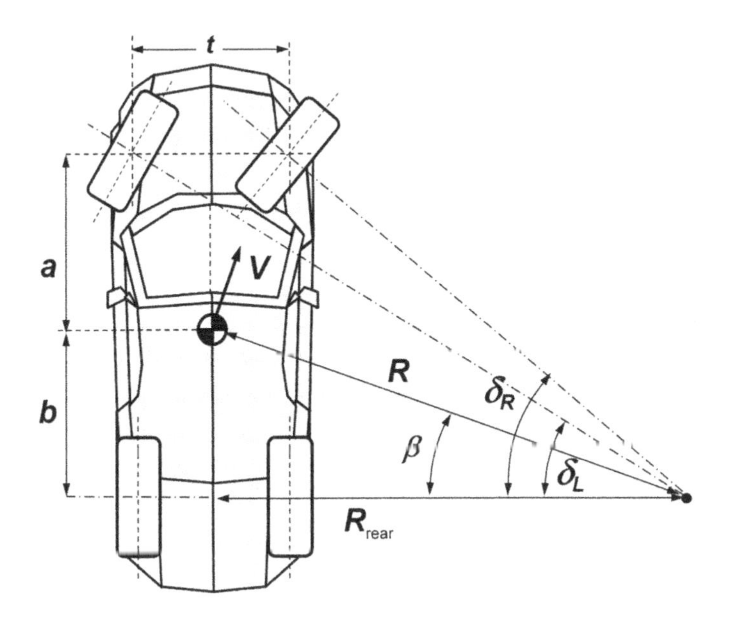

FIGURE 4.3 Low-speed cornering.

From this figure, one easily obtains, for pure rolling of wheels

$$\cot(\delta_L) - \cot(\delta_R) = \frac{t}{a+b} \equiv \frac{t}{L} \tag{4.2}$$

for track width t and wheelbase L. Unfortunately, no practical steering mechanism satisfies Eq. (4.2); some examples are illustrated next, discussed earlier by Genta and Morello in Refs. [10] and [11]. The Ackermann share of a steering mechanism is defined as (see Ref. [16], Chapter 1):

$$\text{Ackermann share:} \quad \frac{\delta_R - \delta_L}{\delta_R - \delta_{L,\mathrm{AM}}} \times 100[\%] \tag{4.3}$$

with inner wheel steering angle δ_R, outer wheel steering angle δ_L, and exact Ackermann outer steering wheel angle $\delta_{L,\mathrm{AM}}$, which satisfies Eq. (4.2).

Pure Ackermann steering means that the inner and outer wheel steering angles are not identical. The reader may easily verify that Eq. (4.2) results in a steering angle difference, as shown in Figure 4.4, for $t = 1.5$ [m] and $L = 2.76$ [m]. One observes a small difference for wheel steering angle, up to 5°, which is the range for normal handling situations. Clearly, applying the same steering angle for both the inner and outer wheels is an acceptable approximation for handling analyses (discussed in Chapter 5). For parking conditions with a large steering angle, one may expect a difference between inner and outer wheel steering angle, up to 10°.

Let us consider the quadrilateral steering mechanism, shown in Figure 4.5. We follow the analysis in Ref. [10]. The mechanism consists of four bars, connected at four rotational joints. The length of the lower bar, f_2, is found from

$$f_2 = f - 2 \cdot d \cdot \sin\gamma$$

The lateral horizontal distance H between the lower two joints follows from the lengths of the remaining bars and the steering angles:

$$H = f - d \cdot \sin(\gamma - \delta_L) - d \cdot \sin(\gamma + \delta_R)$$

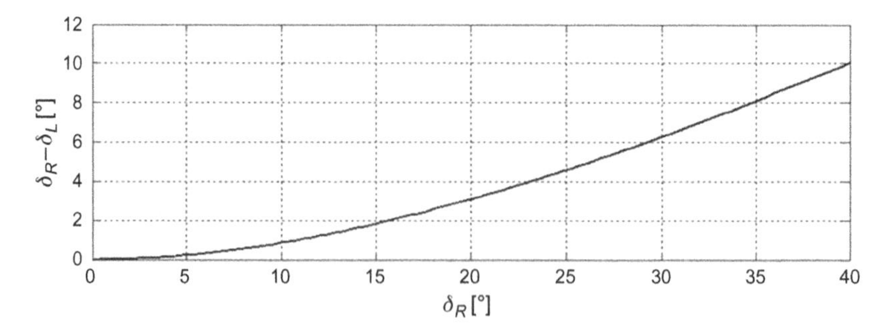

FIGURE 4.4 Difference between inner and outer steering angle, pure Ackermann steering.

Further, H follows from the orientation of the lower bar after steering:

$$H^2 = f_2^2 - [d\cdot\cos(\gamma-\delta_L) - d\cdot\cos(\gamma+\delta_R)]^2$$

These two results are sufficient to determine δ_L from δ_R for this steering mechanism, as we have done. We determined the Ackermann share according to Eq. (4.3) for different inner wheel steering angle and angle γ. Figure 4.6 shows the results for $f = 1.3$ [m] and $d = 0.2$ [m] (right plot). We also determined the difference between the outer wheel steering angle and the optimal Ackermann value according to Eq. (4.2), as shown in the left plot of Figure 4.6.

One observes errors in the outer wheel steering angle, compared to pure Ackermann steering, on the order of 2 to 3 [°], when the angle γ is varied between 16 and 24 [°]. The value γ = 20 [°] appears to be a reasonably good choice. The relative deviation in terms of Ackermann share varies from 60% to 130%, with γ = 22 [°] apparently giving a share closest to 100% over the

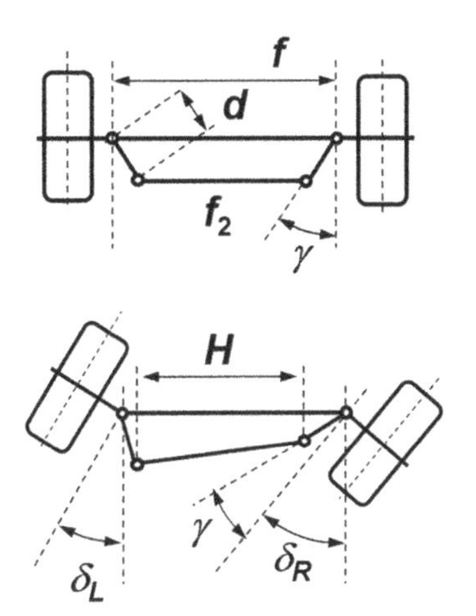

FIGURE 4.5 Example steering mechanism.

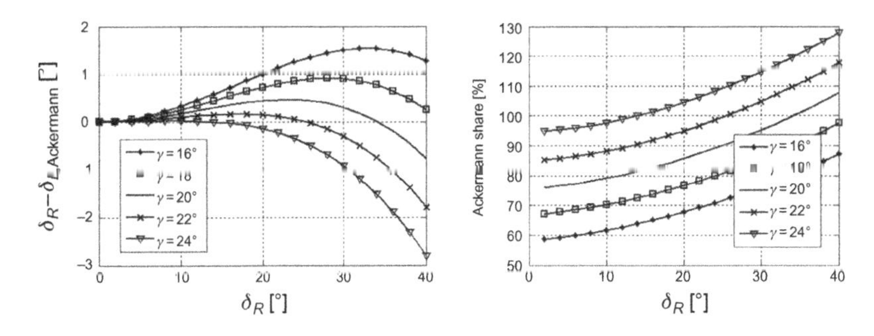

FIGURE 4.6 Deviation of outer wheel steering angle to Ackermann steering (left plot) and Ackermann share (right plot).

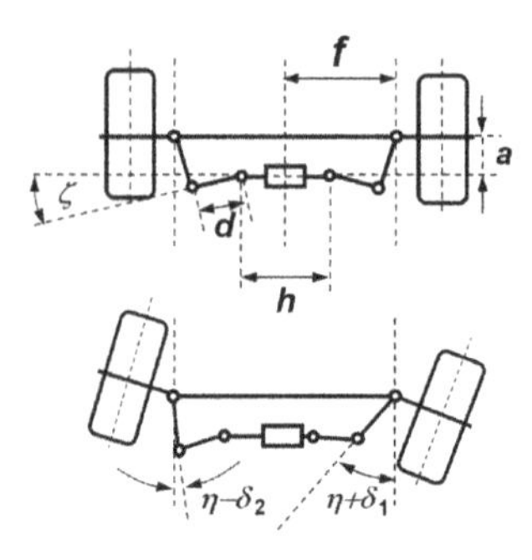

FIGURE 4.7 Rack and pinion system.

full range for the inner wheel steering angle. Note that, for small steering angle and therefore a small absolute deviation from Ackermann steering, the Ackermann share can still be quite different from 100%.

Next, we will consider the rack and pinion system, shown schematically in Figure 4.7. The system is assumed to be in one horizontal plane, defined by the parameters a, h, d, f, and angle ζ.

The rack is positioned at a distance a from the axle. There are six revolute joints, with fixed kingpins on the axle. We will investigate the kinematic properties for variations for these parameters.

The steering angles δ_1 and δ_2 can be determined for different positions of the rack, as indicated in the lower part of Figure 4.7. The end positions of the rack are known, the positions of the kingpins on the axle remain unchanged, and the positions of the other joints are found from the fact that the bar lengths remain unchanged. We completed this analysis for the following reference data:

$$
\begin{aligned}
a &= 0.25 \text{ [m]} \\
h &= 0.50 \text{ [m]} \\
d &= 0.32 \text{ [m]} \\
f &= 0.65 \text{ [m]} \\
\zeta &= 5 \text{ [°]}
\end{aligned}
$$

Results are shown in Figures 4.8 and 4.9. One observes:

- There is an optimal value for d when the other parameters are unchanged. The error, with respect to the Ackermann steering, can be reduced to less than 0.5 [°].
- With the a-value reduced (rack is positioned closer to the axle), the error is reduced as well.
- The same sensitivity is observed when the angle ζ is reduced to zero. Indeed, a quick survey confirms that this angle is chosen small in general, which likely is a consequence of packaging restrictions.
- Finally, the value for rack length h appears to be rather optimal for our choice of the other design parameters.

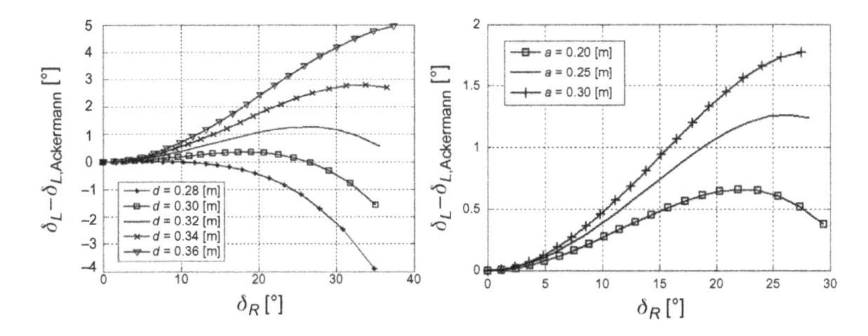

FIGURE 4.8 Deviation of outer wheel steering angle to Ackermann steering, for varying parameters *d* and *a*.

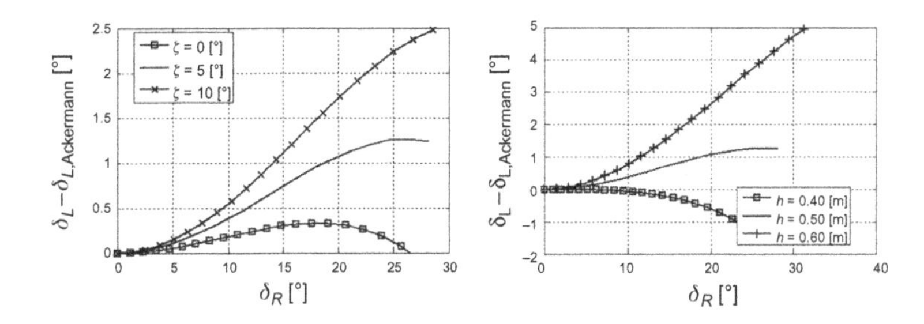

FIGURE 4.9 Deviation of outer wheel steering angle to Ackermann steering, for varying parameters ζ and *h*.

A design for an improved rack and pinion steering system has not been attempted here. The preceding analysis demonstrates the sensitivity of the steering system configuration with respect to certain parameters (it provides a qualitative analysis), with reference to pure Ackermann steering. We conclude that one will not reach the ideal Ackermann performance, but one could get satisfactorily close to it.

Steering is done with the intention to maneuver the vehicle, which means the driver is aiming for a certain trajectory curvature for a given steering angle δ, i.e., for a trajectory curvature gain. We choose the steering angle δ as the average of the left and right steering angle, as follows:

$$\cot\delta=\tfrac{1}{2}\cdot(\cot\delta_L+\cot\delta_R)=\frac{R_{\mathrm{rear}}}{L}$$

which is near the direct average of the steering angles. For kinematic steering, the following *gains* are distinguished (see also Figure 4.3):

$$\text{Trajectory curvature gain:}\quad \frac{1/R}{\delta}=\frac{1}{\delta\cdot\sqrt{b^2+L^2\cdot\cot^2\delta}}\approx\frac{1}{L} \tag{4.4a}$$

$$\text{Body slip angle gain:}\quad \frac{\beta}{\delta}=\frac{1}{\delta}\cdot\arctan\left(\frac{b}{\sqrt{R^2-b^2}}\right)\approx\frac{b}{R\cdot\delta}\approx\frac{b}{L} \tag{4.4b}$$

where the curve radius R is assumed to be large, compared to the wheelbase L. From Eq. (4.4a), the important expression for the Ackermann steering angle δ_{AM} follows as:

$$\delta = \delta_{AM} = \frac{L}{R} \tag{4.5}$$

This expression means that for negligible velocity, and therefore negligible lateral acceleration, the axle steering angle of a vehicle is equal to the ratio of wheelbase and path curve radius. In Chapter 5, we will discuss the relationship between steering angle and lateral acceleration under steady-state conditions. Expression (4.5) provides the first point on that curve, i.e., for $a_y = 0$.

We close this section with some remarks concerning the need for Ackermann steering. As mentioned previously, the situation of pure Ackermann steering is never reached, but can be closely approximated. In addition:

- The condition of negligible vehicle velocity is usually not satisfied.
- There is always side slip because of toe-in (usually present).
- Aligning effects lead to roll-induced steering.
- Suspension compliance and steering compliance lead to additional steering and therefore side slip.

Conversely, too large a deviation from pure Ackermann steering could result in significant tire wear, which in itself will affect the feedback of the steering performance and road conditions to the driver. Turn slip plays an important role in this, and this feedback preferably should not be influenced by side slip response or aligning torque effects.

4.3 THE ARTICULATED VEHICLE

The analysis of low-speed vehicle maneuverability is of interest in determining the amount of space required by the vehicle. In general, the designer's intention is to reduce the requirements for maneuvering space, i.e., to limit this space as much as possible. For a single passenger car, this analysis is straightforward. For articulated vehicles, such as a car–caravan or truck–trailer combination, a limited maneuverability space is not obvious. In general, the different articulations (car, caravan, trailer, etc.) follow different curve radii. As a result, the difference between the minimum inner radius and the maximum outer radius for the vehicle combination during maneuvering can be significantly larger than the width of the vehicle combination. This difference is called the swept path, which should preferably be as small as possible. The optimal situation is when the larger trailer axle is following exactly the same curve as the first vehicle's axle. Consider a car–trailer combination, as schematically shown in Figure 4.10. The trailer is assumed to have one axle that may be steered.

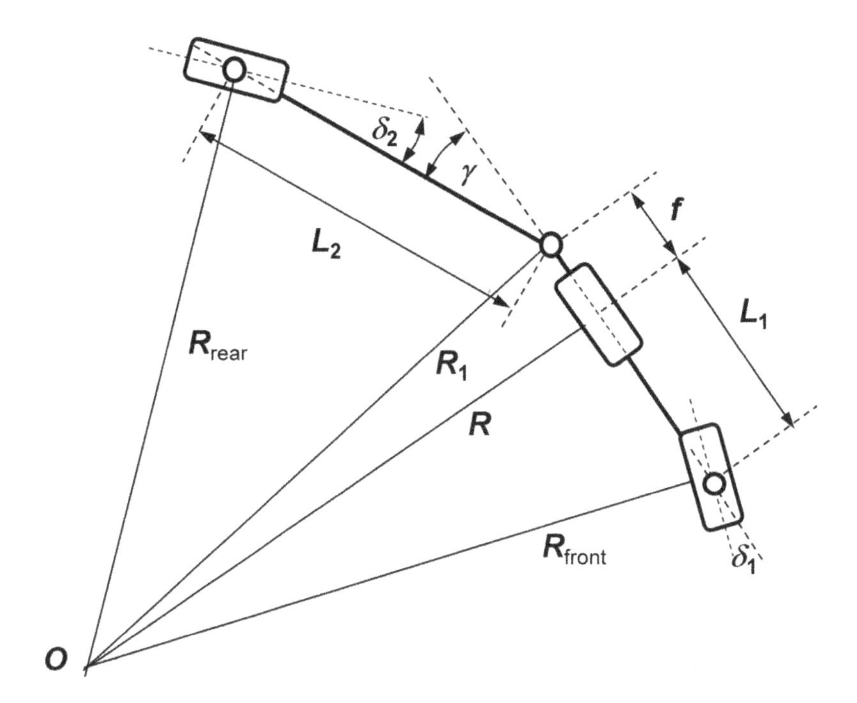

FIGURE 4.10 Car—trailer combination with steerable trailer axle.

The question is how to steer this axle so that the curve radii at vehicle's front axle and the trailer's back axle are identical to create a swept path as minimal as possible.

Let us begin with the situation with the trailer axle as not steered, i.e., $\delta_2 = 0$, and the radius R_{rear} perpendicular to the trailer. Assume that all curve radii are large compared to the dimensions of the vehicle. In that case, one may write

$$\gamma = \pi - \arccos\left(\frac{L_2}{R_1}\right) - \arccos\left(\frac{f}{R_1}\right) \approx \frac{L_2 + f}{R_1}$$

$$R_1 \approx R = \frac{L_1}{\tan(\delta_1)} \approx \frac{L_1}{\delta_1}$$

A combination of these relationships gives the trailer angle gain:

$$\frac{\gamma}{\delta_1} = \frac{L_2 + f}{L_1} \tag{4.6}$$

In other words, a large articulation angle γ is obtained under low-speed conditions, if the coupling overhang f is large or there is a large distance between kingpin and trailer axle.

The radii R_{front} and R_{rear} can be expressed as follows:

$$R_{\text{front}} = \frac{L_1}{\sin\delta_1}$$

$$R_{\text{rear}} = \sqrt{f^2 + \frac{L_1^2}{\tan^2\delta_1} - L_2^2}$$

Eliminating δ_1 leads to

$$R_{\text{rear}}^2 = R_{\text{front}}^2 - L_1^2 - L_2^2 + f^2$$

As illustrated, we have determined the relative off-tracking

$$\frac{R_{\text{front}} - R_{\text{rear}}}{R_{\text{front}}}$$

in percent, for different values of L_2 and f, for $R_{\text{front}} = 10$ [m], and a fixed car wheelbase of 2.76 [m] (Figure 4.11). One observes a dominant effect from the position of the trailer axle, with respect to the kingpin between car and trailer. The parameter f has only a minor effect.

If $\delta_2 \neq 0$, the radius R_{rear} is no longer perpendicular to the trailer. We use the cosine rule in the triangle $R_{\text{rear}}–R_1–L_2$ and find, under the same assumption of large curve radii, that

$$\frac{\gamma}{\delta_1} = \frac{L_2 + f}{L_1} - \frac{\delta_2}{\delta_1} \tag{4.7}$$

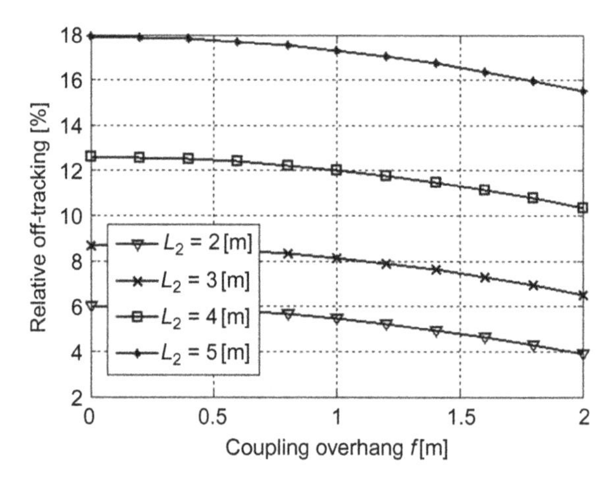

FIGURE 4.11 Off-tracking for different vehicle parameters.

We choose δ_2 such that $R_{\text{rear}} = R$. The cosine rule then results in the following relationship:

$$R^2 + f^2 = R_1^2 = R_{\text{rear}}^2 + L_2^2 - 2 \cdot R_{\text{rear}} \cdot L_2 \cdot \sin \delta_2 \approx R^2 + L_2^2 - 2 \cdot R \cdot L_2 \cdot \delta_2$$

Hence,

$$R = \frac{L_2^2 - f^2}{2 \cdot L_2 \cdot \delta_2}$$

However, we have also

$$R = \frac{L_1}{\delta_1}$$

Consequently,

$$\frac{\delta_2}{\delta_1} = \frac{L_2^2 - f^2}{2 \cdot L_1 \cdot L_2} \quad \text{and} \quad \frac{\delta_2}{\gamma} = \frac{L_2^2 - f^2}{(L_2 + f)^2} \tag{4.8}$$

where we used Eq. (4.7). Expression (4.8) describes how the trailer axle steering angle should be linked to the kingpin angle γ to minimize the swept path at low speed. We plotted this ratio in Figure 4.12 for various values of L_2 and f; this ratio was plotted for the same values for R_{front} and wheelbase L_1 as in Figure 4.11. Clearly, small values of f (such as for compact cars) require the largest trailer axle steering, which is consistent with the largest off-tracking in Figure 4.11. Less steering is required for larger values of f and smaller values of L_2. Note that, for a conventional trailer, the kingpin angle γ increases with L_2 and f, cf. Equation (4.6), and therefore, a smaller gain is required to minimize the off-tracking.

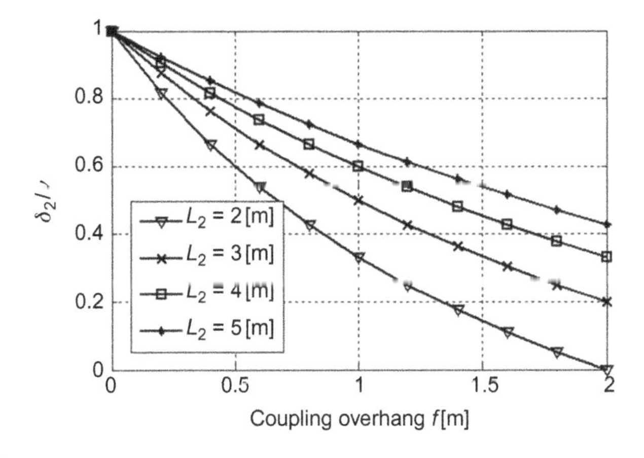

FIGURE 4.12 Optimal trailer axle steering gain with respect to kingpin angle.

Chapter | Five

Vehicle Handling Performance

In this chapter, the handling and stability behavior of vehicles is discussed. Because the contact between vehicle and road is the dominant factor for this behavior, major emphasis is put on the influence of tire properties.

Handling and stability are two properties of the vehicle that are of major importance for the following reasons. First, vehicle behavior should be safe, which means that input from the driver should not lead to an excessive vehicle response that makes the vehicle uncontrollable for the driver. Clearly, a vehicle has certain physical limits. If the friction limits at one of the tires are exceeded, full sliding of this tire may occur. This must be avoided and the main controller of the vehicle to do so is the driver. This means that the driver is constantly monitoring the vehicle behavior with the aim of responding to any vehicle behavior judged to be too critical.

Furthermore, car manufacturers ensure their vehicles have smooth handling behavior, in the sense that it is judged positively by their test drivers and customers. The car should give the driver a feeling that he or she is in control of the car, that the car is responsive to driver inputs, that sudden deviations from a path will damp out quickly, that body roll is limited, etc. This vehicle performance should be experienced over a large range in both longitudinal and lateral acceleration, which may be in conflict with the previous requirement of noticing critical behavior in time in order to perform adequate corrections.

In Section 5.1, we will explain the concept of **good handling** in more depth, with an emphasis on subjective and objective methodology strategies. We will discuss these strategies and how to assess good handling using vehicle tests and simulations. Good handling is strongly related to the experience of the driver in handling the vehicle and how the driver responds to vehicle's behavior under varying circumstances. Will the driver experience an increased workload to maintain a minimum closed-loop performance, and in

Essentials of Vehicle Dynamics.

what way can we assess this workload? These questions are discussed in Chapter 6, where we address the vehicle–driver interface.

In contrast to the kinematic steering case, we will consider vehicle models that allow us to investigate vehicle performance with tire forces considered. Tire forces should be considered in nearly all cases, as they occur when the driver is steering, braking, or accelerating. The simplest vehicle model to address these situations is the **single-track model** (also referred to as the bicycle model). Because of its simplicity, this model is usually used as a first step in vehicle performance analysis for applications such as those involving active steering control. With control measures (such as wheel-by-wheel braking) that have the objective of limiting excessive yaw motion, one must distinguish between left and right tire road contact. The first extension of the bicycle single-track model is to the double-track vehicle model. The single-track model is treated in detail in Section 5.2, with some comments on the extension to the double-track model.

Note that both models may include the full nonlinear tire characteristics. This means that, in case of constant speed, one is faced with a second-order set of **nonlinear differential equations** and the model being used should be treated using the appropriate tools. Consequently, multiple singular points may be expected, depending on the model parameters, with sudden transitions from one to multiple points with a slight change to these parameters (of which the parameters related to the tire will appear to be the dominant ones). Small variations in the initial values do not automatically guarantee small variations in the solution and the local stability analysis is not straightforward.

We explain the steady-state characteristics in Section 5.3, including a discussion on the basic handling curves arising from steady-state circular tests. The steady-state solutions are the singular points for the nonsteady solutions, which are discussed in Section 5.4. The occurrence of unstable motion in yaw (i.e., referred to as **yaw instability**) strongly depends on the tire characteristics, which can be expressed in a **handling diagram**. This handling diagram is directly related to the basic handling curves discussed in Section 5.3 and is specifically suited for studying nonlinear vehicle performance; it will be discussed in detail in Section 5.5. Nonlinear systems are usually primarily used to investigate the qualitative behavior and trends in response to variations in control and system parameters. That means that one is verifying the essential system characteristics, which is, in general, more difficult than fitting a model to test results. If the qualitative match is not satisfactory, the quantitative match will never be possible.

Graphic assessment tools, such as the handling diagram, allow for visualization of these sensitivities. Another way to visualize these sensitivities is to exploit the phase plane approach, which is a common tool for nonlinear dynamical systems in two dimensions. Singular (i.e., steady-state) points and the behavior near these points (i.e., the stability behavior of the vehicle), as well as global stability properties, can be illustrated in an elegant way. Other graphical assessment methods such as the stability diagram, the MMM, and the g-g diagram, will be discussed as well.

5.1 CRITERIA FOR GOOD HANDLING

Different methodologies exist for the assessment of vehicle performance: open-loop and closed-loop tests. With **open-loop tests**, the driver gives specific input to the vehicle and therefore acts more or less as a steering robot. These steering inputs could be a step or ramp input, a fixed steering input with possibly increasing speed, a steering impulse, or a sinusoidal steering input for different frequencies, etc.

For a closed-loop test, the driver responds to the vehicle's behavior to fulfill a specific task. Examples of this include the double lane change maneuver, with the maximum speed as one of the performance metrics, and the slalom test, with cones positioned at a specific distance (usually 18 m).

We refer to the available ISO descriptions for the various tests [62, 63, 64], see also [19]. This distinction refers to the **vehicle control by the driver**, with the driver considered as part of the feedback control loop or not. Methodologies on the assessment of vehicle performance can be done through **judgment by the driver** (based on open- or closed-loop tests) or though measured vehicle behavior (based on a known driver input, which is usually, but not necessarily, based on open-loop tests). These methodologies may be structured as follows:

- **Subjective methodology strategies**
  - *Performance tests* refer to a specific task, such as determining a maximum speed (lane change), minimum lateral deviations, steering motions (straight lane test), etc.
  - *Rating scales* questionnaires based on a scale of some magnitude (5-point, 10-point scale) followed by data reduction.
  - *Open questions* are other data gained that are considered additional to the previous two strategies.
- **Objective methodology strategies**
  - Reference maneuvers with instrumented vehicles.

When the vehicle performance is subjectively assessed, the test driver scores handling based on concepts such as steering feel, controllability, feeling of safety, straight ahead stability, etc. Tests can be selected so that the vehicle behaves more discriminately with respect to these assessment aspects. Examples of this include braking at different road frictions for the left and right tires (μ split), releasing the throttle or braking while cornering, single sinus steer input, releasing the steering wheel while cornering, etc. However, there is very little standardization in subjective handling assessment, meaning that vehicles can be ranked, but a comparison of these assessments completed by various organizations is difficult. Moreover, this approach produces a full vehicle evaluation, meaning the cause of low scores is not immediately clear and requires further investigations.

One should note that not only the ratings themselves are important, but also the deviations among the ratings. These deviations allow a distinction between **individual** assessments by the subjects and assessments with a high level of **consistency**.

Usually, the set of original variables is reduced to a set of principal components or factors that can be regarded as orthogonal (statistically independent) to each of the other components. Principal components are weighted linear combinations of the original measured variables. A next step could then be to reduce this set to new linear combinations with maximum discrimination between two or more clusters (related to maneuverability or stability). This second step is referred to as discriminant function analysis (DFA). Some researchers skip the principal component analysis (PCA) and apply a direct reduction based on the criterion of maximum discrimination, followed by an interpretation toward more independent factors.

Completing a PCA analysis on both open-loop test results and subjective ratings would allow for further correlation studies between the objective and subjective testing procedures.

An alternative approach is the objective reference test. A large number of these tests have been agreed upon within ISO standardization committees. A vehicle is fitted with instruments and given a clearly defined input, for which vehicle behavior is characterized using measured readings such as reaction times, the lag of certain variables in relation to the steering input, etc. An advantage of the objective reference test is that these maneuvers can be imitated in simulation models. In this way, the effect of design changes on vehicle behavior can be assessed at an early stage. However, these measurements only provide a limited picture of the subjective feel of the vehicle for the test driver.

In Table 5.1, some known criteria of good handling are listed that correlate well with driver preferences. We note that most of these criteria can be verified with simple vehicle models. Hence, to identify more refined differences between vehicles, this list is insufficient.

Two ISO tests are described here in some detail:

ISO 4138 : Steady-state circular test

ISO 7401 : Lateral transient response test

5.1.1 ISO 4138: Steady-State Circular Test

The objective of this test is to determine the steady-state vehicle directional control response. A circular path is followed with increasing vehicle speed. The steady-state behavior is usually described in terms of steering wheel angle δ as a function of lateral acceleration a_y (handling curve). This relationship is directly related to the nonlinear axle characteristics, with the shape of the curve describing the important yaw stability properties of the vehicle. We return to this in subsequent sections of this chapter. Other vehicle outputs can be presented versus lateral acceleration, such as

- vehicle side slip angle
- vehicle roll angle

TABLE 5.1 Criteria of Good Handling, an Overview (see also Refs. [33] and [48])

Good Handling Criteria	Clarification
Short time delay	Long delays between command input and vehicle response (yaw rate, lateral acceleration, etc.) should be avoided.
Gains not too large or too small	A large gain is experienced as nervous behavior with only a small input leading to a severe response. On the other hand, a significant driver input to have only a small vehicle response (low gain) is not preferable either.
Compromise between responsiveness and stability	Stability and responsiveness contradict each other. Vehicle instabilities require the driver to provide continuously stabilizing control, which is regarded as disadvantageous. Too high stability leads to delays, which one tries to avoid.
Small body slip angle	This means that the driver is looking in the direction the vehicle is moving. In other words, the heading angle and course angle coincide. On the other hand, a nonzero body slip angle is part of the feedback to the driver.
Response immunity to external disturbances	A vehicle is always subject to external disturbances, such as crosswind gusts, road disturbances, etc.
Small roll response	The best conditions for accurate sensing by the driver of the vehicle environment are obtained for minimum vehicle body roll.
Consistent vehicle behavior	Large changes in vehicle response with speed, loading, road surface conditions, lateral acceleration level, etc. are undesirable from the point of view of building up driving skills. This would lead to consistent vehicle properties over all running conditions and consequently, that the driver is not "warned" when approaching the saturation limits for the tires. Hence, consistent behavior should be limited to nonextreme vehicle behavior.

leading to the following gradients (performance metrics).

Steering wheel gradient:	$\frac{\partial \delta}{\partial a_y}(a_y = 0)$
Side slip gradient:	$\frac{\partial \beta}{\partial a_y}(a_y = 0)$
Roll angle gradient:	$\frac{\partial \varphi}{\partial a_y}(a_y = 0)$

5.1.2 ISO 7401. Lateral Transient Response Test

The primary objective of this standard test is to determine the transient response behavior of a vehicle. Typical characteristic values and functions in the time domain and frequency domain are time lags, response times, gains (lateral acceleration, yaw rate), and overshoot values. An outline of most of the tests in ISO 7401 is given in Table 5.2.

TABLE 5.2 Test Methods, Included in ISO 7401

Test Method	Description	Performance Metrics
Time Domain		
Step input	Rapid change (ramp) of steering angle δ	Response time, peak response time, overshoot values for vehicle motion response (yaw rate a_y)
Sinusoidal input (one period)	One period steering wheel input (0.5 Hz, 1 [Hz] optional) at $a_y = 4$ [m/s^2] (2 or 6 [m/s^2] optional)	First peak values of a_y, yaw rate r, time lags of a_y, r to δ, gains (a_y, r) w.r.t. steering input δ
Frequency Domain		
Random input	Continuous inputs covering a frequency area, for $a_y < 4$ [m/s^2] (2 [m/s^2] recommended)	Steady-state gains (a_y, r) w.r.t. steering input δ, bandwidth, peak ratio, equivalent time
Pulse input	Triangular wave form steering input of 0.3–0.5 s width, with δ_{max} cf. $a_y = 4$ [m/s^2] steady state	Similar to random input
Continuous sinusoidal input	Three successive periods in δ, with increasing frequency and $a_y < 4$ [m/s^2]	Amplitudes (a_y, r), gains (a_y, r) w.r.t. steering input δ, phase shifts between a_y, r, and steering angle δ

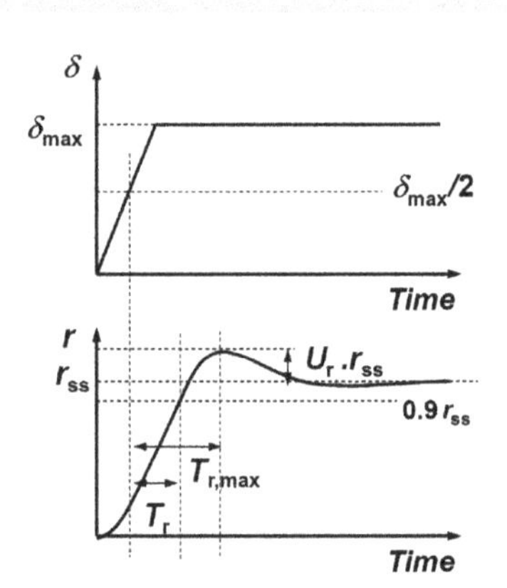

FIGURE 5.1 Ramp steer input and vehicle yaw rate response.

Two tests are discussed in more detail: the step steer input test (also called "J-turn") and the random steer input test.

The input and output for a **J-turn** test are shown in Figure 5.1. The first plot illustrates the steering wheel input and the second plot shows the vehicle response, e.g., the yaw rate. With a constant steering angle input, the vehicle will end up in a steady-state circular curve. The transition from straight ahead

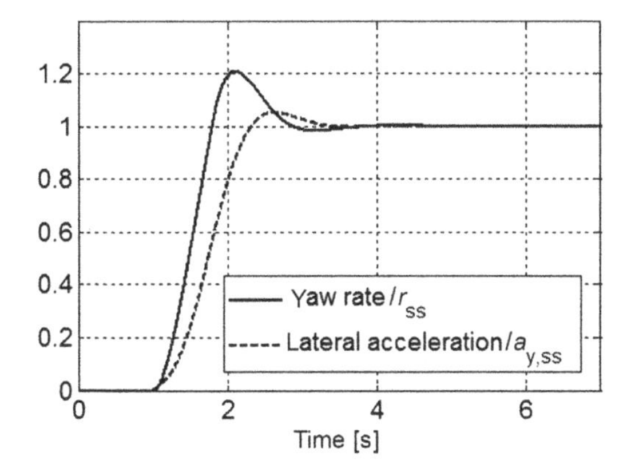

FIGURE 5.2 Yaw rate and lateral acceleration response to ramp steer input.

driving to these steady-state conditions requires a certain response time, and usually leads to overshoot behavior.

Response time T_r is measured from the time that δ reaches 50% of its final value up to the time that 90% of the steady-state value of the vehicle response is reached. Peak response time $T_{r,max}$ is the time required to reach the maximum yaw rate response. Overshoot U_r is defined as the relative difference between the peak value and the yaw rate steady-state value r_{ss}. Other performance metrics considered include the lag time between yaw rate and lateral acceleration and the TB factor. This factor is the product of the steady-state side slip angle and the yaw rate peak time, determined from the step steer test. Xia and Willis refer to this parameter as a vehicle characteristic [33,58]. The TB factor was shown to correlate well with subjective ratings, meaning that a driver is not able to distinguish properly between not being oriented in the driving direction, and a lag in yaw rate response (especially with excessive cornering). When a vehicle experiences a step or ramp steer input, the front tire side forces build up first, and the vehicle begins to yaw. Next, the rear tire side forces build up and, likewise, the lateral speed (drifting). This drifting is therefore delayed with respect to the yaw rate. Drifting (lateral sliding) contributes to the lateral acceleration. Consequently, the lateral acceleration will be delayed with respect to the yaw rate. This means that the driver will experience yaw before the vehicle side force. This is illustrated in Figure 5.2, where the yaw rate and lateral acceleration responses (divided by the steady-state values r_{ss} and $a_{y,ss}$, respectively) are depicted for a ramp steer input. One observes a larger lag for the lateral acceleration.

The random steer test is the experimental assessment of the vehicle frequency performance with lateral acceleration up to 4 [m/s^2]. This behavior can be expressed graphically in Bode diagrams of gain and phase (see Appendix 4). A diagram of yaw rate and some performance metrics is shown in Figure 5.3 for a single-track vehicle model. The upper plot shows the gain, which is the ratio of yaw rate amplitude and steering angle amplitude for a

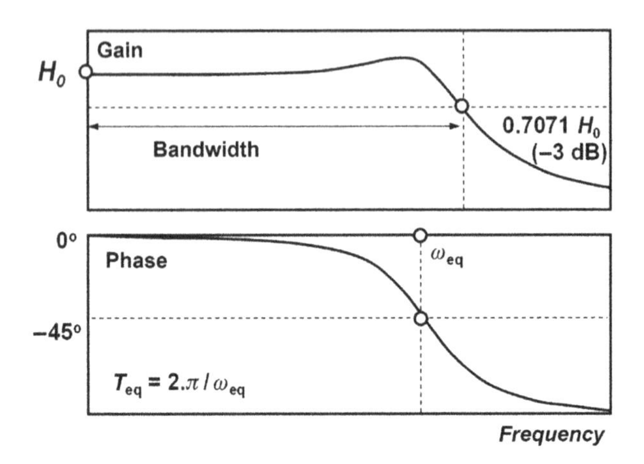

FIGURE 5.3 Random steer test.

sinusoidal steering input. The lower plot shows the phase difference between the yaw rate and the steering angle. Note the resonance peak in the gain and the phase moving to -90 [°] for large frequency. This frequency behavior will be discussed in more detail in subsequent sections [54].

Bandwidth ω_b	:	range of frequencies for which the gain equals at least $0.707 = 1/\sqrt{2}$ of the steady-state gain H_0
Equivalent time T_{eq}	:	$2.\pi/\omega_{eq}$ where ω_{eq} corresponds to a phase lag of 45[°]

Weir and Dimarco suggested plots for the steady-state yaw rate gain H_0 (ratio of yaw rate and steering angle) versus the equivalent time T_{eq} and indicated areas of preferred vehicle behavior for American cars around 1978 [57]. A large value for H_0 means that small changes in the steering angle lead to large effects in response, which are conditions only an experienced driver could handle. A small value for H_0 means that a large steering input is required to have a sufficient vehicle response, which is not preferable either.

Consequently, the optimal range in steady-state gain is a band between a maximum and minimum value. For an expert driver, this band covers larger values for H_0 than for the average driver. The equivalent time expresses the dynamic response of the vehicle to sudden changes in the steering angle (such as avoiding unexpected obstacles). The smaller T_{eq}, the better the vehicle is able to follow the driver input to avoid accidents. In Figure 5.4, some (H_0, T_{eq}) values are shown for European cars that were tested around 1995.

Two bands, originated from Weir and Dimarco [57], have been added to this figure: one for the expert driver and one for the typical driver. Various clusters of cars have been included, and one observes the dependency of vehicle cluster on T_{eq}. Larger cars tend to show more time lag for quick

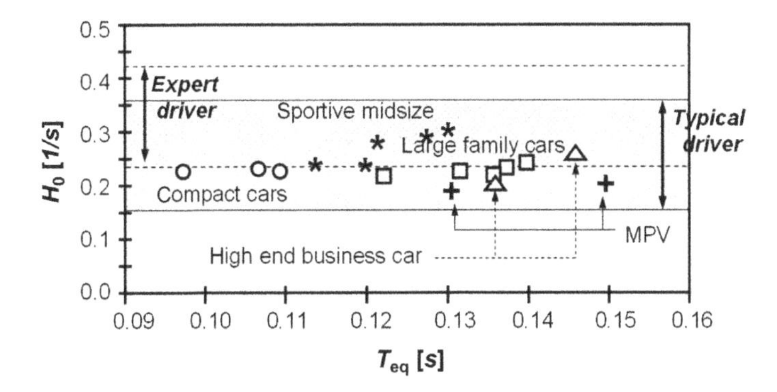

FIGURE 5.4 Weir–Dimarco plot for European cars.

driver input. Remarkably, all cars seem to have steady-state gain close to the lower boundary for the expert driver, which is in the lower half of the area for a typical driver. The more sporty cars show a larger yaw rate gain.

5.2 SINGLE-TRACK VEHICLE MODELING

5.2.1 The Single-Track Model

In this section, the more fundamental aspects of possible vehicle motions are addressed. Instead of discussing simulation runs of complicated vehicle models, we use the simplest automobile model, which runs at a constant speed over an even horizontal road surface.

The steer and slip angles are restricted to relative small values. The driving force required to keep the speed constant is assumed to remain small with respect to the lateral forces acting on the tires. Brake and drive forces are neglected. The vehicle body roll and possible pitch behavior are neglected too. This assumption holds if the height of the center of gravity (CoG) of the vehicle is small compared to the track width. Another situation when roll and pitch motions are negligible is when friction coefficients between tires and road are not too large.

Using these assumptions, the front wheels can be considered one system with an overall lateral force, aligning torque, and a slip angle obtained from the combined effect of left and right wheel. A similar simplification holds for the rear wheels. That means that tire characteristics are replaced by axle characteristics, which finally leads to a vehicle model as shown in Figure 5.5. Observe that the pole of the rotation has moved forward compared to kinematic steering (see Figure 4.3) and that, therefore, the driver must turn his or her head slightly to the left (i.e., in outer direction), to see where he or she is going.

The slip angles for the front and rear axle are denoted by α_1 and α_2, respectively. The total lateral force and aligning torque for front and rear axles are denoted by F_{y1} and F_{y2}, and M_{z1} and M_{z2}, respectively. The aligning torques will be disregarded in the forthcoming analysis.

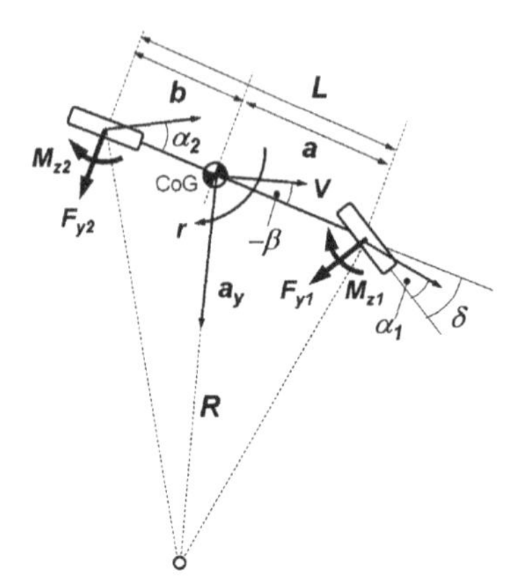

FIGURE 5.5 One-track vehicle handling model.

The vehicle CoG is positioned between both axles, at distances a and b from front and rear axle, respectively. The horizontal behavior of the vehicle is described using a local lateral velocity v at the CoG, a local forward velocity u, yaw rate r, and body slip angle β. Because the lateral velocity will be small compared to the forward velocity u, this body slip angle follows from

$$\beta = \arctan\left(\frac{v}{u}\right) \approx \frac{v}{u} \tag{5.1}$$

The total vehicle velocity is denoted as V. For small body slip angle, u is equal to V up to second-order; therefore, we replace u with V when appropriate.

The equations of motion follow from the statements that equilibrium must hold in longitudinal, lateral, and yaw direction. The longitudinal and lateral vehicle accelerations must be balanced by the horizontal tire forces. The yaw moment acting on the vehicle must be balanced by the tire forces moment (where aligning torques are neglected). In the global coordinate system (Figure 5.6), this means that the following equations hold:

$$\begin{aligned} m \cdot \ddot{X} &= F_X \\ m \cdot \ddot{Y} &= F_Y \\ J_z \cdot \ddot{\psi} &= M_Z \end{aligned} \tag{5.2}$$

where m is the vehicle mass and J_z is the polar moment of inertia in z direction (yaw moment of inertia), and F_X, F_Y, and M_Z are the total forces and moments.

Moving from a global to a local coordinate system, we can define the locally defined forces and moments as

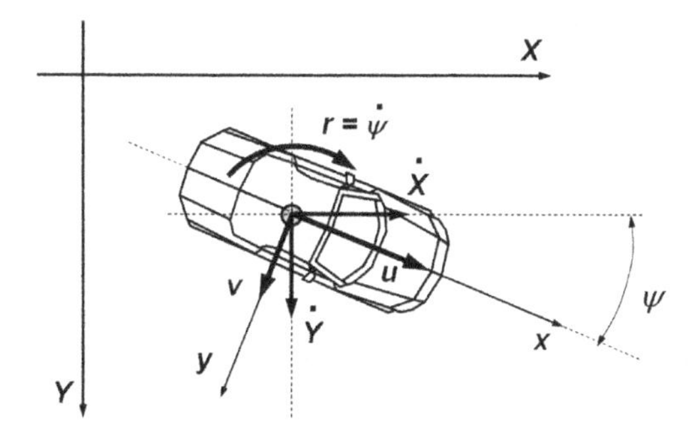

FIGURE 5.6 Vehicle behavior for local and global coordinates.

$$
\begin{aligned}
F_x &= F_X \cdot \cos\psi + F_Y \cdot \sin\psi \\
F_y &= -F_X \cdot \sin\psi + F_Y \cdot \cos\psi \\
M_z &= M_Z
\end{aligned}
\tag{5.3}
$$

The velocity vector is transformed as follows:

$$
\begin{pmatrix} \dot{X} \\ \dot{Y} \end{pmatrix} = \begin{pmatrix} \cos\psi & -\sin\psi \\ \sin\psi & \cos\psi \end{pmatrix} \cdot \begin{pmatrix} u \\ v \end{pmatrix}
\tag{5.4}
$$

and therefore, for the accelerations

$$
\ddot{X} = \dot{u} \cdot \cos\psi - \dot{v} \cdot \sin\psi - \dot{\psi} \cdot (u \cdot \sin\psi + v \cdot \cos\psi)
$$

$$
\ddot{Y} = \dot{u} \cdot \sin\psi + \dot{v} \cdot \cos\psi + \dot{\psi} \cdot (u \cdot \cos\psi - v \cdot \sin\psi)
$$

In combination with Eqs. (5.2) and (5.3), one arrives at

$$
\begin{aligned}
m \cdot (\dot{u} - v \cdot r) &= F_x \\
m \cdot (\dot{v} + u \cdot r) &= F_y \\
J_z \cdot \dot{r} &= M_z
\end{aligned}
\tag{5.5}
$$

We substitute the tire forces according to Figure 5.5 and account for possible external force F_{ye} and moment M_{ze}.

Then, the following handling equations arise for the bicycle one-track vehicle handling model.

$$
m \cdot (\dot{v} + u \cdot r) = F_{y1} + F_{y2} + F_{ye}
\tag{5.6a}
$$

$$
J_z \cdot \dot{r} = a \cdot F_{y1} - b \cdot F_{y2} + M_{ze}
\tag{5.6b}
$$

To be more accurate, the pneumatic trails front and aft should have been accounted for in the distances a and b in Eq. (5.6b), but these effects will be disregarded.

The first equation in (5.5) describes the forward behavior of the vehicle, i.e., where F_x is the total longitudinal force (drive, brake, resistances). Remember that we assume the vehicle speed V to be constant, which means that a drive force should exist to counteract rolling resistance, aerodynamic drag, slope resistance, etc. In addition, we note that the second term in the left-hand side of this equation is of higher order (both lateral speed and yaw rate assumed small).

REMARKS REGARDING FORCES ACTING ON THE VEHICLE

1. The external force and moment in expressions (5.6a) and (5.6b) could be from an aerodynamic source. Considering only the planar motion of the vehicle, a side force $F_{y,\text{aer}}$ and a yaw moment $M_{z,\text{aer}}$ might be included in expressions (5.6a) and (5.6b), of the form

$$F_{ye} = F_{y,\text{aer}} = \frac{1}{2} \cdot \rho_{\text{air}} \cdot S \cdot C_y \cdot V_{\text{r}}^2 \quad M_{ze} = M_{z,\text{aer}} = \frac{1}{2} \cdot \rho_{\text{air}} \cdot S \cdot L \cdot C_z \cdot V_{\text{r}}^2$$

   with air density ρ_{air}, vehicle frontal area S, relative wind speed V_{r}, wheelbase L, lateral force coefficient C_y, and yawing moment coefficient C_z. The last two coefficients are functions of the relative wind direction angle (see Ref. [1] and Chapter 21 in Ref. [11]). These additional loads may be due to crosswind disturbance.

   Similarly, an aerodynamic vehicle lift or down force may be accounted for, as well as aerodynamically induced roll and pitch moments. These forces will result in larger or smaller wheel loads, and therefore, affect the tire–road contact characteristics.

2. When brake and/or driving forces are not neglected, they contribute to the side force at each steered axle. For the vehicle in Figure 5.5, this leads to modified model equations for the single-track model:

$$m \cdot (\dot{v} + u \cdot r) = F_{y1} \cdot \cos\delta + F_{y2} + \sum_{\text{front wheels}} F_x \cdot \sin\delta$$

$$J_z \cdot \dot{r} = a \cdot F_{y1} \cdot \cos\delta - b \cdot F_{y2} + \sum_{\text{front wheels}} a \cdot F_x \cdot \sin\delta$$

3. We assume a flat road. A road with a transversal slope ζ will initiate an additional side force of $F_{ye} = m \cdot g \cdot \sin(\zeta)$.

Expressions (5.6a) and (5.6b) can also be derived from Lagrange's equations (see Appendix 5). We follow the approach in Ref. [32], but neglect roll. The kinetic energy T of the vehicle is due to forward speed, lateral speed, and yaw rate. The lateral speed and yaw rate contribute to the cornering energy. In total, we have

$$T = \frac{1}{2} \cdot m \cdot \left[v^2 + u^2\right] + \frac{1}{2} \cdot J_z \cdot r^2 \tag{5.7}$$

The potential energy U is built up in the springs (suspension, roll stiffness). With only the planar motion accounted for, these contributions are neglected and the Lagrangian equation is taken as the kinetic energy T. With generalized velocities $\dot{X}$, $\dot{Y}$, and $r = \dot{\psi}$ (global coordinates X, Y) the Lagrangian equations (with U omitted) become

$$\frac{\mathrm{d}}{\mathrm{d}t}\frac{\partial T}{\partial \dot{q}_i} - \frac{\partial T}{\partial q_i} = \frac{\mathrm{d}}{\mathrm{d}t} \cdot \frac{\partial T}{\partial \dot{q}_i} = Q_i$$

With (cf. Eq. (5.4))

$$u = \dot{X} \cdot \cos \psi + \dot{Y} \cdot \sin \psi \quad v = -\dot{X} \cdot \sin \psi + \dot{Y} \cdot \cos \psi$$

and the transformation between global generalized force and local lateral force (see Eq. (5.3))

$$Q_Y \cdot \cos \psi - Q_X \cdot \sin \psi = F_{y1} + F_{y2}$$

the equation follows as

$$\frac{\mathrm{d}}{\mathrm{d}t} \cdot \frac{\partial T}{\partial v} + r \cdot \frac{\partial T}{\partial u} = F_{y1} + F_{y2}$$

With T from expression (5.7), one easily derives Eq. (5.6a). The derivation of Eq. (5.6b) is found using the Lagrange equations for generalized rotational velocity r.

Sometimes, the yaw moment of inertia is written in terms of the radius of gyration r_g:

$$J_z = m \cdot r_g^2 \tag{5.8}$$

The axle loads front and aft follow from equilibrium in the (x, z) plane:

$$F_{z1} = m \cdot g \cdot \frac{b}{a+b} = m \cdot g \cdot \frac{b}{L} \quad F_{z2} = m \cdot g \cdot \frac{a}{a+b} = m \cdot g \cdot \frac{a}{L} \tag{5.9}$$

for acceleration of gravity g and wheelbase L. To obtain simpler equations, it is convenient to approximate the yaw moment of inertia using the value obtained with the mass consisting of two separate parts, concentrated at the front and rear axle:

$$J_z \approx J_{z,\mathrm{conc}} = a^2 \cdot \frac{m \cdot b}{L} + b^2 \cdot \frac{m \cdot a}{L} = m \cdot a \cdot b, \text{ i.e., } r_g = \sqrt{a \cdot b} \tag{5.10}$$

If the vehicle is not too small, this is not a bad approximation. Moreover, this approximation does not affect the qualitative aspects of the vehicle handling performance, which is the subject of the current analysis.

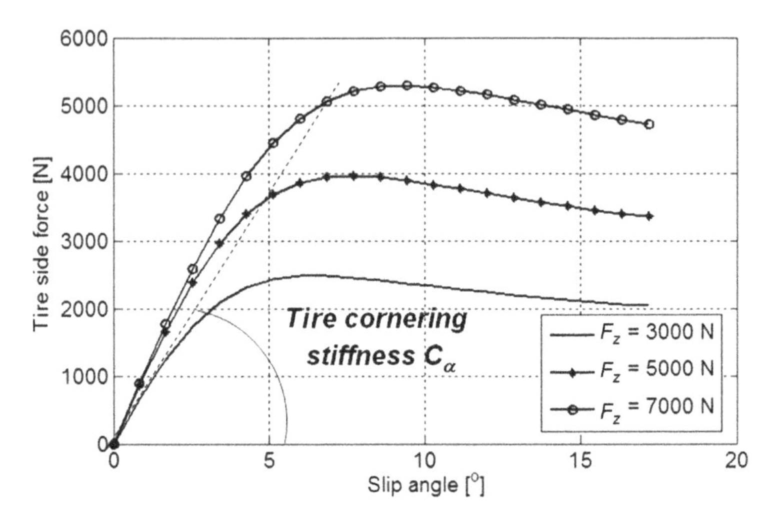

FIGURE 5.7 Lateral tire force versus slip angle.

Expressions (5.6a) and (5.6b) is formulated in terms of two dependent variables, v and r. The lateral axle forces F_{y1} and F_{y2} depend on the slip angles α_1 and α_2:

$$F_{y1} = F_{y1}(\alpha_1) \qquad F_{y2} = F_{y2}(\alpha_2) \tag{5.11}$$

where we recall that these functions are highly nonlinear. In Figure 5.7, we depicted the tire side force for the tire data, given in Appendix 6, and certain road friction (with the sum of left and right side force being the lateral axle force).

The tire slip angle is defined as the angle between the symmetry plane of the wheel and the local velocity vector. The local velocity vector is a result of the velocity of the vehicle CoG and the rotation of the vehicle around this point.

In general dynamic terms, if a body is moving with a speed vector $\underline{V}_{\text{CoG}}$ and rotating with a rotation velocity vector $\underline{\omega}$, then the global speed of any point P of this body is found from

$$\underline{V}_{\text{P}} = \underline{V}_{\text{CoG}} + \underline{\omega} \times (\underline{X}_{\text{P}} - \underline{X}_{\text{CoG}})$$

The rotation velocity is given by the yaw rate r. Consider the right front wheel in Figure 5.8 with steering wheel angle δ and half-track width d at the front axle. The slip angle α_1 is then found from

$$\alpha_1 = \delta - \arctan\left(\frac{v + a \cdot r}{u - d \cdot r}\right) \approx \delta - \frac{v + a \cdot r}{u - d \cdot r} \tag{5.12}$$

In most cases, the forward speed u exceeds $d \cdot r$ by far, which means that this final term can be neglected. A similar approach can be followed for the

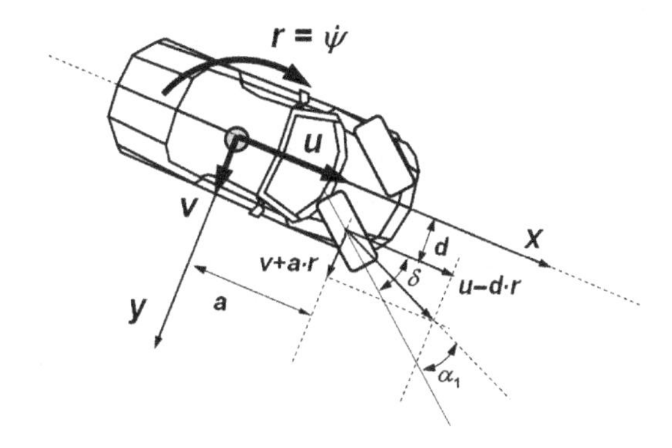

FIGURE 5.8 Kinematic description of tire slip angle.

rear axle, which, accounting for the approximation of small track width times yaw rate with respect to forward speed u, leads to

$$\alpha_1 = \delta - \frac{v + a \cdot r}{u} \qquad \alpha_2 = -\frac{v - b \cdot r}{u} \tag{5.13}$$

with steering angle δ. Equation (5.13) can be used to eliminate the slip angles in Eqs. (5.6a) and (5.6b), leaving a set of equations with lateral speed v and yaw rate r as the only states.

Equations (5.13) can also be used to eliminate lateral speed and yaw rate in Eqs. (5.6a) and (5.6b). When we replace the state variables v and r with the slip angles α_1 and α_2 and neglect F_{ye} and M_{ze}, we obtain

$$\begin{pmatrix} \dot{\alpha}_1 \\ \dot{\alpha}_2 \end{pmatrix} = \frac{u}{L} \begin{pmatrix} -1 & 1 \\ -1 & 1 \end{pmatrix} \cdot \begin{pmatrix} \alpha_1 \\ \alpha_2 \end{pmatrix} - \frac{1}{u} \cdot \begin{pmatrix} \frac{1}{m} + \frac{a^2}{J_z} & \frac{1}{m} - \frac{a \cdot b}{J_z} \\ \frac{1}{m} - \frac{a \cdot b}{J_z} & \frac{1}{m} + \frac{b^2}{J_z} \end{pmatrix} \cdot \begin{pmatrix} F_{y1} \\ F_{y2} \end{pmatrix}$$
$$+ \frac{u \cdot \delta}{L} \cdot \begin{pmatrix} 1 \\ 1 \end{pmatrix} + \begin{pmatrix} \dot{\delta} \\ 0 \end{pmatrix} \tag{5.14}$$

Using expression (5.9), and under the assumption (5.10), these equations simplify to

$$\begin{pmatrix} \dot{\alpha}_1 \\ \dot{\alpha}_2 \end{pmatrix} = \frac{u}{L} \cdot (\alpha_2 - \alpha_1 + \delta) \cdot \begin{pmatrix} 1 \\ 1 \end{pmatrix} - \frac{g}{u} \cdot \begin{pmatrix} f_{y1}(\alpha_1) \\ f_{y2}(\alpha_2) \end{pmatrix} + \begin{pmatrix} \dot{\delta} \\ 0 \end{pmatrix} \tag{5.15}$$

with normalized axle side forces (lateral friction coefficient) f_{y1} and f_{y2} for front and rear axles, respectively, defined as the ratio of axle side force and axle load. Consequently, we derived two sets of Eqs. (5.6a), (5.6b) and (5.14), which are completely identical in terms of system performance; the only difference is that we use either (v, r) or (α_1, α_2) as dependent variables (states).

The lateral forces can be taken in their full nonlinear setting or they can be approximated using **linear relationships** $C_{\alpha i} \cdot \alpha$ with cornering stiffness $C_{\alpha i}$ for front and rear axles (with $i = 1, 2$, respectively) for small slip angle α. This cornering stiffness is indicated in Figure 5.7. Note that the cornering stiffness $C_{\alpha i}$ refers to the axle characteristics. Expressing the right-hand side of expressions (5.6a) and (5.6b) in terms of the tire cornering stiffness, a factor of 2 must be included if load transfer is not considered.

When we introduce the "**derivatives of stability**" Y_β, Y_r, N_β, and N_r:

$$Y_\beta = -(C_{\alpha 1} + C_{\alpha 2}) \quad Y_r = -\frac{a \cdot C_{\alpha 1} - b \cdot C_{\alpha 2}}{V} \tag{5.16a}$$

$$N_\beta = -(a \cdot C_{\alpha 1} - b \cdot C_{\alpha 2}) \quad N_r = -\frac{a^2 \cdot C_{\alpha 1} + b^2 \cdot C_{\alpha 2}}{V} \tag{5.16b}$$

expressions (5.6a) and (5.6b) simplifies, in case of linear axle characteristics and constant vehicle speed V, to the linear equations

$$m \cdot V(\dot{\beta} + r) = Y_\beta \cdot \beta + Y_r \cdot r + C_{\alpha 1} \cdot \delta + F_{ye} \tag{5.17a}$$

$$J_z \cdot \dot{r} = N_\beta \cdot \beta + N_r \cdot r + a \cdot C_{\alpha 1} \cdot \delta + M_{ze} \tag{5.17b}$$

where we have replaced v with β using expression (5.1), and used the fact that $u \approx V$. For nonconstant speed, one must add a term $m \cdot \dot{V} \cdot \beta$ to the left-hand side of Eq. (5.17a).

Note that we neglected the axle aligning torques in expressions (5.6a) and (5.6b). If we account for the torques, equation (5.16b) would be adjusted as follows (see also Ref. [11]):

$$N_\beta = -(a \cdot C_{\alpha 1} - b \cdot C_{\alpha 2}) + \frac{\partial M_{z1}}{\partial \alpha_1}(\alpha_1 = 0) + \frac{\partial M_{z2}}{\partial \alpha_2}(\alpha_2 = 0)$$

$$N_r = -\frac{a^2 \cdot C_{\alpha 1} + b^2 \cdot C_{\alpha 2} + a \cdot \frac{\partial M_{z1}}{\partial \alpha_1}(\alpha_1 = 0) + b \cdot \frac{\partial M_{z2}}{\partial \alpha_2}(\alpha_2 = 0)}{V}$$

The expressions (5.16a) and (5.16b) are referred to as derivatives of stability because they are the derivatives of the right-hand sides of expressions (5.6a) and (5.6b) to the body slip angle β and yaw rate r, for linear axles. Note that $-Y_\beta$ and $-N_r$ act as damping coefficients in Eqs. (5.17a) and (5.17b).

We close this section with comments about the validity of the one-track vehicle model. This model was introduced under a number of assumptions. These assumptions may be relaxed (or corrected), thus yielding a much larger area for application of this model.

5.2.2 Effect of Body Roll and Lateral Load Transfer

The lateral tire force F_y depends on the tire load F_z in a nonlinear fashion. An example is shown in Figure 5.9 in which the tire force F_y depicts versus tire load F_z for different slip angles. Tire data is taken from Appendix 6. One observes the decreasing absolute slope of these curves, meaning that under load transfer of for example 1500 [N] (being the increase, decrease of the tire load at outer and inner tire, respectively) and with an axle slip angle of 3 [°], the total lateral axle force (being the sum of the side forces at inner and outer wheel) is reduced.

Consequently, the entire lateral axle characteristics are changed. Note that load transfer changes with increasing lateral acceleration and thus with increasing slip angle. The load transfer can be different for both axles and it depends on the roll stiffness distribution over both axles. Most load transfer will occur at the axle with the largest roll stiffness. We assume steady-state behavior, i.e., roll damping is neglected. The roll stiffness at axle i ($i = 1, 2$) is denoted as $K_{\varphi,i}$. Using a vehicle CoG height h_{CoG} and a track width t_i ($i = 1, 2$), the load transfers ΔF_{z1} and ΔF_{z2} from inner to outer wheels satisfy

$$t_1 \cdot \Delta F_{z1} + t_2 \cdot \Delta F_{z2} = h_{\mathrm{CoG}} \cdot F_y \tag{5.18}$$

where F_y is the total vehicle side force. This is a consequence of the equilibrium in roll moment around the vertical projection of the vehicle's CoG on the road. Lateral force times CoG height must be balanced by vertical reactions at the wheels. The contribution $h_{\mathrm{CoG}} \cdot m \cdot g \cdot \varphi$ (mass, acceleration of gravity) in this roll moment, due to the lateral deviation of the CoG over the

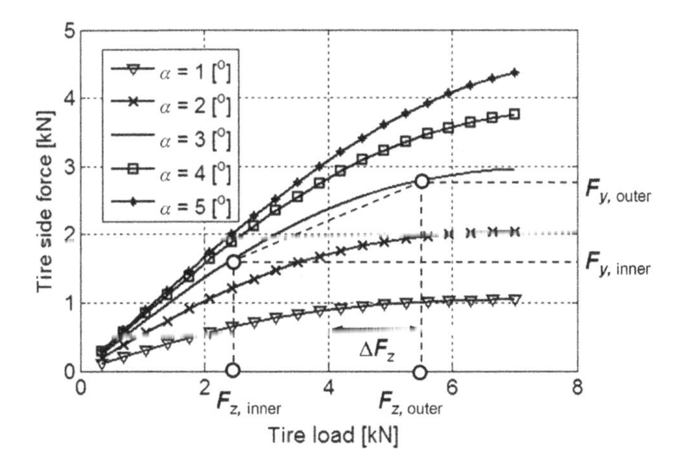

FIGURE 5.9 Tire side force versus tire load, load transfer.

roll angle φ, has been neglected. The roll angle φ satisfies the following relationships:

$$\varphi = \frac{t_1 \cdot \Delta F_{z1}}{K_{\varphi 1}} = \frac{t_2 \cdot \Delta F_{z2}}{K_{\varphi 2}} = \frac{\sum_i t_i \cdot \Delta F_{zi}}{\sum_i K_{\varphi i}} = \frac{h_{\text{CoG}} \cdot F_y}{\sum_i K_{\varphi i}} \tag{5.19}$$

and therefore

$$\Delta F_{z1} = \frac{K_{\varphi 1}}{\sum_i K_{\varphi i}} \cdot \frac{h_{\text{CoG}} \cdot F_y}{t_1} \quad \Delta F_{z2} = \frac{K_{\varphi 2}}{\sum_i K_{\varphi i}} \cdot \frac{h_{\text{CoG}} \cdot F_y}{t_2} \tag{5.20}$$

From Eq. (5.20), it is clear that load transfer is directly related to the distribution of axle roll stiffnesses. Assuming steady-state conditions, we will show in Section 5.2.3:

$$\frac{F_{y1}}{F_{z1}} = \frac{F_{y2}}{F_{z2}} = \frac{F_y}{m \cdot g}$$

In other words, constants M_i, $i = 1, 2$, exist such that

$$\Delta F_{zi} = M_i \cdot F_{yi}$$

Applying this in the description of the lateral tire characteristics, denoted as $F_{\text{lateral}}(\alpha, F_z)$, one may write for the front axle:

$$F_{y1} = F_{\text{lateral}}\left(\alpha, \frac{1}{2} \cdot F_{z1} + M_1 \cdot F_{y1}\right) + F_{\text{lateral}}\left(\alpha, \frac{1}{2} \cdot F_{z1} - M_1 . F_{y1}\right) \tag{5.21}$$

This is a nonlinear equation in F_{y1} that can be solved by iteration. We solved it, assuming that 75% of the roll stiffness is covered by the front axle. In this way, we distinguished between the noncorrected axle side force characteristic (twice the tire side force) and the solution of expression (5.21). Figure 5.10 shows the results.

Observe that the side force shows a significant reduction. Also, in spite of the fact that the slope of this curve at $\alpha = 0$ doesn't change, a lower effective cornering stiffness (linearization over the linear slip angle range 0–4 [°]) is observed.

This load transfer could have been determined using a **double-track vehicle model**. This is the straightforward extension of the single-track model, with distinction of considering separate wheels instead of the axles. Slip angles and steering angles are assumed the same for left and right wheels. Using this model means that one must determine the individual **wheel ground contact forces** under varying conditions of changing lateral acceleration a_y. If we allow the vehicle speed to change, one must account for changing longitudinal acceleration a_x as well, which means that we include the

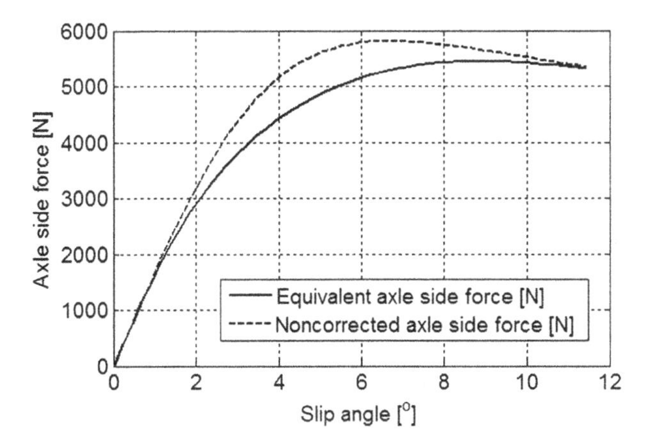

FIGURE 5.10 Effect of load transfer on lateral axle characteristics.

longitudinal load transfer in this model. There are different approaches to determine these contact forces (Genta and Morello [11], Kiencke and Nielsen [20], and Venhovens and van der Knaap [55]). Starting with the first approach (extending expression (5.20)) provides the following.

CONTACT FORCES ACCORDING TO GENTA AND MORELLO

$$\frac{1}{m}\cdot F_{z1L} = \frac{b}{2\cdot L}\cdot g - \frac{h_{CoG}\cdot K_{\varphi 1}}{t_1\cdot \sum_i K_{\varphi i}}\cdot a_y - \frac{h_{CoG}}{2\cdot L}\cdot a_x \tag{5.22a}$$

$$\frac{1}{m}\cdot F_{z1R} = \frac{b}{2\cdot L}\cdot g + \frac{h_{CoG}\cdot K_{\varphi 1}}{t_1\cdot \sum_i K_{\varphi i}}\cdot a_y - \frac{h_{CoG}}{2\cdot L}\cdot a_x \tag{5.22b}$$

$$\frac{1}{m}\cdot F_{z2L} = \frac{a}{2\cdot L}\cdot g - \frac{h_{CoG}\cdot K_{\varphi 2}}{t_2\cdot \sum_i K_{\varphi i}}\cdot a_y + \frac{h_{CoG}}{2\cdot L}\cdot a_x \tag{5.22c}$$

$$\frac{1}{m}\cdot F_{z2R} = \frac{a}{2\cdot L}\cdot g + \frac{h_{CoG}\cdot K_{\varphi 2}}{t_2\cdot \sum_i K_{\varphi i}}\cdot a_y + \frac{h_{CoG}}{2\cdot L}\cdot a_x \tag{5.22d}$$

with wheel load F_{zij} with $i = 1,2$ (front and rear axle) and $j = \text{L, R}$ (left, right wheel). Kiencke and Nielsen introduce virtual masses at front and wheel axles:

$$m_1^* = \frac{F_{z1}}{g},\quad m_2^* = \frac{F_{z2}}{g}$$

to obtain, for the left and right wheel loads

$$F_{zij} = \frac{1}{2} \cdot F_{zi} \pm m_i^* \cdot \frac{h_{\mathrm{CoG}}}{t_i} \cdot a_y = \left(\frac{1}{2} \pm \frac{h_{\mathrm{CoG}}}{t_i} \cdot \frac{a_y}{g}\right) \cdot F_{zi}$$

with the axle loads being changed by longitudinal load transfer. Together, this leads to

CONTACT FORCES ACCORDING TO KIENCKE AND NIELSEN

$$\frac{1}{m} \cdot F_{z1\mathrm{L}} = \left(\frac{b}{L} \cdot g - \frac{h_{\mathrm{CoG}}}{L} \cdot a_x\right) \cdot \left(\frac{1}{2} - \frac{h_{\mathrm{CoG}}}{t_1} \cdot \frac{a_y}{g}\right) \tag{5.23a}$$

$$\frac{1}{m} \cdot F_{z1\mathrm{R}} = \left(\frac{b}{L} \cdot g - \frac{h_{\mathrm{CoG}}}{L} \cdot a_x\right) \cdot \left(\frac{1}{2} + \frac{h_{\mathrm{CoG}}}{t_1} \cdot \frac{a_y}{g}\right) \tag{5.23b}$$

$$\frac{1}{m} \cdot F_{z2\mathrm{L}} = \left(\frac{a}{L} \cdot g + \frac{h_{\mathrm{CoG}}}{L} \cdot a_x\right) \cdot \left(\frac{1}{2} - \frac{h_{\mathrm{CoG}}}{t_2} \cdot \frac{a_y}{g}\right) \tag{5.23c}$$

$$\frac{1}{m} \cdot F_{z2\mathrm{R}} = \left(\frac{a}{L} \cdot g + \frac{h_{\mathrm{CoG}}}{L} \cdot a_x\right) \cdot \left(\frac{1}{2} + \frac{h_{\mathrm{CoG}}}{t_2} \cdot \frac{a_y}{g}\right) \tag{5.23d}$$

Venhovens and van der Knaap extended the approach of Kiencke and Nielsen, allowing the inertia J_{xz} to not be equal to zero. This provides additional contributions from yaw acceleration to the wheel loads. However, they took the track width as identical for front and rear axles. Their approach is identical to expressions (5.23a)–(5.23d), if $J_{xz} = 0$. For identical track widths, one can show that expressions (5.23a)–(5.23d) implies that the lateral load transfer per axle is proportional to the instantaneous axle load:

$$\frac{F_{z1\mathrm{R}} - F_{z1\mathrm{L}}}{F_{z1}} = \frac{F_{z2\mathrm{R}} - F_{z2\mathrm{L}}}{F_{z2}}$$

Venhovens and van de Knaap [55] indicated it as a design of good handling when this proportionality was fulfilled.

We compared both approaches for the rear axle wheel loading using the vehicle data from Appendix 6 and for a roll stiffness distribution between front and rear axles of 2:3. The longitudinal acceleration a_x has been varied from 0 to 0.6 [g] (with the upper bound being too large for driving but acceptable for braking) and the lateral acceleration from 0 to 0.8 [g].

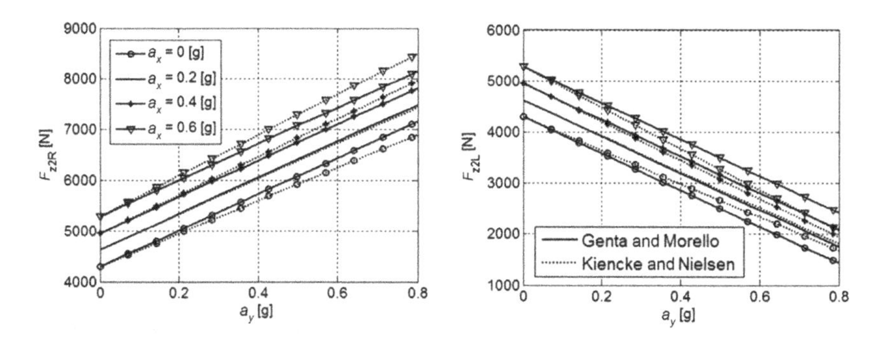

FIGURE 5.11 Wheel loads for different a_x and a_y according to Eqs. (5.22a)–(5.22d) and (5.23a)–(5.23d).

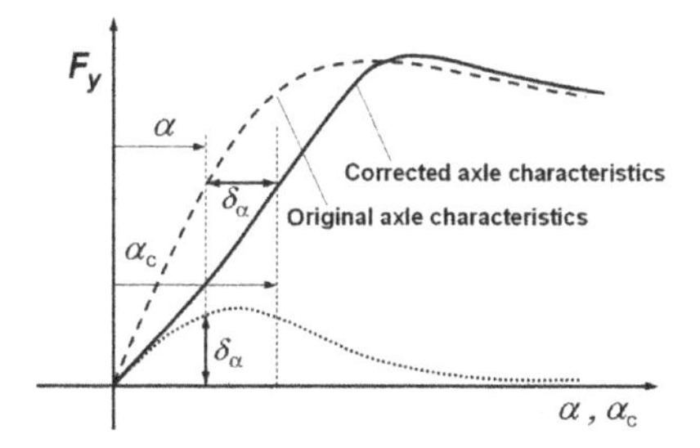

FIGURE 5.12 Effective axle characteristics in case of steering or suspension compliance.

The curves in Figure 5.11 illustrate that the $a_x \cdot a_y$ term in expression (5.23a)–(5.23d) tends to reduce the effect of a_x for the inner wheel and increase it for the outer wheel. Note that a more extreme distribution of roll stiffness will increase the difference between the two approaches.

5.2.3 Alignment and Compliance Effects

Pacejka [32] discusses the determination of effective axle characteristics in the presence of steering compliance and suspension kinematics. Additional steering angles may arise because of

- Roll steer.
- Steering compliance due to a torque on the wheels, which is the side force times the trail (pneumatic, caster).
- Suspension compliance.

Similar effects may be observed with respect to additional camber angles. Both effects will change the axle characteristics. If we restrict ourselves to additional steering angles, this means that points on the axle characteristic curve (F_y versus α) are shifted horizontally along the α-axis (from α to α_c, Figure 5.12). It is assumed that additional steering angles $\delta_{\alpha L}$ and $\delta_{\alpha R}$ arise at left and right wheels from the preceding compliances and suspension kinematics, respectively. These steering angles depend on the side force and we obtain (see also expression (5.21)):

$$F_y = F_{\text{lateral}}(\alpha + \delta_{\alpha \text{L}}(F_y)) + F_{\text{lateral}}(\alpha + \delta_{\alpha \text{R}}(F_y))$$

which is a nonlinear equation from which $F_y(\alpha)$ can be determined. One might combine this with the load transfer, described by expression (5.21). The graphical interpretation is given in Figure 5.12 (taken from Ref. [32]). Because the disturbance slip angle δ_α depends on F_y, we may also write it as a function of the undisturbed angle α. This relationship is then used to modify the axle side force behavior, as indicated in Figure 5.12.

We mention two sources where explicit data is given on the impact of compliance and alignment effects on the **cornering compliance** CC_α, which is defined as follows:

$$CC_{\alpha i} = \frac{F_{zi}}{C_{\alpha i}} \tag{5.24}$$

i.e., being the ratio of axle load and the axle cornering stiffness for axle i, $i = 1,2$. Using the given information about the cornering compliance, and accounting for some of the preceding effects and known axle load, one is able to derive the effective axle cornering stiffness.

The first source is from Bickerstaff [3], and uses a pickup truck. In Table 5.3, we list the causes for modified cornering compliance in the first column. In the next two columns, we provide the values for the cornering compliance, taking into account the effects up to that row, including all the potential causes in this row and previous rows. The last two columns list the cornering stiffnesses. Based on Ref. [3], we assumed a front axle load of 3000 [lb] = 6675 [N], and a rear axle load of 6000 [lb] = 13350 [N].

TABLE 5.3 Effective Axle Cornering Stiffness for a Pickup Truck [3]

	$CC_{\alpha 1}$	$CC_{\alpha 2}$	$C_{\alpha 1}$ (N/rad)	$C_{\alpha 2}$ (N/rad)
Tire response, load transfer	0.103	0.131	64822	101986
+aligning stiffness	0.117	0.106	57082	125393
+roll steer	0.134	0.080	49669	166282
+camber thrust	0.138	0.080	48411	166282
+lateral compliance	0.155	0.080	42972	166282
+steering compliance	0.173	0.080	38631	166282

We observe that the front axle cornering stiffness can be reduced to 60% of the original value, whereas the rear axle cornering stiffness may increase over 60%. Note that the reference is from the 1970s and is therefore not current. That means a little out of date. To discuss this table in more detail, we first address the axle stiffness based on only the tire data, including the effect of load transfer. The aligning torque, which is usually neglected, has an effect on the vehicle steering response that can be accounted for by modifying the axle cornering stiffnesses. Roll steer means that steering arises from body roll, which is experienced as a modified side force for the same slip angle, i.e., as a different axle cornering stiffness. A similar effect is observed for camber. Finally, the front axle cornering stiffness is affected by lateral (depending on the position of the steering linkage) and steering compliance.

The second source [5] is more recent. It refers to a passenger car, with mass of 1200 [kg], and front and rear axle load given as 6474 [N] and 5297 [N], respectively. Results are presented in Table 5.4 where, as in Table 5.3, the data has been transferred to Newtons and radians instead of pounds and degrees.

Again, observe the significant effect, specifically on the front axle cornering stiffness (a reduction of 30%). In more detail, the modified CoG position is due to a possible different loading balance of the vehicle. This leads to different axle loads and might require different tire pressures (these effects have been included). Roll steer and camber have been discussed previously in Table 5.3. They lead to a different side force for the same slip angle and therefore, a different effective cornering stiffness. Aligning torques have a similar effect, especially for the front axle in combination with the steering compliance. In addition, they move the tire side forces rearward; this movement is referred to as aligning torque on a rigid body.

5.2.4 Effect of Combined Slip

For combined slip, when a specific brake or drive torque is applied, we can account for that torque in the lateral axle forces, as described in Section 2.6.

TABLE 5.4 Effective Axle Cornering Stiffness for a Passenger Car [5]

	$CC_{\alpha 1}$	$CC_{\alpha 2}$	$C_{\alpha 1}$ (N/rad)	$C_{\alpha 2}$ (N/rad)
Tire response, load transfer	0.105	0.105	61828	50586
+modified CoG position	0.110	0.099	58884	53249
+roll steer	0.119	0.091	54554	58369
+roll camber effect on steer	0.122	0.099	52995	53249
+lateral compliance	0.126	0.103	51523	51444
+aligning torque compliance	0.145	0.105	44695	50586
+aligning torque on rigid body	0.147	0.103	44163	51444

This can be completed by applying the combined slip Magic Formula description for the front or rear axle, or by using one of the approximations introduced in Section 2.6. One might think of the general situation when a certain drive force is required to overcome various resistances (rolling resistance, aerodynamic drag, slopes). Genta and Morello [11] discussed this case, showing the impact on the neutral steer point (see Section 5.3.3). For wheel-by-wheel braking or driving, one requires at least a double-track vehicle model.

5.3 STEADY-STATE ANALYSIS

5.3.1 Steady-State Solutions

In this section, we shall discuss the steady-state solutions of expressions (5.6a) and (5.6b). We consider vehicle behavior as described by the steady-state circular test ISO 4138. Stability of vehicle behavior can be described as the return of the vehicle to steady-state behavior following a disturbance from some deviation from the original path. Following the dynamic behavior in terms of states $v(t)$ and $r(t)$, the steady-state solutions can be considered as the final limits of the vehicle states. In other words, the steady-state solutions correspond to the singular points for the full dynamic system.

Under steady-state conditions, and neglecting external loads F_{ye} and M_{ze}, time derivatives vanish and expressions (5.6a) and (5.6b) reduces to

$$m \cdot u \cdot r = F_{y1} + F_{y2} = F_y \tag{5.25a}$$

$$a \cdot F_{y1} = b \cdot F_{y2} \tag{5.25b}$$

In terms of the axle loads as given by Eq. (5.9), one finds from Eq. (5.25b)

$$\frac{F_{y1}(\alpha_1)}{F_{z1}} = \frac{F_{y2}(\alpha_2)}{F_{z2}} \tag{5.26}$$

Hence, the normalized axle characteristics or lateral friction coefficients $f_i(\alpha_i)$ coincide. Substituting Eq. (5.26) in Eq. (5.25a), one finds

$$f_{y1}(\alpha_1) = f_{y2}(\alpha_2) = \frac{u \cdot r}{g} = \frac{u^2}{g \cdot R} = \frac{F_y}{m \cdot g} (= a_y \text{ in } g's) \tag{5.27}$$

for radius of a vehicle path R. The lateral acceleration F_y/mg depends on the relative path curvature L/R (wheelbase L) in a linear sense:

$$\frac{F_y}{m \cdot g} = \left(\frac{u^2}{g \cdot L}\right) \cdot \frac{L}{R} \tag{5.28}$$

In terms of this same relative path curvature, it follows from Eq. (5.13), between slip angles and state variables (v, r), that

$$\delta = \frac{L}{R} + (\alpha_1 - \alpha_2) \tag{5.29}$$

See also expression (4.5). The effect of finite velocity compared to the situation of kinematic steering is that the axle steering angle is changed with $\alpha_1 - \alpha_2$. Equality of normalized axle side forces (see expression (5.26)) follows very easy from the equations of motion, in terms of the slip angles (see expression (5.15)). In addition, Eq. (5.27) follows directly from expression (5.15), in combination with Eq. (5.29). Likewise, expression (5.14) (without the simplification $J_z = m \cdot a \cdot b$) can be used to obtain the same results.

REMARK

Note that different sets of (α_1, α_2) may satisfy Eqs. (5.26) and (5.29) due to the nonlinear behavior of f_{y1} and f_{y2}. This has been illustrated in Figure 5.13, in which two normalized axle curves are shown. One is looking for sets (α_1, α_2), with a difference equal to $\delta - L/R$ and with equal values of f_{yi}. Three pairs of solutions are indicated, one corresponding to both the parts of both curves with positive slope, and two corresponding when the axle slip for a maximum side force is exceeded ("over the top").

In summary, one observes, from the previous expressions, that the centrifugal force F_y can be described in terms of both the slip angles through the normalized axle curves (these relationships are highly nonlinear) and in terms of the path curvature where slip angles and path curvature are related through expression (5.29).

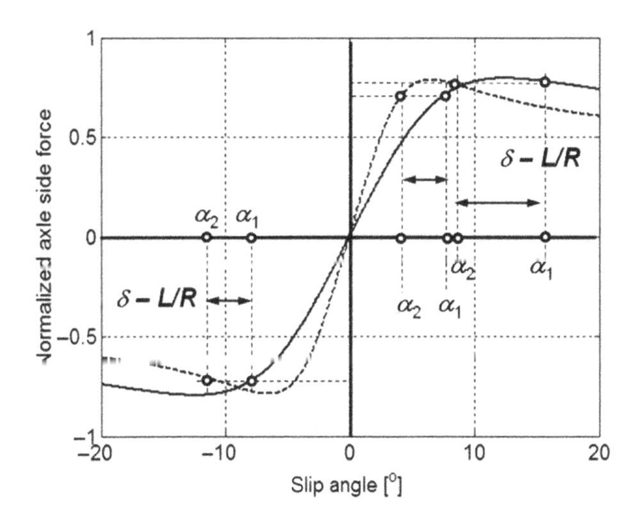

FIGURE 5.13 Different steady-state solutions.

5.3.2 Understeer and Oversteer

Let us examine Eqs. (5.25a), (5.25b) and (5.26) in more detail, starting with **linear tire behavior**. One finds from expressions (5.26) that

$$\frac{u^2}{g \cdot R} = \frac{C_{\alpha 1} \cdot \alpha_1}{F_{z1}} = \frac{C_{\alpha_2} \cdot \alpha_2}{F_{z2}} \tag{5.30}$$

which means that

$$\left[\frac{F_{z1}}{C_{\alpha 1}} - \frac{F_{z2}}{C_{\alpha 2}}\right] \cdot \frac{u^2}{g \cdot R} \equiv \eta \cdot \frac{u^2}{g \cdot R} = \eta \cdot a_y(g) = \alpha_1 - \alpha_2 = \delta - \frac{L}{R} \tag{5.31}$$

where η is the coefficient, referred to as **understeer gradient**. This relationship (5.31) expresses clearly the dependency of the vehicle cornering performance on tire characteristics. This can be explained as follows. For very low speed u, the steering angle necessary to negotiate a curve is equal to the relative curvature L/R (Ackermann angle). Increasing speed over the same curve (a circle with radius R), creates a necessary change in required steering angle that depends on the understeer gradient and hence, on the tire characteristics. For a positive understeer gradient η, the steering angle δ must increase, whereas for a negative value of η, the opposite is true and the steering angle δ must be reduced. In the second case, without this change in δ, the vehicle would end up on a curve with a small curve radius. Increasing speed means a larger lateral acceleration and hence, more extreme conditions. Without corrective action from the driver, and for negative η, these conditions lead to a smaller curve radius which further increases the lateral acceleration. Hence, the vehicle exhibits a self-reinforcing effect, making things worse for understeer gradient $\eta < 0$. It will be demonstrated that, beyond a certain speed, this case will lead to instability.

As one observes, the understeer gradient is written as the difference of two terms, related to the front and the rear axles, and was introduced in Section 5.3.1 as the cornering compliance of the front axle $CC_{\alpha 1}$ and rear axle $CC_{\alpha 2}$ (see expression (5.24)). These terms include contributions from the tire characteristics (cornering stiffness), as well as from other effects such as aligning torque, compliance, camber effects, etc. The data from Tables 5.3 and 5.4 can be used to show the impact on the understeer gradient, simply by subtracting $CC_{\alpha 2}$ from $CC_{\alpha 1}$. The resulting understeer gradient is given in Table 5.5.

Consequently, if one is restricted to tire response only, the pickup truck has a negative understeer gradient. The other effects result in a positive value for η. In a relative sense, this change in η is much larger than the impact on the individual axle contributions ($CC_{\alpha i}$) to η. The same can be said for the passenger car. The impact on the separate axles was significant, but the relative effect of compliances, suspension kinematics, and alignment is more

TABLE 5.5 Understeer Gradient, Accounting for Different Compliance and Alignment Effects

Pickup Truck [3]	η	Passenger Car [5]	η
Tire response, load transfer	−0.028	Tire response, load transfer	0.000
+aligning stiffness	0.010	+modified CoG position	0.010
+roll steer	0.054	+roll steer	0.028
+camber thrust	0.058	+roll camber effect on steer	0.023
+lateral compliance	0.075	+lateral compliance	0.023
+steering compliance	0.093	+aligning torque	0.040
		+aligning torque on rigid body	0.044

pronounced. In Section 5.4, we discuss how these changes in η crucially affect the vehicle handling performance and yaw stability.

We provide definitions for **understeer and oversteer behavior**, following from Eq. (5.31).

DEFINITION 1

A vehicle is **understeered** if the axle steering angle must be increased for an increasing vehicle forward speed to negotiate the same curve. A vehicle is **oversteered** if the opposite is true, i.e., the steering angle must be decreased for increasing vehicle forward speed to negotiate the same curve. We call a vehicle **neutrally steered** if no adjustment of δ is required.

DEFINITION 2

A vehicle is understeered if the front axle slip angle exceeds the rear axle slip angle under steady-state conditions: $\alpha_1 > \alpha_2$ (or, in more general terms, $|\alpha_1| > |\alpha_2|$). The vehicle is oversteered if the opposite is true ($\alpha_1 < \alpha_2$).

DEFINITION 3

A vehicle is understeered if the understeer gradient $\eta > 0$, i.e., the front axle normalized axle cornering stiffness is exceeded by the rear axle normalized axle cornering stiffness. Note that

$$\eta \cdot Y_r > 0; \quad \eta \cdot N_\beta > 0 \tag{5.32}$$

in case that $\eta \neq 0$, and with Y_r and N_β defined in Eqs. (5.16a) and (5.16b). If $\eta < 0$, we define the vehicle as being oversteered.

Note that the understeer property is directly related to the **steering wheel gradient**, which is related to the steady-state circular behavior (see Section 5.1). This means that the following alternative definition of understeer may be introduced.

DEFINITION 4
A vehicle is understeered if the steering wheel gradient $\partial\delta/\partial a_y$ ($a_y = 0$) is positive. A vehicle is oversteered if this gradient is negative.

We conclude that, for linear tire characteristics, all **four definitions are identical**. Definition 1 is general and applies for larger lateral acceleration where nonlinear axle behavior cannot be neglected. Definitions 2–4 are related to tire behavior for small lateral acceleration, i.e., for the range of a_y for which linear tire behavior is a realistic approximation.

Consider Figure 5.14, in which we depicted possible relationships between axle steering angle δ and lateral acceleration a_y (handling curves), as a result of a stationary vehicle situation, i.e., driving a circle with a constant radius R. The left part of the curves corresponds to linear tire behavior, cf. expression (5.31). It is clear that, with increasing speed and lateral acceleration, this lateral acceleration cannot be increased forever. There is a moment when the tires saturate and the vehicle will skid out, either at the front axle or at the rear axle. That means that the curve, which begins as a linear relationship for small a_y, will have to bend off upward or downward. Compare this with Figure 5.13 where the normalized axle characteristics were assumed to intersect at a certain slip angle. This means a steady-state situation with the same slip angles at front and rear axles and thus, $\delta = L/R$. This is only possible if the curve in Figure 5.14 is bending off downward, which means that there will be an area in terms of a_y where the vehicle behaves as oversteered (according to definition 1). Observe that the rear axle is saturated at this intersection point, i.e., the vehicle is expected to skid out at the rear axle. If the normalized axle characteristics do not intersect, the situation of the upper curve occurs, where the vehicle remains understeered until it skids out at the front axle.

Now let us assume again linear axle behavior and examine the vehicle steady-state steering behavior more closely. Remember that we derived expressions for the trajectory gain and the body slip angle gain in Section 4.1 under

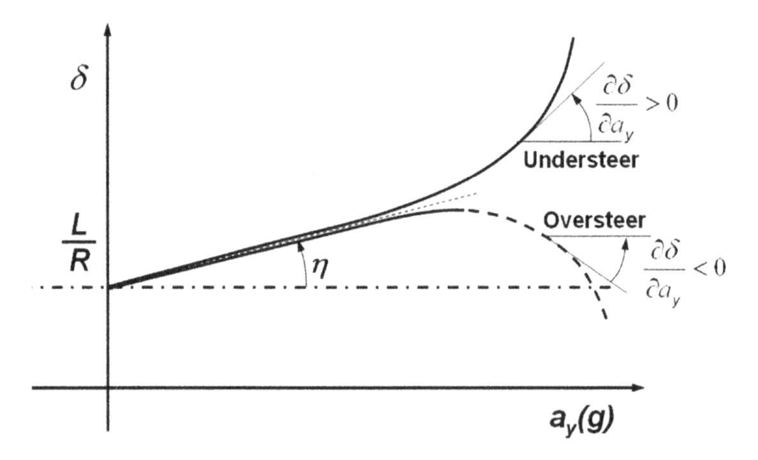

FIGURE 5.14 Stationary steering performance.

conditions of kinematic steering. Under steady-state conditions, one can derive from Eqs. (5.25a) and (5.25b), in combination with expression (5.13), that

Trajectory curvature gain:

$$\left(\frac{1/R}{\delta}\right)=\frac{L\cdot C_{\alpha1}\cdot C_{\alpha2}}{L^2\cdot C_{\alpha1}\cdot C_{\alpha2}+m\cdot u^2\cdot N_\beta} \tag{5.33a}$$

Body slip angle gain:

$$\left(\frac{\beta}{\delta}\right)=\frac{b\cdot L\cdot C_{\alpha1}\cdot C_{\alpha2}-a\cdot C_{\alpha1}\cdot m\cdot u^2}{L^2\cdot C_{\alpha1}\cdot C_{\alpha2}+m\cdot u^2\cdot N_\beta} \tag{5.33b}$$

Yaw rate gain:

$$\left(\frac{r}{\delta}\right)=\frac{u\cdot L\cdot C_{\alpha1}\cdot C_{\alpha2}}{L^2\cdot C_{\alpha1}\cdot C_{\alpha2}+m\cdot u^2\cdot N_\beta} \tag{5.33c}$$

Lateral acceleration gain:

$$\left(\frac{a_y}{\delta}\right)=\frac{u^2\cdot L\cdot C_{\alpha1}\cdot C_{\alpha2}}{L^2\cdot C_{\alpha1}\cdot C_{\alpha2}+m\cdot u^2\cdot N_\beta} \tag{5.33d}$$

with N_β defined in Eqs. (5.16a) and (5.16b). The reader may verify that Eqs. (5.33a) and (5.33b) correspond to the values for Ackermann steering if $u\to 0$.

Define the **stability factor** K_s:

$$K_s=\frac{m\cdot N_\beta}{L^2\cdot C_{\alpha1}\cdot C_{\alpha2}}=\frac{1}{g\cdot L}\cdot\eta \tag{5.34}$$

The yaw rate gain and body slip angle gain can now be written as

$$\left(\frac{r}{\delta}\right)=\frac{u}{L\cdot(1+K_s\cdot u^2)} \tag{5.35a}$$

$$\left(\frac{\beta}{\delta}\right)=\frac{b-B\cdot u^2}{L\cdot(1+K_s\cdot u^2)}\quad\text{where } B=\frac{a\cdot m}{L\cdot C_{\alpha2}} \tag{5.35b}$$

From these expressions, one can draw two important conclusions:

1. If the vehicle is understeered (i.e., $K_s>0$), the yaw rate gain and body slip angle gain are bounded. A speed u exists for which the yaw rate gain has a maximum value. This speed is called the **characteristic speed** u_{ch}.
2. If the vehicle is oversteered (i.e., $K_s<0$), these gains will become unbounded for a certain speed, which is referred to as the **critical speed** u_{cr}.

The following expressions hold for the characteristic and critical speeds:

$$u_{\mathrm{ch}}^2 = \frac{1}{K_\mathrm{s}} = \frac{g \cdot L}{\eta} \quad \eta > 0 \tag{5.36}$$

$$u_{\mathrm{cr}}^2 = -\frac{1}{K_\mathrm{s}} = -\frac{g \cdot L}{\eta} \quad \eta < 0 \tag{5.37}$$

We evaluated the relationships (5.35a) and (5.35b) for the vehicle parameters of Appendix 6 for understeer ($\eta = 0.042$), neutral steer ($\eta = 0$), and oversteer vehicle ($\eta = -0.02$). This leads to

$$u_{\mathrm{ch}} = 89.3 \text{ [km/h]}$$

$$u_{\mathrm{cr}} = 132.5 \text{ [km/h]}$$

Results are shown in Figure 5.15. Observe the increasing gains (in absolute sense) for oversteer, the linear yaw rate gain for neutral steer, and the maximum yaw rate gain for $u = u_{\mathrm{ch}}$.

The next step is to extend Eq. (5.31) for nonlinear axles, which means that we want to describe the curve of δ versus a_y cf. Figure 5.14 for the full nonlinear regime.

The first step is to invert Eq. (5.27), where we denote

$$g_i(a_y) = \mathrm{inv}\{f_{yi}(\alpha_i)\}(a_y) \quad i = 1, 2 \tag{5.38}$$

with inv{f} indicating the inverse of a function f. The functions g_i are now multivalued functions in the lateral acceleration a_y (in g's, i.e., $F_y/(mg)$), where both single-valued branches may be treated separately.

It follows using Eq. (5.29) that

$$\delta - \frac{L}{R} = \alpha_1 - \alpha_2 = g_1(a_y) - g_2(a_y) \equiv h(a_y) \tag{5.39}$$

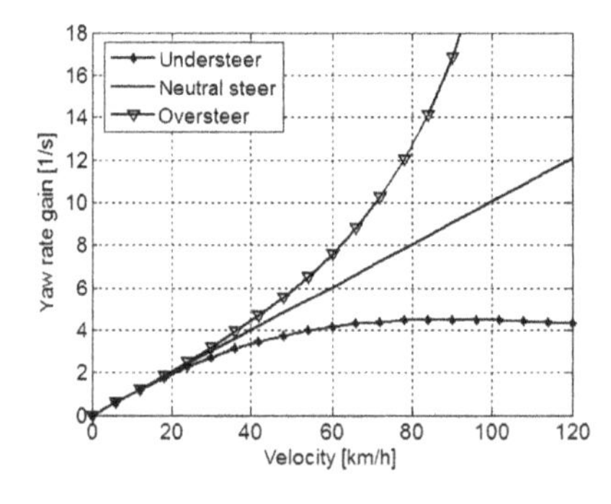

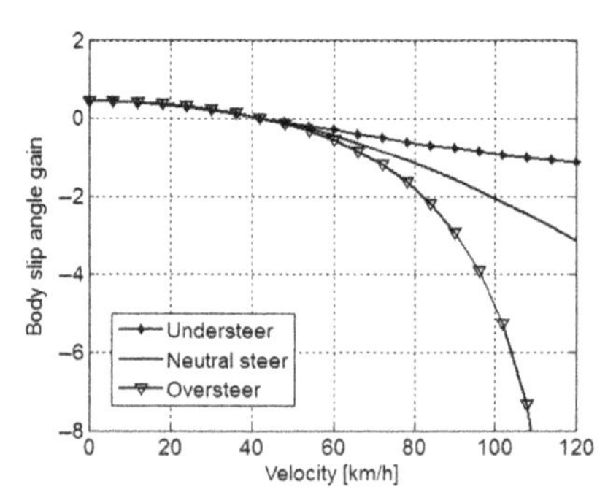

FIGURE 5.15 Vehicle yaw rate and body slip angle gain.

This relationship is the nonlinear extension of Eq. (5.31), where the linear term $\eta \cdot a_y(g)$ is replaced by $h(a_y)$.

It follows that the curve of $\delta = L/R + \alpha_1 - \alpha_2$ versus a_y/g, as shown in Figure 5.14, can be obtained using vertical subtraction of g_1 and g_2 in the (a_y, δ) plane, i.e., using horizontal subtraction of the original normalized axle characteristic curves in the (α_i, f_{yi}) plane. Because the functions g_i are multi-valued, there are different ways to carry out this subtraction, i.e., different branches of the two functions g_1 and g_2 can be combined.

Branches can only exist when the a_y/g value is below the lowest maximum (of the two maximum values of both normalized axle characteristics, respectively). Consider the axle characteristics, depicted in Figure 5.13. They are also shown in the left-hand plot of Figure 5.16. The dashed line in Figure 5.16 corresponds to the function $h(a_y) = \alpha_1 - \alpha_2$, described in Eq. (5.39). Because of the intersection of f_1 and f_2, the function $h(a_y)$ intersects the line $\alpha_1 - \alpha_2 = \delta - L/R = 0$. This is clear from the right-hand figure, with $\delta - L/R$ plotted against $a_y(g)$. This plot is the representation of the stationary steering performance for the axle characteristics from Figure 5.13. The dashed line in the left-hand image of Figure 5.16 is usually referred to as **handling curve**. Consequently, according to *definition 1*, the vehicle behaves as understeered up to the maximum of this handling curve. Beyond this point, the vehicle behaves as oversteered. Note that, even in case of oversteer, we may still obtain $\alpha_1 > \alpha_2$. This implies that *definition 2* does not hold for the nonlinear case.

In the same way, one may use axle characteristics that do not intersect, as indicated in Figure 5.17 (left-hand plot). In this case, the nonlinear handling curve in the right-hand plot is continuously increasing, in contrast to Figure 5.16, which means that the vehicle behaves as understeered for the full a_y range. The maximum of f_{y1} is now decisive for the a_y/g versus $\alpha_1 - \alpha_2$ branch in the left-hand image in Figure 5.17.

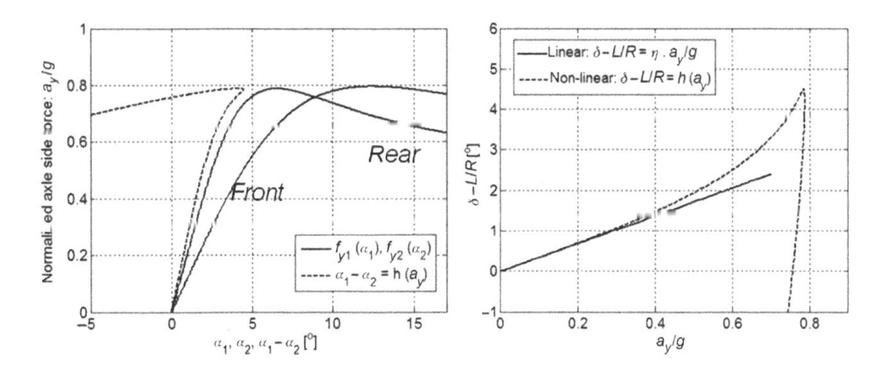

FIGURE 5.16 Handling curve, example 1.

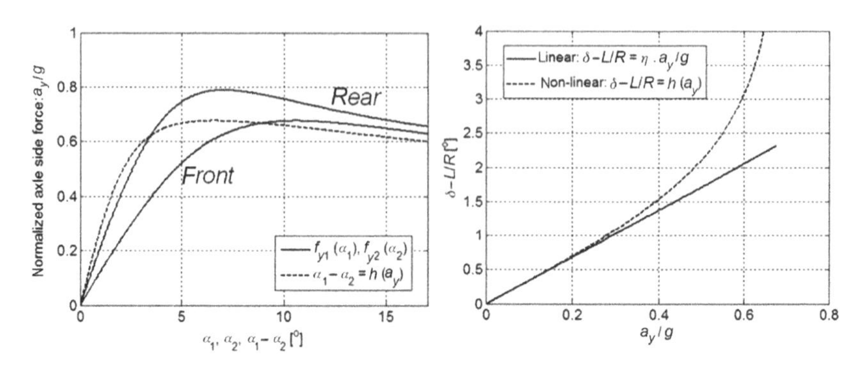

FIGURE 5.17 Handling curve, example 2.

These figures clearly show that the vehicle steering characteristics depend directly on the axle characteristics and thus the tire characteristics. The understeer region corresponds to the situation in which

$$\frac{\partial f_{y1}(\alpha_1)}{\partial \alpha_1} < \frac{\partial f_{y2}(\alpha_2)}{\partial \alpha_2} \tag{5.40}$$

showing that *definition 3* still holds if η is replaced by

$$\eta_{\text{nonlinear}} = \frac{1}{(\partial f_{y1}(\alpha_1))/(\partial \alpha_1)} - \frac{1}{(\partial f_{y2}(\alpha_2))/(\partial \alpha_2)} = \frac{\partial h(a_y)}{\partial a_y} \tag{5.41}$$

where a_y is expressed in terms of g. The extension of *definition 4* to nonlinear axle characteristics is straightforward from Eq. (5.41).

5.3.3 Neutral Steer Point

Another approach to understeer is to consider a vehicle under a lateral force and no steering. With the force acting at a point at the front of the vehicle, the vehicle is expected to experience a yaw motion. The same is true if the force is acting at a point at the rear of the vehicle, but with the yaw direction opposite to the previous situation. One may expect the existence of a point, in the (x, z) plane of symmetry, such that any side force acting at that point does not cause a steady-state yaw motion, i.e., the vehicle will only drift sideways with body slip angle (β). This point P_{NS} is referred to as the **neutral steer point** (Figure 5.18) and has an x position relative to the vehicle CoG, denoted as x_{NS}. The **static margin** M_s is defined as

$$M_s = \frac{x_{NS}}{L} \tag{5.42}$$

with wheelbase L. Refer also to Genta and Morello [11] and Dukkipati et al. [6] for further discussion of the neutral steer point and the static margin.

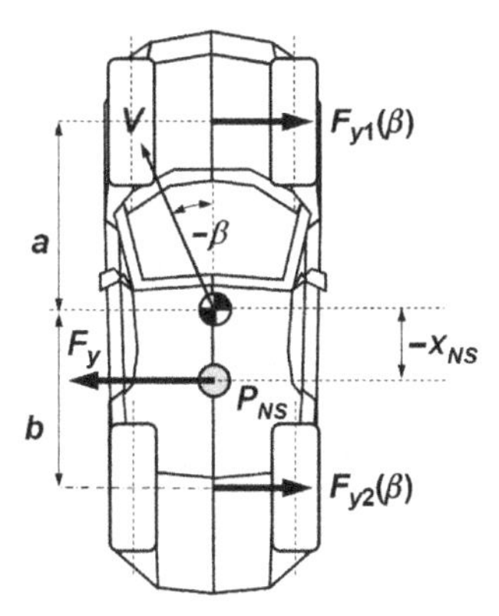

FIGURE 5.18 Vehicle under lateral force F_y, acting at the Neutral Steer Point.

In Ref. [6], a different sign convention is used, in the sense that x_{NS} indicates the vehicle CoG with respect to the neutral steer point, instead of the inverse. The consequence is that x_{NS}, as well as M_s, will have a different sign compared to the analysis here. The total force F_y must be equal to the sum of both axle side forces. Each axle experiences a slip angle equal to β, because the yaw rate $r = 0$.

Consequently,

$$F_y = F_{y1}(\beta) + F_{y2}(\beta)$$

which can be described, for linear axle characteristics, as

$$F_y = Y_\beta \cdot \beta = -(C_{\alpha 1} + C_{\alpha 2}) \cdot \beta$$

The value of x_{NS} follows from equilibrium of moments around the vehicle CoG:

$$x_{\mathrm{NS}} = \frac{a \cdot F_{y1}(\beta) - b \cdot F_{y2}(\beta)}{F_{y1}(\beta) + F_{y2}(\beta)}$$

In case of linear axle characteristics,

$$x_{\mathrm{NS}} = \frac{a \cdot C_{\alpha 1} - b \cdot C_{\alpha 2}}{C_{\alpha 1} + C_{\alpha 2}} = \frac{N_\beta}{Y_\beta} \tag{5.43}$$

Let us investigate what occurs for linear axles if the neutral steer point is located behind the CoG, i.e., $x_{\mathrm{NS}} < 0$.

A force, acting on the vehicle at the CoG, will result in a yaw motion as the vehicle moves away from this force. The slip angle at the front axle will therefore increase (with $a \cdot r/V$) and the slip angle at the rear axle will

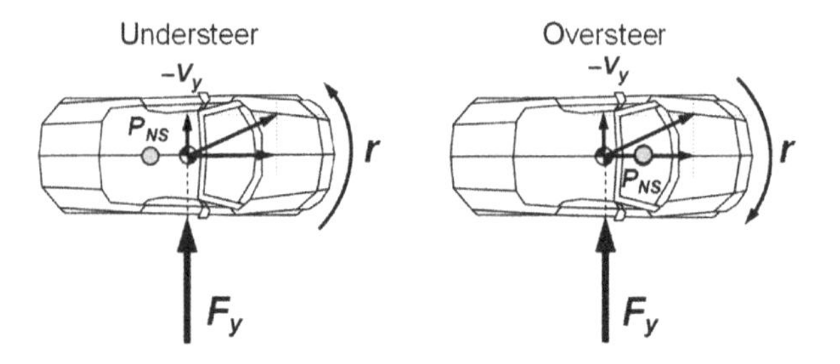

FIGURE 5.19 Vehicle response to external force in absence of steering.

TABLE 5.6 Steering Performance and Neutral Steer Point

Behavior	η	$\alpha_1-\alpha_2$	x_{NS}	M_s
Understeer	>0	>0	<0	<0
Oversteer	<0	<0	>0	>0

decrease. Consequently, $\alpha_1-\alpha_2$ will increase, which means that the vehicle behaves as understeered.

The opposite is true if the neutral steer point is ahead of the CoG. When the vehicle experiences a force acting on vehicle CoG (e.g., a crosswind), drifting will move the vehicle away from the source of this force. However, the resulting yaw motion will give the vehicle the tendency to move toward this force, i.e., to act against the loading.

In this case, $\alpha_1-\alpha_2$ will decrease, which means that the vehicle behaves as oversteered. Also compare Eq. (5.43) with expression (5.31), from which it follows that x_{NS} has the same sign as $-\eta$.

Both cases (understeer and oversteer) are schematically shown in Figure 5.19 and are summarized in Table 5.6.

5.4 NONSTEADY-STATE ANALYSIS

5.4.1 Yaw Stability

In this section, we discuss the solutions of the equations of motion (5.6a) and (5.6b) where we neglect F_{ye} and M_{ze}. In Section 5.3, the steady-state solutions of equations (5.6a) and (5.6b) have been discussed and we observed the relevance of the understeer and oversteer properties, with respect to the vehicle's steady-state behavior.

Let us now discuss **stability** of these steady-state solutions. We will linearize around the steady-state solutions and discuss the nontrivial solutions of the resulting set of homogeneous equations. We use the theory of system dynamics as discussed in Appendix 2.

Let the slope of the axle characteristics front and rear for steady-state slip angles α_1 and α_2 be denoted by A_1 and A_2, respectively (Figure 5.20). Note

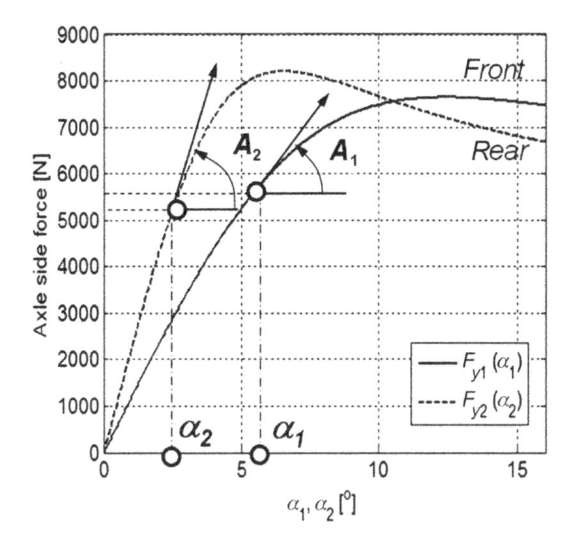

FIGURE 5.20 Steady-state solutions.

that these slopes are the same as the cornering stiffnesses $C_{\alpha 1}$ and $C_{\alpha 2}$ if linear axle characteristics are assumed. Linearization around the steady-state solution leads to equations in $d_v = v - v_s$ and $d_r = r - r_s$, with an index s indicating steady-state behavior.

Let these equations have solutions of the form (see Appendix 2)

$$\mathrm{d}_v = K_v \cdot \mathrm{e}^{\lambda \cdot t} \quad d_r = K_r \cdot \mathrm{e}^{\lambda \cdot t}$$

resulting in the equation

$$\begin{pmatrix} m \cdot \lambda + \dfrac{A_1 + A_2}{u} & m \cdot u + \dfrac{a \cdot A_1 - b \cdot A_2}{u} \\ \dfrac{a \cdot A_1 - b \cdot A_2}{u} & J_z \cdot \lambda + \dfrac{a^2 \cdot A_1 + b^2 \cdot A_2}{u} \end{pmatrix} \cdot \begin{pmatrix} K_v \\ K_r \end{pmatrix} = \begin{pmatrix} 0 \\ 0 \end{pmatrix} \tag{5.44}$$

The characteristic equation follows from stating that the determinant of the coefficient matrix vanishes:

$$\begin{aligned} & m \cdot J_z \cdot u^2 \cdot \lambda^2 + u \cdot \lambda \cdot (J_z \cdot (A_1 + A_2) + m \cdot a^2 \cdot A_1 + m \cdot b^2 \cdot A_2) \\ & + A_1 \cdot A_2 \cdot J \cdot L^2 - m \cdot (a \cdot A_1 - b \cdot A_2) = 0 \end{aligned} \tag{5.45}$$

The local stability of the vehicle around the steady-state point is determined by the resulting eigenvalues. In Appendix 2, we discuss this stability and the type of local behavior (oscillatory, monotonous).

First, we first assume **linear axle characteristics**. The extension to nonlinear axle characteristics is treated in detail in Section 5.5.2. The same equation as expression (5.45) is then derived, with A_1 and A_2 replaced by $C_{\alpha 1}$ and $C_{\alpha 2}$, respectively. This equation, and all equations derived from Eq. (5.45) are simpler if we take $J_z = m \cdot a \cdot b$, as suggested in expression (5.10). In that case, Eq. (5.45) becomes

$$\begin{aligned} &a \cdot b \cdot m^2 \cdot u^2 \cdot \lambda^2 + m \cdot u \cdot \lambda \cdot L \cdot (a \cdot C_{\alpha 1} + b \cdot C_{\alpha 2}) \\ &+ C_{\alpha 1} \cdot C_{\alpha 2} \cdot L^2 - m \cdot u^2 \cdot (a \cdot C_{\alpha 1} - b \cdot C_{\alpha 2}) = 0 \end{aligned} \tag{5.45a}$$

This is no real restriction in the sense that stability properties are qualitatively the same as for arbitrary J_z. From Eq. (5.34), we find for the understeer gradient:

$$\eta = \frac{m \cdot g}{L \cdot C_{\alpha 1} \cdot C_{\alpha 2}} \cdot (b \cdot C_{\alpha 2} - a \cdot C_{\alpha 1}) \tag{5.46}$$

resulting, after substituting in Eq. (5.45a), in

$$\lambda^2 + \frac{L \cdot (a \cdot C_{\alpha 1} + b \cdot C_{\alpha 2})}{m \cdot u \cdot a \cdot b} \cdot \lambda + \frac{C_{\alpha 1} \cdot C_{\alpha 2} \cdot L \cdot \eta}{m^2 \cdot a \cdot b \cdot g} \cdot \left(1 + \frac{g \cdot L}{\eta \cdot u^2}\right) = 0 \tag{5.47}$$

There are in general two solutions λ_{12}, written as

$$\lambda_{12} = -\zeta \cdot \omega_0 \pm \omega_0 \cdot \sqrt{\zeta^2 - 1} \quad \text{if } \zeta > 1 \text{ (overdamped)}$$

$$\lambda_{12} = -\omega_0 \quad \text{if } \zeta = 1 \text{ (critically damped)}$$

$$\lambda_{12} = -\zeta \cdot \omega_0 \pm \mathrm{i} \cdot \omega_0 \cdot \sqrt{1 - \zeta^2} \quad \text{if } \zeta < 1 \text{ (underdamped)}$$

with

$$\omega_0^2 = \frac{C_{\alpha 1} \cdot C_{\alpha 2} \cdot L \cdot \eta}{m^2 \cdot a \cdot b \cdot g} \cdot \left(1 + \frac{g \cdot L}{\eta \cdot u^2}\right) \tag{5.48}$$

$$\zeta = \frac{L \cdot (a \cdot C_{\alpha 1} + b \cdot C_{\alpha 2})}{2 \cdot m \cdot a \cdot b \cdot u \cdot \omega_0} \tag{5.49}$$

For positive ζ and ω_0^2, these eigenvalues have a negative real part (or are real and negative), which implies asymptotic stability. We indicated overdamped because any deviation from the steady-state solution will behave nonoscillatory in this case (the eigenvalue is real and assumed to be negative). This corresponds to the two-sided node, described in Appendix 2. Underdamped means that deviations will oscillate while decaying to zero (the eigenvalue has a nonzero imaginary part). This corresponds to the focus or spiral point (again, see Appendix 2).

Let us consider these eigenvalues for different values of η. Remember that the vehicle is understeered if $\eta > 0$ and oversteered if $\eta < 0$.

From expression (5.48), it can easily be concluded that

$$\begin{aligned}&\text{If } \eta > 0 \text{ then } \omega_0^2 > 0 \text{ for all velocities } u\\&\text{If } \eta < 0 \text{ then } \omega_0^2 > 0 \text{ if and only if } u < u_{cr}\end{aligned}$$

with u_{cr} given by expression (5.37). Furthermore, using Eqs. (5.48) and (5.49), one is able to prove that

$$\text{If } \eta > 0 \text{ then } \frac{d\zeta}{du} < 0 \text{ for all velocities } u > 0$$

$$\text{If } \eta < 0 \text{ then } \frac{d\zeta}{du} > 0 \text{ for all velocities satisfying } 0 < u < u_{cr}$$

We used the fact that the cornering stiffnesses $C_{\alpha 1}$ and $C_{\alpha 2}$ are positive. Following the same approach for nonlinear axles, one or two of the slopes A_1 and A_2 may become negative, which changes the situation. We return to this in Section 5.5.2.

For an understeered (linear) vehicle, for large values of u, ω_0 approaches a finite nonzero value, which implies that ζ will become arbitrarily small, i.e., <1. For small values of velocity u the value of ζ approaches

$$\zeta \rightarrow \frac{a \cdot C_{\alpha 1} + b \cdot C_{\alpha 2}}{2 \cdot \sqrt{a \cdot b \cdot C_{\alpha 1} \cdot C_{\alpha 2}}} > 1 \quad \text{if} \quad a \cdot C_{\alpha 1} \neq b \cdot C_{\alpha 2}$$

Hence, the understeered vehicle will respond as overdamped to deviations from steady-state behavior up to a certain speed, and will respond as underdamped beyond that speed.

For the oversteered vehicle, the velocity must be smaller than the critical velocity u_{cr}. At this critical speed, the steady-state behavior becomes unbounded, as shown in Figure 5.15. However, a steady-state solution exists if $u > u_{cr}$ for an oversteered vehicle. If $u > u_{cr}$ and $\eta < 0$, then $\omega_0^2 < 0$ (this follows from expression (5.48)). Consequently, as following from expression (5.47), the resulting eigenvalues are real: one positive and one negative. According to Appendix 2, this stationary (equilibrium) point is an (unstable) saddle point in the phase plane.

We determined the relationship between ω_0 and ζ versus velocity u, with the result plotted in Figures 5.21 and 5.22 for an understeered vehicle ($\eta > 0$), an oversteered vehicle ($\eta < 0$), and a neutrally steered vehicle ($\eta = 0$).

Observe the behavior of ω_0 near the critical speed u_{cr} for an oversteered vehicle. No eigenfrequency can be found for $u > u_{cr}$. The damping ratio ζ satisfies $\zeta > 1$ for all velocities; hence; the vehicle responds as overdamped,

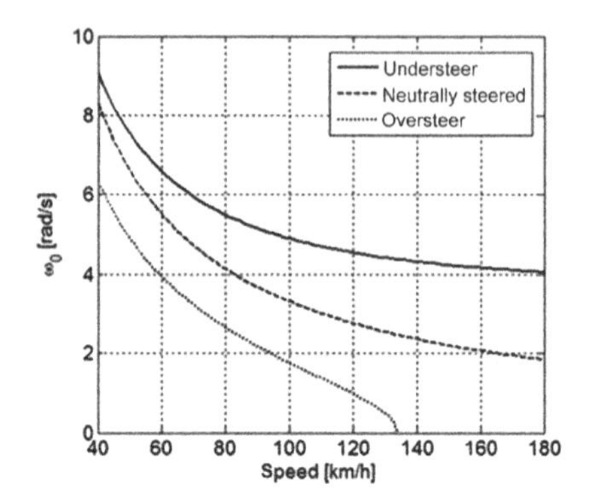

FIGURE 5.21 Parameter ω_0 versus speed u.

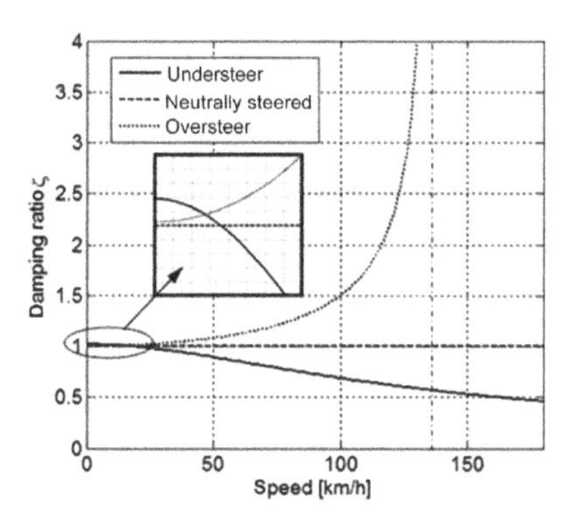

FIGURE 5.22 Damping ratio ζ versus speed u.

in contrast to the understeered vehicle for not too small speed. For small speed, the damping ratio ζ exceeds 1, cf. Figure 5.22).

We have chosen the vehicle data from Appendix 6 for the understeered vehicle. For the oversteered vehicle, we have chosen $\eta = -0.02$.

What would occur if the driver suddenly changed the steering angle with a finite step, resulting in the **ramp or step steer response** (or **J-turn**) as discussed in Section 5.1? Assuming a step change in the steering angle, the front axle slip angle α_1 will experience a step change as well according to expression (5.13). As a result, there is an instantaneous change in the front axle side force and in the lateral acceleration (see Appendix 1). The yaw rate now begins to change, resulting in a side force at the rear axle, with the lateral acceleration increasing further. Finally, the vehicle is reaching the steady-state behavior after a possible overshoot in the yaw rate. This overshoot indicates underdamped oscillatory behavior near the steady-state yaw rate value, hence a focus or spiral point in the phase plane (see Appendix 2). When the maximum yaw rate is reached, the rear axle drift is still increasing. Hence, the lateral acceleration (and therefore, the vehicle lateral speed) lags behind the yaw rate. Consequently, applying a step (or ramp) steer input, a vehicle will first yaw and then drift (see Figure 5.2 for a ramp steer). We have used

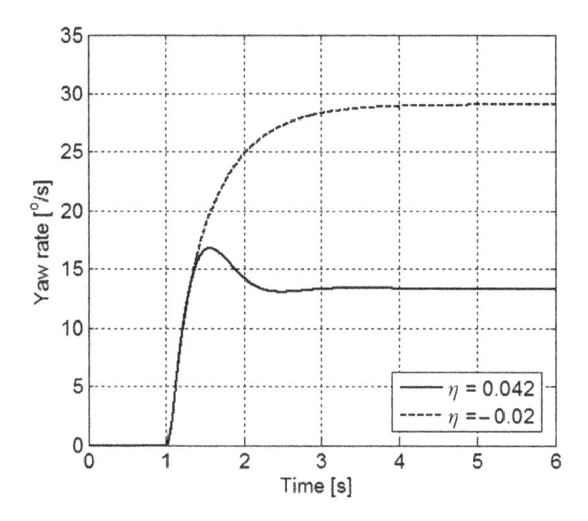

FIGURE 5.23 Yaw rate response to a ramp steer for an understeered ($\eta = 0.042$) and oversteered ($\eta = -0.02$) vehicle.

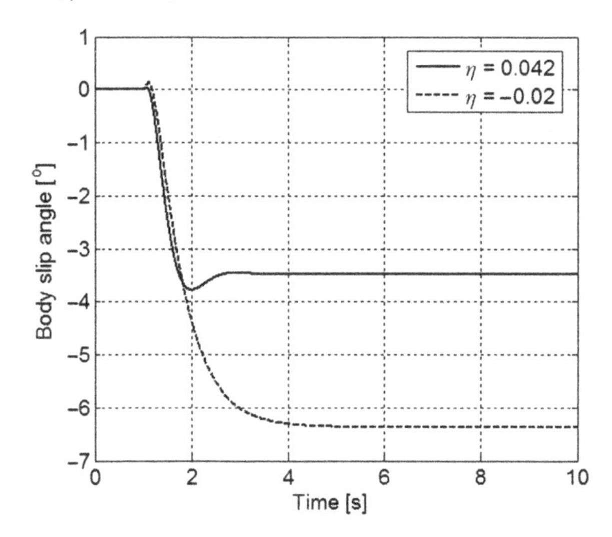

FIGURE 5.24 Body slip angle response to a ramp steer for an understeered ($\eta = 0.042$) and oversteered ($\eta = -0.02$) vehicle.

the state space model from Appendix 1 for two different values of the understeer gradient, and determined the yaw rate response to a ramp steer (120 km/h, maximum axle steering angle of 3°), Figure 5.23. For $\eta > 0$, the plot shows overshoot behavior in contrast to the case of $\eta < 0$. In addition, one observes a much larger yaw rate steady-state value (i.e., the yaw rate gain) for the oversteered vehicle compared to the understeered vehicle. This was shown previously in Figure 5.15, in which we depicted the yaw rate gain versus vehicle speed for varying vehicle velocity. We also depicted the body slip angle for the same ramp steer input (Figure 5.24).

Another step input refers to the situation where a side force is applied on a vehicle, driving straight ahead, i.e., with a steering angle of zero. Such a force could be a result of a sudden change in the lateral slope of the road or a crosswind loading. In Section 5.3.3, we introduced the neutral steer point, which is the point in the (x, z) plane of symmetry such that any side force acting at that point does not cause a yaw motion. For an understeered vehicle, this point lies behind the vehicle CoG. For an oversteered vehicle, the neutral steer point lies ahead of the CoG (see Table 5.6). As a result, if a sudden (ramp input) lateral force F_e is applied at the vehicle CoG, the ultimate vehicle response will only include drifting, i.e., a constant body slip angle and no yaw if the vehicle is neutrally steered. If the vehicle is understeered, the lateral force F_e acts in front of the neutral steer point. The vehicle will have a yaw response, making the vehicle move away from the source of the side force F_e. A similar qualitative behavior is observed for any vehicle subjected to a lateral force acting in front of the neutral steer point and in absence of an external yaw moment. An oversteered vehicle (or a vehicle with the external force acting behind the neutral steer point) will move toward this source. The vehicle response to a ramp input for an external force F_e, in case of understeer, neutral steer, and oversteer is shown in Figure 5.25.

The following discussion is taken from Milliken and Milliken [26]. Consider a road with a lateral slope that induces a lateral force acting on the neutrally steered vehicle. The understeered vehicle will move away from the force, i.e., run down the slope, in contrast to the oversteered vehicle, which will run up the slope. This is an efficient way to determine the understeer tendency of a vehicle.

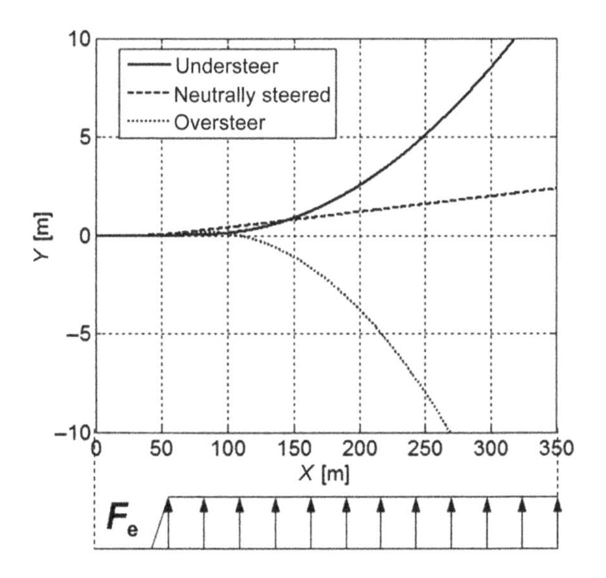

FIGURE 5.25 Vehicle response to a ramp input for a lateral force, acting at the CoG.

It is of interest to consider the vehicle stability properties in terms of the normalized slip stiffnesses, denoted as $c_{\alpha i}$, $i = 1, 2$ and defined by

$$c_{\alpha i} = \frac{\partial f_{yi}(\alpha_i = 0)}{\partial \alpha_i} = \frac{C_{\alpha i}}{F_{zi}} \quad i = 1, 2$$

Here, we indicate the different stability areas in the $(c_{\alpha 2}, c_{\alpha 1})$ plane, which is also denoted as the **stability diagram** (see Section 5.5). The following transitions between stability areas are relevant:

i. The transition of understeer to oversteer.
ii. The transition of stable oversteer to unstable oversteer.
iii. The transition of oscillatory behavior to nonoscillatory behavior near the steady-state solution for an understeered vehicle at low speed.

AD (i)
From *definition 3* of understeer (and expression (5.31)), it is clear that this transition is given by

$$c_{\alpha 2} - c_{\alpha 1} = 0$$

AD (ii)
This is the curve where ω_0 vanishes for negative η. Using Eq. (5.48) and the definition of η according to expression (5.31), one finds

$$-\eta = \frac{1}{c_{\alpha 2}} - \frac{1}{c_{\alpha 1}} = \frac{g \cdot L}{u^2}$$

Hence, we arrive at a curve in the stability diagram passing the origin under an angle of 45° and with a vertical asymptote for

$$c_{\alpha 2} = \frac{u^2}{g \cdot L}$$

AD (iii)
This is the curve where $\zeta = 1$ for positive η. Using Eqs. (5.48) and (5.49), together with expression (5.31), we find

$$c_{\alpha 2} - c_{\alpha 1} = \frac{4 \cdot u^2}{g \cdot L}$$

These curves are plotted in the $(c_{\alpha 2}, c_{\alpha 1})$ plane in Figure 5.26 for two different speeds, u_1 and $u_2 > u_1$.

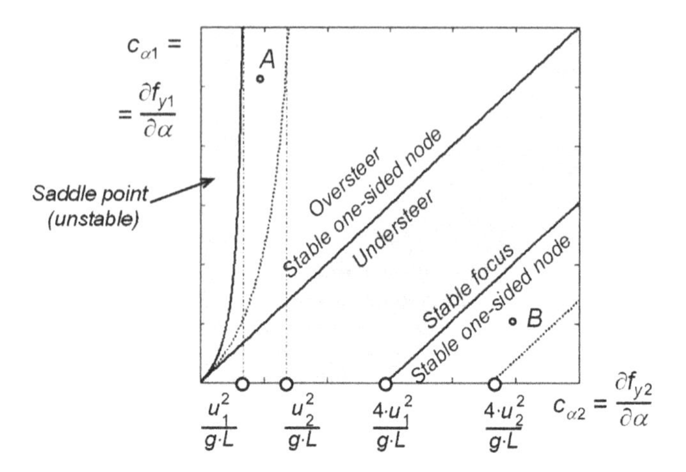

FIGURE 5.26 Stability of a linear one-track vehicle model in terms of the normalized slip stiffnesses.

For speed u_1, the vehicle is stable for all combinations of the normalized slip stiffnesses, lying to the right of the solid hyperbolic curve. Left of that, the vehicle is oversteered and unstable, with the steady-state solution being a saddle point in the phase plane (see Appendix 2). The stable area is divided into three areas, one for an oversteered but still stable vehicle, one for the understeered vehicle with oscillatory behavior near the steady-state point in the phase plane (focus), and one for the stable understeered vehicle with local nonoscillatory behavior (and hence, no overshoot for a ramp steer input).

Consider point A in the plot. For speed u_1, this point corresponds to a stable vehicle. For speed u_2, the stability has been lost. In the graph, the asymptote has shifted to the right and passed this point.

Such behavior is valid for every point above the line $c_{\alpha 1} = c_{\alpha 2}$. There is always a forward speed u such that the asymptote for $c_{\alpha 2} = u^2/(g \cdot L)$ passes this point, corresponding to loss of stability. In other words, if the speed is high enough, stability is lost for an oversteered vehicle, as concluded previously.

Next, we consider point B. For speed u_1, a vehicle with these normalized slip stiffnesses, would respond monotonously to a step or ramp steer input. With increased speed u_2, this behavior changes, and an overshoot will be observed (as shown in Figure 5.23 for $\eta > 0$. In other words, assuming linear axle characteristics, a vehicle is expected to show overshoot behavior in response to sudden steering changes, but only if the speed is not too low.

We close this section with the root locus visualization of the eigenvalues of the linear single-track vehicle model, i.e., the solutions of the characteristic Eq. (5.45). This visualization shows the eigenvalues in the complex domain for variation of a specific vehicle parameter, e.g., vehicle velocity. The importance of a root locus plot is that the speed sensitivity of parameters, such as the damped eigenfrequency (imaginary part) and the damping ratio (angle between eigenvalue and imaginary axis), for a specific eigenmode can be read immediately from the plot. We refer to Appendix 3 for more

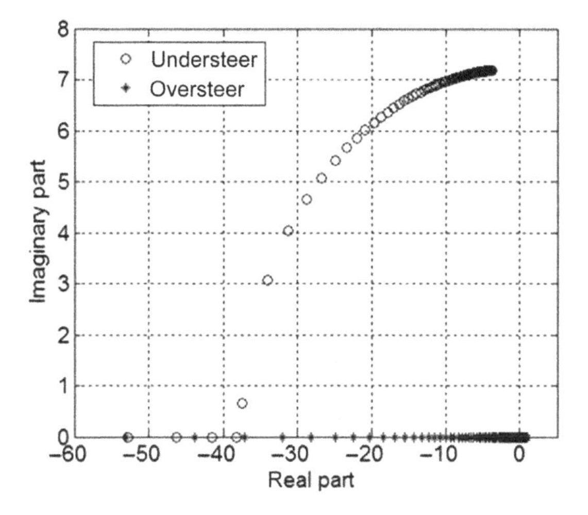

FIGURE 5.27 Root locus plot (understeer and oversteer).

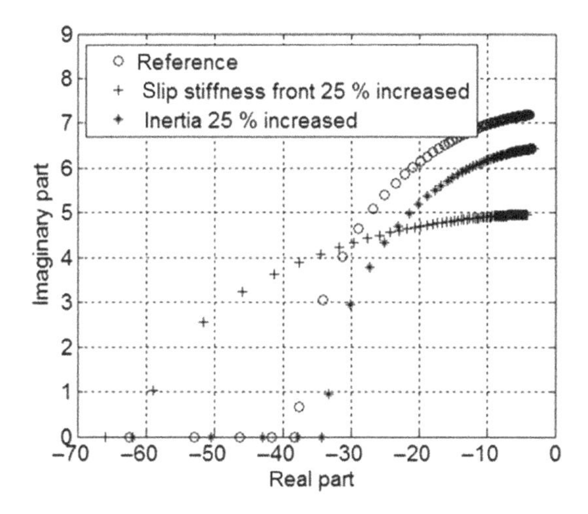

FIGURE 5.28 Root locus plot, understeer, varying vehicle parameters.

information. This improves our qualitative understanding of vehicle performance, which is especially relevant if more degrees of freedom (i.e., more eigenmodes and more eigenvalues in one root locus plot) must be considered. We varied the vehicle velocity and plotted one of the eigenvalues for both the understeer and oversteer situations (Figure 5.27).

Observe the increasing damped eigenfrequency and decreasing damping ratio for increasing velocity and compare this with Figure 5.22. The oversteered vehicle becomes unstable beyond a certain critical speed, which is indicated by passing the imaginary axis into the right half plane. In Figure 5.28, we have plotted the eigenvalue for modified slip stiffness of the front axle and for increased inertia (mass, yaw moment of inertia). As expected, larger inertia reduces the eigenfrequency. Larger slip stiffness at the front axle also leads to lower eigenfrequencies, but to a larger damping ratio for a lower speed. This is consistent with the fact that a smaller stiffness results in less understeer, and that oversteer corresponds to a damping ratio exceeding 1.

5.4.2 Frequency Response

In this section, we consider the linear vehicle response to an oscillatory steering input:

$$\delta(t) = A_\delta \cdot e^{i \cdot \Omega \cdot t} \tag{5.50}$$

Here, we assume linear axle characteristics with the equations of motions given in Eqs. (5.17a) and (5.17b). Substituting expression (5.50) into Eqs. (5.17a) and (5.17b) for $F_{ye} = 0$ and $M_{ze} = 0$, and solving for yaw rate and body slip angle, we arrive at the following expressions:

$$r(t) = A_r \cdot e^{i \cdot \varphi_r} \cdot e^{i \cdot \Omega \cdot t} \tag{5.51}$$

$$\beta(t) = A_\beta \cdot e^{i \cdot \varphi_\beta} \cdot e^{i \cdot \Omega \cdot t} \tag{5.52}$$

In Appendix 4, we discuss Bode diagrams, in which a graphical representation of the frequency transfer function $G(i \cdot \Omega)$ is defined as

$$x = G(i \cdot \Omega) \cdot u$$

for input u (steering angle, amplitude A_δ), and state x (yaw rate or body slip angle). With the notation of Eqs. (5.50)–(5.52), we obtain

$$G_r(i \cdot \Omega) = \frac{A_r}{A_\delta} \cdot e^{i \cdot \varphi_r}$$

$$G_\beta(i \cdot \Omega) = \frac{A_\beta}{A_\delta} \cdot e^{i \cdot \varphi_\beta}$$

Using Eqs. (5.17a) and (5.17b), the following expressions for these frequency transfer functions may be obtained:

$$G_r(i \cdot \Omega) = \frac{C_{\alpha 1}}{m \cdot J_z \cdot V} \cdot \frac{L \cdot C_{\alpha 2} + i \cdot a \cdot m \cdot V \cdot \Omega}{\omega_0^2 + 2 \cdot i \cdot \zeta \cdot \omega_0 \cdot \Omega - \Omega^2} \tag{5.53}$$

$$G_\beta(i \cdot \Omega) = \frac{C_{\alpha 1}}{m \cdot J_z \cdot V^2} \cdot \frac{b \cdot L \cdot C_{\alpha 2} - a \cdot m \cdot V^2 + i \cdot J_z \cdot V \cdot \Omega}{\omega_0^2 + 2 \cdot i \cdot \zeta \cdot \omega_0 \cdot \Omega - \Omega^2} \tag{5.54}$$

with ω_0 and ζ introduced in Eqs. (5.48) and (5.49). That means that we use the simplification (5.10).

We plotted the frequency response in terms of Bode diagrams (see Appendix 4), for the yaw rate and body slip angle in Figures 5.29 and 5.30, respectively, for different speeds for an understeered vehicle (vehicle data taken from Appendix 6).

The following observations can be made:

- The yaw rate damping is reduced with increased speed, which is consistent with Figure 5.22. The same is true for the body slip angle damping (drifting).
- The damped eigenfrequency shifts to large values with increasing speed, which is consistent with Figure 5.28.
- The steady-state gain is not monotonous in speed V. Apparently, 150 [km/h] is beyond the vehicle's characteristic speed (see Figure 5.15).
- A slight phase lead is observed for high speed, for low frequency below the yaw resonance frequency.
- A small steady-state value in the body slip angle magnitude plot is because the steady-state body slip angle is quite small for $V = 30$ [km/h] (see Figure 5.15).
- The steady-state body slip angle changes sign around 40 [km/h] (see again Figure 5.15). For that reason, the phase is changed with 180 [°] for increasing speed, as indicated in Figure 5.30.

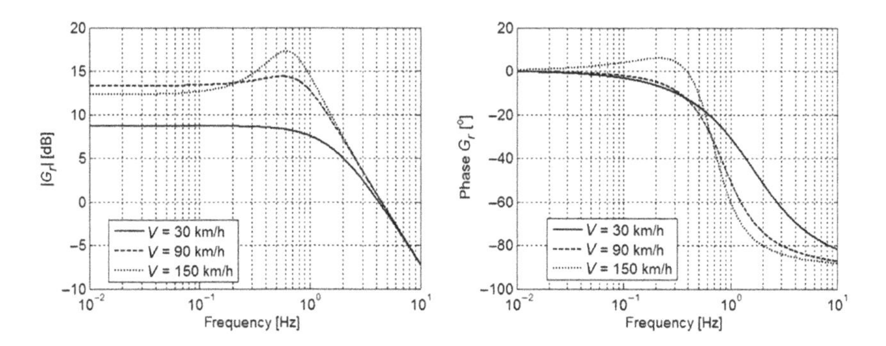

FIGURE 5.29 Bode diagrams for yaw rate frequency transfer.

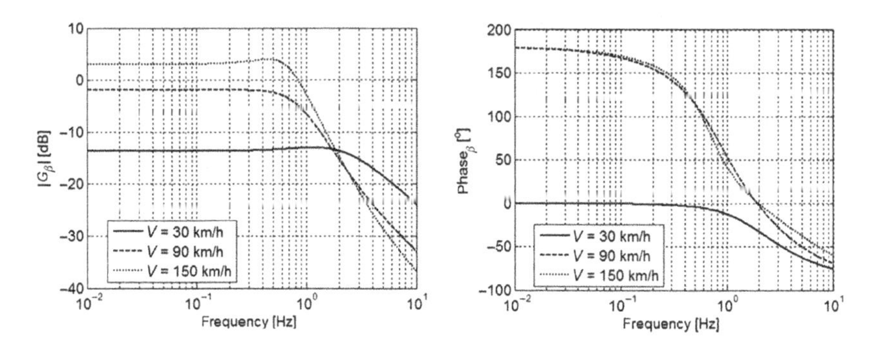

FIGURE 5.30 Bode diagrams for body slip angle frequency transfer.

5.5 GRAPHICAL ASSESSMENT METHODS

5.5.1 Phase Plane Analysis

The Eqs. (5.6a) and (5.6b) are a second-order system, with vehicle states (v, r) of (β, r), where $v = \beta \cdot V$ and with vehicle total speed V, body slip angle β, yaw rate r, and lateral speed v. This set of equations can also be described in terms of vehicle states (α_1, α_2) through Eq. (5.15), where α_1 and α_2 are the slip angles at the vehicle's front and rear axles. Assume the steering angle δ and vehicle speed V are constant. That means that the vehicle will be steady state on a curve with radius R, if

$$\delta = \frac{L}{R} + h\left(\frac{V^2}{g \cdot R}\right) \tag{5.55}$$

according to expression (5.39), where the function h is the difference of the inverse functions of the normalized axle side force according to Eq. (5.38). In case of linear axle characteristics, Eq. (5.55) can be replaced by Eq. (5.31), depending on the understeer gradient η. If expression (5.55) is not satisfied, the vehicle will not be in steady state and the states will change in time. Suppose a vehicle following a steady-state circle suddenly experiences a disturbance such that expression (5.55) is no longer satisfied. If the driver does not change the steering angle or the speed, two things can occur. First, the vehicle returns to the steady-state curve, i.e., it supports the driver to reach stable cornering conditions. Second, the vehicle is lost in spinning or drifting with extreme yaw rate or body slip. Under extreme driving conditions (large lateral acceleration), a third option exists, which is limit cycle behavior.

Obviously, the first type of behavior is preferred. As discussed in Section 5.4.1 for linear axle characteristics, this behavior is related to the stability of the steady-state solution, which depends on the normalized axle slip stiffness, as indicated in Figure 5.26. This stability is global, meaning that, regardless of the size of the disturbance, the vehicle will always return to the steady-state situation. In case of nonlinear axle characteristics, stability depends on the eigenvalues of the equations, obtained using linearizations of these equations near the steady-state solutions. We refer to Appendix 2 for further details. See Section 5.5.2, in which the *stability diagram* for linear equations (Figure 5.26) is extended to nonlinear axle characteristics. Demonstrating stability in that way does not guarantee global stability.

To understand global stability (what happens to large disturbances?), we take advantage of the fact that the bicycle model is a second-order system. We can draw solution curves in the (β, r)- or (α_1, α_2) plane. We can visualize the global behavior near the steady-state points (also called critical points) in these planes. A plot of all possible solution curves for a fixed input is called a **phase plane**, and the solution curves are called **trajectories**. Smakman [50] applied the phase plane in terms of β and $\dot{\beta}$ in the design of a wheel load controller to improve the vehicle handling properties. For that, he derived an area in this phase plane of acceptable combinations of these vehicle states, for which smooth handling is fulfilled.

An interesting and elegant approach was proposed by Guo [15], selecting the following nondimensional states:

$$x_{\mathrm{Guo}} = \beta; \quad y_{\mathrm{Guo}} = \frac{r_{\mathrm{g}} \cdot r}{V} \tag{5.56}$$

where r_{g} is the radius of gyration (see expression (5.8)). He denoted the phase plane for these states as the **energy phase plane**. Further, he observed that the sum of squares of these states is equal to the ratio of cornering kinetic energy T_{c} and translational kinetic energy T_{k}:

$$x_{\mathrm{Guo}}^2 + y_{\mathrm{Guo}}^2 = \frac{T_{\mathrm{c}}}{T_{\mathrm{k}}} \tag{5.57}$$

where

$$T_{\mathrm{c}} \equiv \frac{1}{2} \cdot m \cdot V^2 \cdot \beta^2 + \frac{1}{2} \cdot J_z \cdot r^2, \quad T_{\mathrm{k}} \equiv \frac{1}{2} \cdot m \cdot V^2$$

Consequently, the following interpretation for points in the energy phase plane holds:

Interpretation 1

The square of the distance of points on trajectories in the energy phase plane to the origin is directly proportional to the cornering energy of the vehicle.

Let us draw the trajectories for three different cases, for the vehicle and axle data corresponding to the handling curves of Figure 5.16:

i. $V = 70$ [km/h], $\delta = 2$ [°]
ii. $V = 90$ [km/h], $\delta = 2$ [°]
iii. $V = 70$ [km/h], $\delta = 3$ [°]

Results are shown in Figure 5.31. We highlighted the specific trajectory passing through the origin (which corresponds to a step steer input response) in bold.

The following observations can be made:

- A critical stationary point exists that attracts the trajectories near this point in an oscillatory way. According to Appendix 2, this stationary point is a stable focus.
- Following the solution curve passing (0, 0), the body slip angle is first positive and then becomes negative. Compare this with Figure 5.15, in which

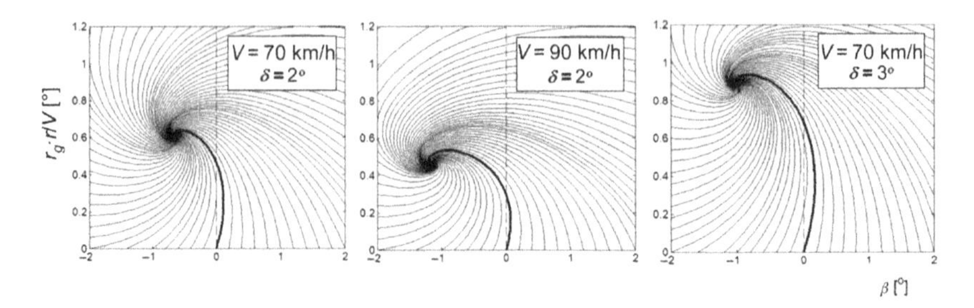

FIGURE 5.31 Energy phase plane for the one-track vehicle mode, axle characteristics (cf. Figure 5.13) for $(V, \delta) = (70, 2)$, $(90, 2)$, and $(70, 3)$ ([km/h], [°]).

the steady-state body slip angle gain is plotted versus speed V, with $\beta > 0$ for small speed and $\beta < 0$ for large slip. Starting at $(\beta, r) = (0, 0)$, the rear axle slip angle needs time to build up to a certain value, i.e., it follows the steering angle with some lag. With the rear slip angle not fully developed, the lateral speed still points into the curve, i.e., with positive β. Beyond a certain speed, the steady-state body slip angle will be negative, which means that β will change sign at a certain time, as shown in Figure 5.31.

- Considering the same curve, the vehicle shows yawing before it shows drifting (lateral sliding). Consequently, the yaw rate will respond faster to changes in steering angle than the body slip angle.
- Increasing speed or steering angle will increase the cornering energy. Note that x and y scales are not the same in Figure 5.31.
- Increasing speed leads to a smaller y_{Guo} for the final stationary solution. This is because speed has a nonproportional effect on the yaw rate gain (see Figure 5.15).The state y_{Guo} has an additional term V in the denominator, pushing y_{Guo} down with increasing speed.
- This is different for changing steering angle δ, as it pushes the yaw rate r upward (see Eq. (5.33c)].
- In both cases, increasing speed V and increasing steering angle δ, the steady-state body slip angle is increased as well.

Let us next consider tire slip angles α_1 and α for front and rear axles. These angles can be expressed in terms of x_{Guo} and y_{Guo} as follows:

$$\alpha_1 = \delta - x_{\mathrm{Guo}} - \frac{a}{r_{\mathrm{g}}} \cdot y_{\mathrm{Guo}}; \quad \alpha_2 = -x_{\mathrm{Guo}} + \frac{b}{r_{\mathrm{g}}} \cdot y_{\mathrm{Guo}} \tag{5.58}$$

These expressions indicate that constant slip angles correspond to fixed straight lines in the energy phase plane. In other words:

Interpretation 2

Families of straight lines $x_{\mathrm{Guo}} + (a/r_g)\, y_{Guo} = constant$ and $x_{Guo} - (b/r_g) y_{Guo} = constant$ correspond to constant values of $\alpha_1 - \delta$ and α_2.

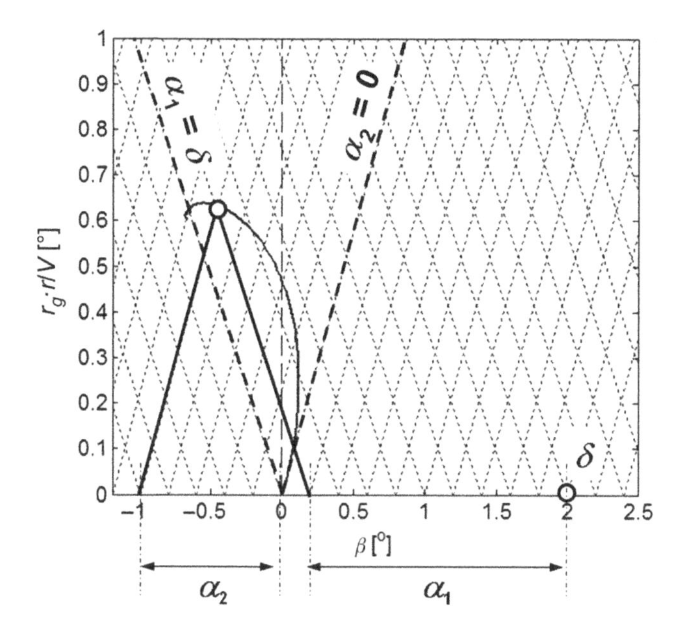

FIGURE 5.32 Wheel slip angles in the energy phase plane.

For $V = 70$ [km/h] and $\delta = 2$ [°], these lines are shown in Figure 5.32, in combination with the trajectory through the origin (response of the step steer input). Each point of this trajectory (and any other trajectory) falls upon two straight lines that correspond to a certain slip angle for the front and rear axles, respectively. The rear slip angle is obtained by taking the distance between the origin and the intersection of the line for which α_2 is constant with the β-axis ($y_{Guo} = 0$). A positive value for α_2 corresponds to an intersection at the left side of the origin. The difference between the steering angle and the front slip angle follows from the distance of the origin and the intersection of the line for which α_1 is constant with the β-axis. A positive value of $\alpha_1 - \delta$ again corresponds to an intersection at the left side of the origin.

Following the trajectory from the origin to the final steady-state point, the rear axle slip angle α_2 appears to grow from 0 to approximately 1.2 [°]. The front axle slip angle starts at the value of δ, then slightly decreases up to about 1.5 [°], after which it increases again to a value, slightly exceeding the steering angle δ.

Note that the constant $\alpha_1 = \delta$ and $\alpha_2 = 0$ lines cut the two quadrants in the energy phase plane in two parts for positive yaw rate corresponding to a specific vehicle behavior in terms of slip angles and body slip angle, as schematically depicted in Figure 5.33.

Starting on the right, all local speeds at the axles and at the vehicle's CoG are pointing to the right of the vehicle, where the front axle slip angle may be negative. Consequently, the pole of the vehicle motion must lie behind the rear axle and the driver is looking outside of the curve. Passing the line $\alpha_2 = 0$ means that the rear axle slip angle changes sign and indicates

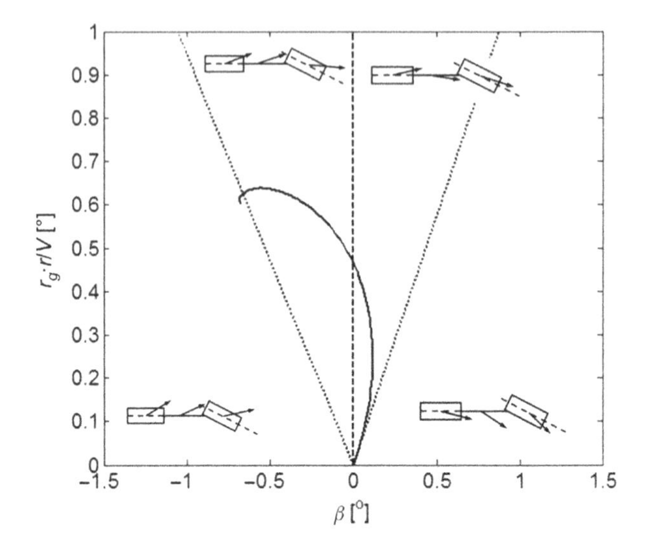

FIGURE 5.33 Typical vehicle behavior in terms of body slip angle and wheel slip angles.

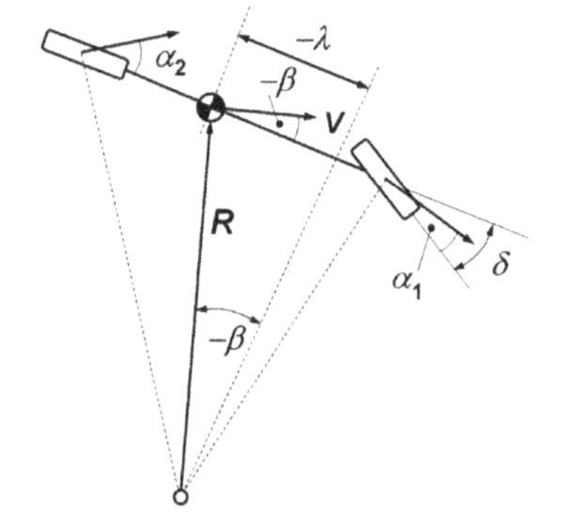

FIGURE 5.34 Rotating length (λ) and curvature radius (R).

that the pole of the vehicle's motion is shifted in forward direction (relative to the vehicle's direction). Passing the line $\beta = 0$ means that the body slip angle changes sign. Therefore, the pole of the motion has shifted further in forward direction and the driver now looks into the curve. Finally, passing the line $\alpha_1 = \delta$ means that α_1 exceeds δ and the local speed at the front axle points to the left, relative to the vehicle's orientation (port side). The pole of the vehicle motion will now be ahead of the front axle position.

The distance between the pole and the vehicle CoG, projected on the vehicle symmetry plane, is called the *rotating length* λ, Figure 5.34. This length is positive if the pole lies behind the vehicle's CoG. According to Figure 5.34, the rotating length will exceed b for large positive β and finite yaw rate r. The pole of rotation will then lie behind the rear axle. Similarly, the rotating length will exceed, in an absolute sense, the distance a between vehicle CoG and front axle for large negative β and finite yaw rate.

The path curvature and the radius of curvature are related to the course angle η, which is shown in Figure 4.2 and is the sum of the yaw angle ψ and the body slip angle β. From standard kinematics for a planar motion of a particle, one can show that

$$\frac{1}{R} \approx \frac{\dot{\eta}}{V} = \frac{\dot{\beta} + r}{V} = \frac{a_y}{V^2}$$

for small angles. For small changes in the body slip angle (e.g., when the vehicle behavior is close to steady state), the curve radius can be approximated by

$$R \approx \frac{r_g}{y_{Guo}} \tag{5.59}$$

Consequently, large values of y_{Guo} correspond to large values of path curvature. Clearly, this will be true if changes in the body slip angle are not small, as well.

In other words:

Interpretation 3

Large values of y_{Guo} correspond to small values of the curve radius R, i.e., large curvature, in case that the vehicle behavior is close to steady state.

If Eq. (5.59) holds, the rotating length λ can be expressed in x_{Guo} and y_{Guo} as follows:

$$\lambda = R \sin\beta \approx r_g \cdot \frac{\sin x_{Guo}}{y_{Guo}} \approx r_g \cdot \frac{x_{Guo}}{y_{Guo}} \tag{5.60}$$

This expression confirms the change in pole position, indicated previously. For large positive value of x_{Guo}/y_{Guo}, the pole will move to the rear. If x_{Guo}/y_{Guo} is large in negative direction, the pole will move to the front, with respect to the vehicle orientation.

Interpretation 4

The rotating length, which describes the pole position during handling, varies with x_{Guo}/y_{Guo}. This means that points in the energy phase plane close to the y-axis for finite y_{Guo} will have a pole close to the vehicle CoG position, whereas (for a positive yaw rate) moving these points to large $|x_{Guo}|$ will shift the pole to the front ($x_{Guo} < 0$) or to the rear ($x_{Guo} > 0$). For a negative yaw rate, the steering orientation is reversed. Therefore, the pole will, in that case, move to the front if $x_{Guo} > 0$ and to the rear if $x_{Guo} < 0$.

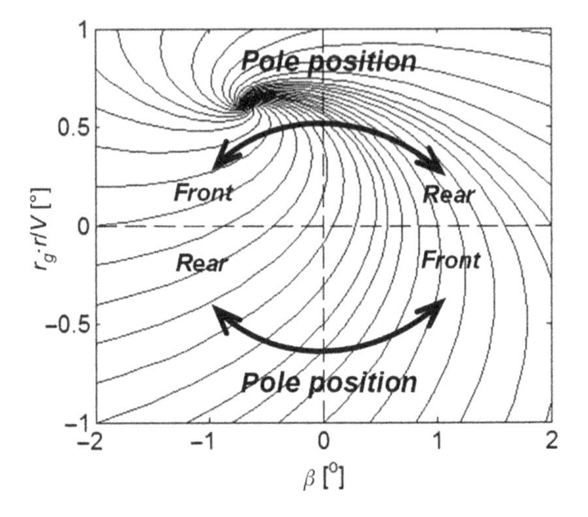

FIGURE 5.35 Variation of pole position in the phase plane.

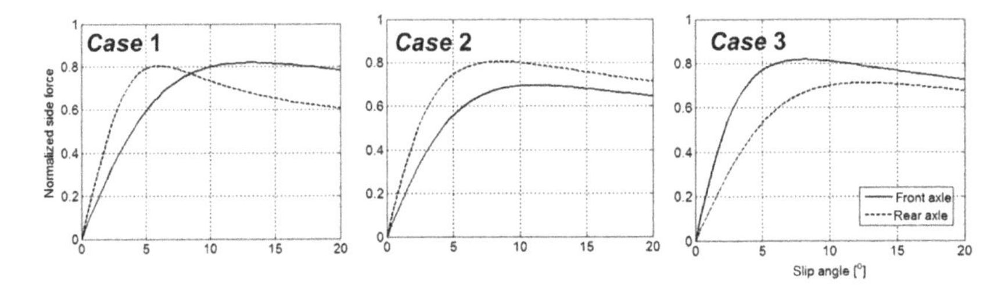

FIGURE 5.36 Three different sets of normalized axle characteristics.

This interpretation is schematically shown in Figure 5.35.

Thus far, we considered one steady-state point in the phase plane. However, Figure 5.13 in Section 5.3.1 indicates that more than one steady-state point may exist for a large axle side force. Therefore, more than one critical point may be expected in the energy phase plane. We consider three different sets of axle characteristics (Figure 5.36). If we consider only the linear range of these sets, then the first two cases correspond to an understeered vehicle and the last case corresponds to an oversteered vehicle.

A steady-state solution exists with identical nontrivial slip angles for case 1 but not for case 2. Compare this with the handling curves in Figures 5.16 and 5.17, corresponding with case 1 and case 2, respectively.

Let us examine case 1 closer. If we look for steady-state solutions with a fixed lateral acceleration (and therefore also a fixed normalized axle force) between 0.6 and 0.8 g, two different values of α_2 are found for only one choice of α_1. For one case, $\alpha_1 < \alpha_2$ and for the other case, $\alpha_1 > \alpha_2$. In case of a fixed steering angle δ and speed V, we show, in Section 5.5.3, that three steady-state solutions exist for case 1. These steady-state solutions correspond

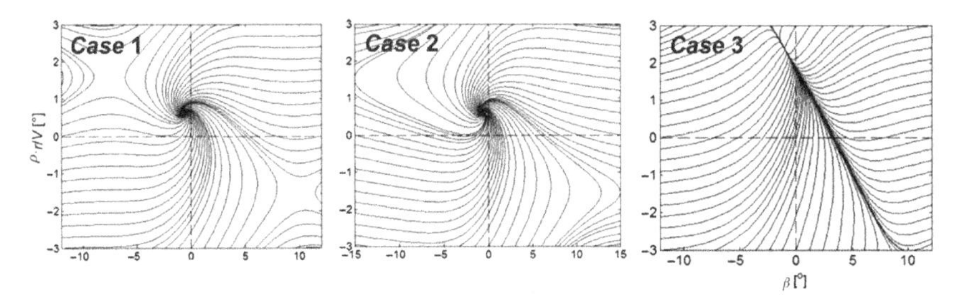

FIGURE 5.37 Phase plane plots for the sets of axle characteristics of Figure 5.36.

to different lateral accelerations. Two of these solutions are lost in case 2. Case 3 also corresponds to one steady-state solution.

We determined the energy phase planes for $V = 70$ [km/h] and $\delta = 2$ [°] for cases 1 and 2, and for $V = 40$ [km/h] and $\delta = 3$ [°] for case 3, respectively (using the vehicle data from Appendix 6). Figure 5.37 illustrates these phase plane plots.

For case 1, three steady-state points are recognized: one being a stable focus and the other two unstable saddle points. See Appendix 2 for a more detailed discussion of the possible types of steady-state (critical) points. With the intersection between the axle characteristics gone, the saddle points disappear as well (see the second plot in Figure 5.37 for case 2). Locally, near the focus point, nothing has truly changed. However, as some trajectories would never reach this focus point for case 1, all possible trajectories tend to move to this point for case 2. Hence, the local stability has not changed, but the global stability has. For case 3, no focus point appears; however, one recognizes a stable two-sided node. Clearly, the vehicle is behaving stable near this point, which is because the speed has been chosen as low. Oversteered vehicles are only stable if the speed is not too high, as we have seen in Section 5.3.

The *interpretations 1* to *4* are not exclusive for solutions with constant input, but also hold for nonstationary solutions of our one-track vehicle model (5.6a) and (5.6b) with varying steering input. Let us consider a lane change, which will be discussed in Section 6.4. The vehicle response is shown in Figure 5.38 for 90 [km/h]. The corresponding Guo states (x_{Guo}, y_{Guo}) are plotted in Figure 5.38. The lines for vanishing $\alpha_1 - \delta$ and α_2 are shown as dashed lines. Following the previous interpretations, we observe that, during the lane change, the cornering energy varies significantly as expected, that the slip angles at front and rear axles vary qualitatively similarly to the Guo states themselves, and the pole of rotation moves from front to rear.

We conclude that the phase plane representation is a powerful tool to visualize the local and global behavior of solution curves (trajectories) near critical points, as well as to interpret the performance along these solution curves in terms of position of the pole of rotation, of the slip at front and rear axles, and of the curvature. Nonstationary solutions with varying input can be visualized and interpreted in a similar way.

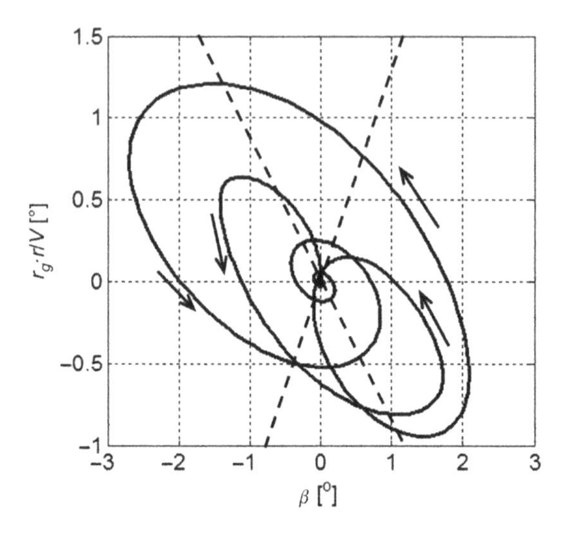

FIGURE 5.38 Solution of a lane change in terms of Guo states (x_{Guo}, y_{Guo}).

5.5.2 Stability Diagram

Let us return to Eq. (5.45) describing the eigenvalues of the linearized bicycle model. As in Section 5.4, we will take $J_z = m \cdot a \cdot b$. We will extend the **stability diagram**, shown in Figure 5.26 to nonlinear axle characteristics. This diagram was first introduced by Pauwelussen in Ref. [37]. Writing

$$a_{\alpha i} = \frac{\partial f_{yi}(\alpha_i)}{\partial \alpha_i} = \frac{A_i}{F_{zi}} \quad i = 1, 2 \tag{5.61}$$

we may conclude (as in Section 5.4) that for *positive* slope $a_{\alpha i}$ of the normalized axle side forces, the steady-state solution is stable if

$$a_{\alpha 2} > \frac{a_{\alpha 1}}{1 + a_{\alpha 1} \cdot \frac{g \cdot L}{u^2}} \tag{5.62}$$

Similar to Figure 5.26, one may distinguish between areas of oscillatory stability (stable focus in the phase plane), nonoscillatory stability (stable one-sided node), and loss of stability (saddle point). These types of stationary (critical) points are further explained in Appendix 2. See also Section 5.5.1 on phase plane analysis. For nonlinear axle characteristics, one or both of the slopes may be negative. Let us, as an illustration, take the vehicle from Appendix 6 with the normalized axle characteristics shown in Figure 5.39. Let us assume a ramp steer input, with a rise time of 0.2 [s], with the axle steering angle changing from 0 to 10 [°]. Results are shown in Figure 5.40. One observes a stable response, but with the saturated front axle. This indicates that stability can be obtained, even under adverse conditions.

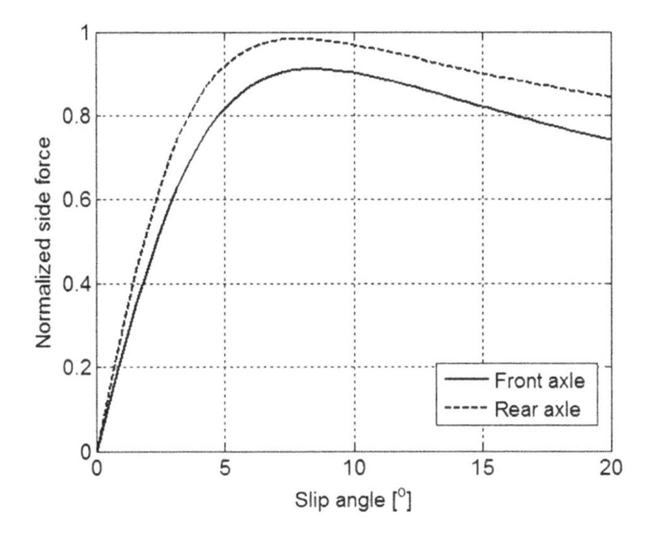

FIGURE 5.39 Normalized axle characteristics.

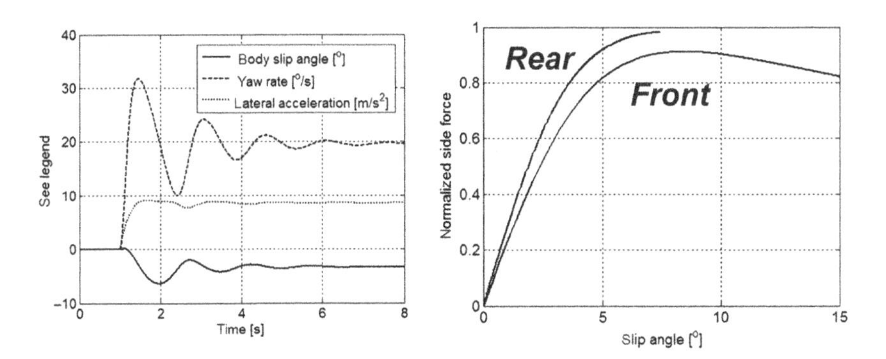

FIGURE 5.40 Ramp steer response for large steering angle.

Using (5.45) and substituting (5.61), the characteristic equation can be written as follows:

$$\lambda^2 + \frac{g}{u} \cdot (a_{\alpha 1} + a_{\alpha 2}) \cdot \lambda + \frac{g}{L} \cdot (a_{\alpha 2} - a_{\alpha 1}) + \frac{a_{\alpha 1} \cdot a_{\alpha 2} \cdot g^2}{u^2} = 0 \tag{5.63}$$

A solution with only negative real parts requires

$$a_{\alpha 2} + a_{\alpha 1} > 0 \tag{5.64}$$

Oscillatory solutions (which are also stable if expression (5.64) is satisfied) are obtained if the damping ratio is bounded by 1, which is equivalent to the condition

$$(a_{\alpha 1} - a_{\alpha 2}) \cdot \left(1 + \frac{g \cdot L}{4 \cdot u^2} \cdot (a_{\alpha 1} - a_{\alpha 2})\right) < 0$$

This condition is satisfied, if and only if

$$a_{\alpha 2} - \frac{4 \cdot u^2}{g \cdot L} < a_{\alpha 1} < a_{\alpha 2} \tag{5.65}$$

This is a strip with constant width in the $(a_{\alpha 2}, a_{\alpha 1})$ plane, just as the area in Figure 5.26 indicated as related to a stable focus.

Nonoscillatory stable solutions (two-sided node) are obtained if, in combination with Eq. (5.64), the constant term in Eq. (5.63) is positive:

$$\frac{g}{L} \cdot (a_{\alpha 2} - a_{\alpha 1}) + \frac{a_{\alpha 1} \cdot a_{\alpha 2} \cdot g^2}{u^2} > 0$$

which is identical to

$$a_{\alpha 1} < \frac{a_{\alpha 2}}{1 - (g \cdot L)/u^2 \cdot a_{\alpha 2}} \quad \text{if} \quad a_{\alpha 2} < \frac{u^2}{g \cdot L} \tag{5.66a}$$

$$a_{\alpha 1} > \frac{a_{\alpha 2}}{1 - (g \cdot L)/u^2 \cdot a_{\alpha 2}} \quad \text{if} \quad a_{\alpha 2} > \frac{u^2}{g \cdot L} \tag{5.66b}$$

When we combine the preceding conditions and allow for nonpositive values of $a_{\alpha 1}$ and $a_{\alpha 2}$, we arrive at the full *stability diagram* (shown in Figure 5.41).

Figure 5.41 is the visualization of the extension of the stability analysis of Section 5.4.1, with the first quadrant corresponding to Figure 5.26. We list some conclusions that can be drawn from Figure 5.41.

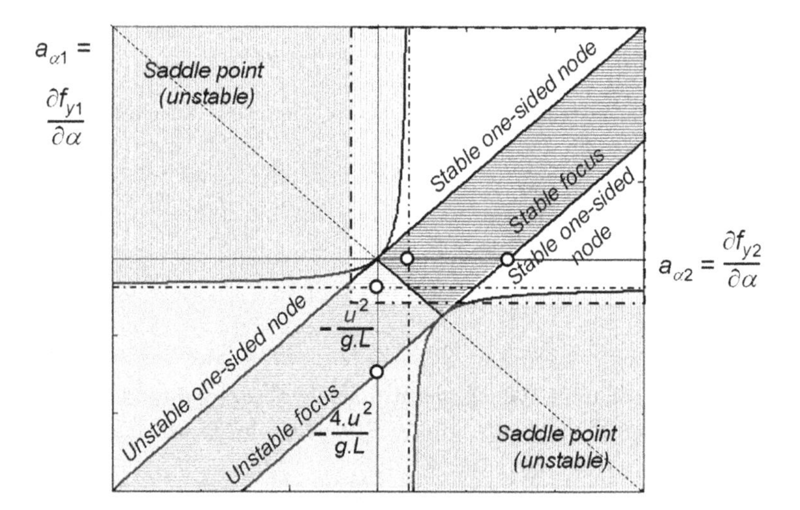

FIGURE 5.41 Stability of the steady-state solution of a nonlinear one-track vehicle model in terms of the slopes of the normalized axle side forces at this solution.

1. Stability of the stationary solution is lost if $a_{\alpha 1} + a_{\alpha 2} < 0$.
2. Two areas exist bounded by hyperbolic curves, and extending to large absolute values of $a_{\alpha 1} + a_{\alpha 2}$ that consist of stationary points with real eigenvalues with opposite signs, i.e., corresponding to saddle points in the phase plane.
3. A strip exists between two lines under 45 [°], passing through the origin and the point $(0, -4u^2/(g \cdot L))$, respectively, such that the behavior near stationary points, within this strip, behave oscillatory. Depending on the sign of $a_{\alpha 1} + a_{\alpha 2}$, these stationary solutions can be stable or unstable.
4. There exist nonoscillatory stable and unstable steady-state solutions (two-sided node). The areas for $(a_{\alpha 2}, a_{\alpha 1})$ for which such nodes exist are bounded by the saddle point areas and the strip of focus points. Again, the stability depends on the sign of $a_{\alpha 1} + a_{\alpha 2}$.
5. Consequently, stable steady-state behavior is possible for a negative slope for the front normalized axle curve. This is referred to as *excessive understeer*. This point could be a focus (oscillatory behavior) or a two-sided node (nonoscillatory behavior).
6. The area of stable two-sided nodes for positive front normalized axle characteristics slope $a_{\alpha 1}$ corresponds to stable oversteer behavior. This area is reduced for increasing vehicle speed u. Consequently, stability is obtained if the speed is bounded by a certain maximum value (the critical speed).
7. The area of stable stationary points in case of excessive understeer is increased for increasing speed u. Consequently, stability of the stationary solution is obtained if the speed exceeds a minimum value.

In practice, large negative values of $a_{\alpha 1}$ and $a_{\alpha 2}$ will not occur. We have indicated the area of practical interest using a dash-dotted box in Figure 5.41 with the remaining part of this figure being shaded.

5.5.3 The Handling Diagram

When a vehicle is under steady-state conditions, it follows a circular path with a specific curve radius R, with a certain speed V for which a steering angle δ is required. These parameters cannot be chosen independently. Section 5.5.1 shows that more than one steady-state solution may exist, and these solutions determine the global stability properties of the stable steady-state solution (−s). Hence, one is faced with the following questions:

1. If, for a nonlinear vehicle (i.e., allowing nonlinear axle characteristics), two of the three parameters (speed, curve radius, and/or steering angle) are known, how does the third parameter depend on these two, and how is this related to the axle characteristics?
2. In what way does the number of steady-state solutions depend on the axle characteristics and the selection of the mutually dependent parameters speed, curve radius, and steering angle?

The answers to these questions can be derived based on the **handling diagram**, which has been first introduced and discussed extensively by Pacejka [32].

Let us consider Figure 5.16, in which the left plot shows both the normalized axle characteristics and the curve derived from horizontal subtraction of these characteristics (referred to as **handling curve**). This handling curve describes the relationship between $\alpha_1 - \alpha_2$ and a_y/g under steady-state conditions, and visualizes the vehicle's handling performance. Increasing the vehicle speed slowly, such that steady-state behavior is maintained, this curve shows to what extent the vehicle is understeered or oversteered in relationship to vehicle speed. This curve does not depend on the steering angle or curve radius, but only on the axle characteristics.

Let us consider this dashed handling curve from Figure 5.16 closer, and combine the situations where the vehicle may turn right or turn left (i.e., positive and negative values of a_y). Using Eq. (5.39), i.e.,

$$\alpha_1 - \alpha_2 = \delta - \frac{L}{R}$$

the handling curve in Figure 5.16 (case 1 in Figure 5.36) can be depicted as shown in Figure 5.42. Steady-state solutions must coincide with points of this handling curve. We also indicated the understeer gradient η for the linear approximation, as defined in Eq. (5.31).

As previously mentioned in Section 5.3.2, more branches should be included because the function $h(a_y)$ in expression (5.39) is multivalued. For case 1, these extra branches correspond to values of f_{y1} with negative slope. For case 2 (cf. Figure 5.17), extra branches correspond to the values of f_{y2} with negative slope. In this section, we restrict ourselves to the branches that are passing through the origin and being continuously extended for larger lateral acceleration. For more information, including these additional branches, see Ref. [32].

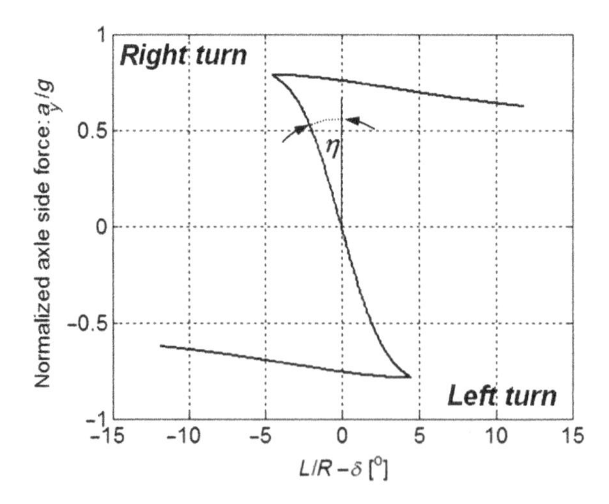

FIGURE 5.42 Handling curve for case 1 (see Figure 5.36).

Next, we combine this handling curve with an expression of the lateral acceleration in terms of curvature. For steady-state behavior, the following expression holds:

$$\frac{a_y}{g} = \frac{V^2}{R \cdot g} = \frac{V^2}{L \cdot g} \cdot \frac{L}{R} \tag{5.67}$$

for speed V, wheelbase L, and curve radius R. In other words, the normalized lateral acceleration is a linear function in the nondimensional curvature L/R with the slope of linear relationship increasing with V^2.

This is indicated in Figure 5.43, and it is an obvious consequence of Eq. (5.17a) under steady-state conditions, for which the yaw rate r equals V/R.

We now have two curves (a nonlinear handling curve in Figure 5.42 and a linear relationship in Figure 5.43, referred to as the **speed curve**) for which, in both cases, steady-state solutions must coincide with points of each of these curves. Figure 5.42 shows a_y/g versus $L/R - \delta = \alpha_1 - \alpha_2$, whereas Figure 5.43 shows a_y/g versus L/R. Consequently, when we shift the linear curve in Figure 5.43 to the left (in negative direction) over a distance δ, the abscissa of both figures coincide, and we can plot the curves in one plot, as shown in Figure 5.44, where we have selected $V = 70$ [km/h]. The combination of these curves is referred to as the **handling diagram**.

Because steady-state solutions correspond to points of both curves, these steady-state solutions are found from the points of intersection after shifting over the steering angle δ, which is taken here as 4 [°], and indicated in Figure 5.44 with numbers 1, 2, and 3. Comparing these three steady-state solutions for case 1 of Section 5.5.1 with Figure 5.37, one observes solution 1 to be stable and solutions 2 and 3 to be unstable.

The handling diagram shows all the three parameters R, V, and δ in one image. Speed V corresponds to the slope of the linear curve, which is

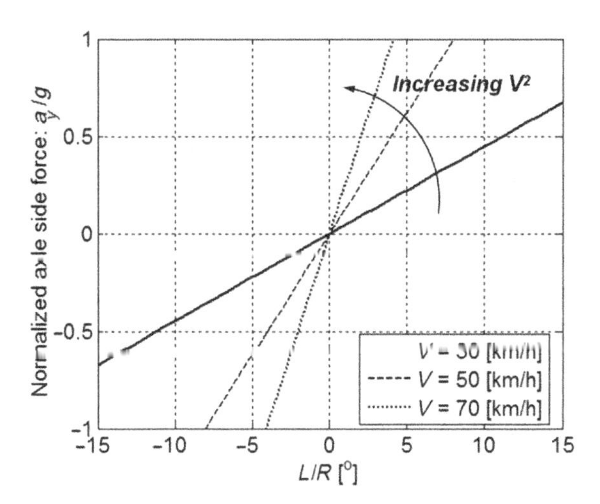

FIGURE 5.43 Lateral acceleration versus nondimensional curvature L/R.

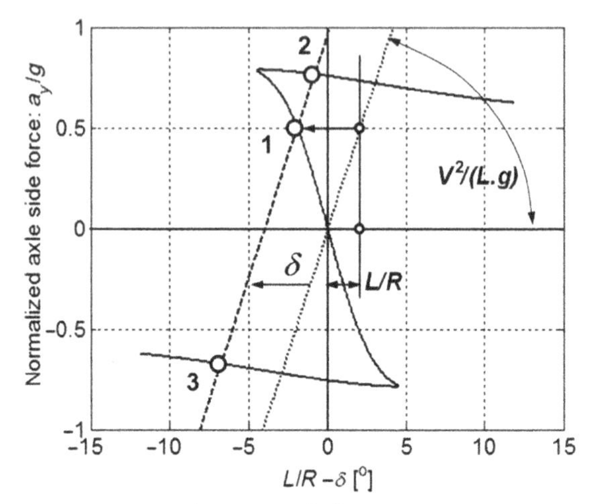

FIGURE 5.44 Handling diagram.

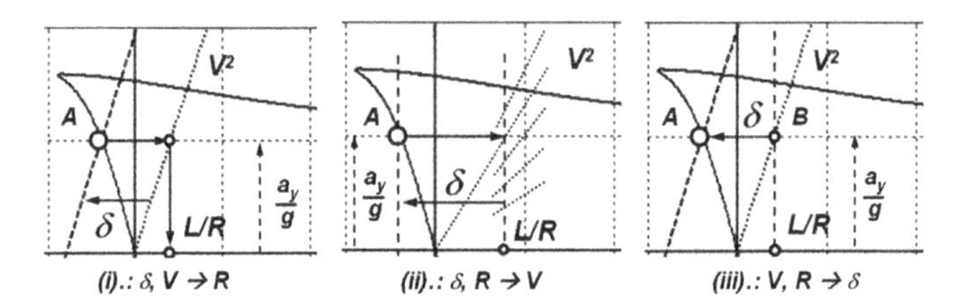

FIGURE 5.45 How to derive steady-state parameters from the handling diagram.

described by Eq. (5.67), δ is the shift of this linear curve to obtain the steady-state solutions from the intersections of the shifted curve with the handling curve, and the curvature L/R follows from the original linear curve and the lateral acceleration for the steady-state solution.

Let us return to the first question at the beginning of this section, and assume two of these three parameters known. How can we determine the third parameter? Figure 5.45 shows a schematic indication of the procedures explained next.

i. *Steering angle δ and speed V are known*

The speed V determines the linear speed curve. Shift this curve over a distance δ, to find the intersection A with the handling curve. This leads to the lateral acceleration a_y/g, which in turn determines L/R.

ii. *Steering angle δ and curvature L/R are known*

The curvature L/R corresponds to a vertical line passing through the horizontal axis at position L/R. Shift this line over a distance δ, to find the intersection A with the handling curve. This leads to the lateral acceleration a_y/g. The speed curve is the line that passes through the origin and the point (L/R, a_y/g). The slope of this curve determines V^2 and therefore, the speed V.

iii. *Speed V and curvature L/R are known*

The speed curve and the vertical line passing through the horizontal axis at position L/R, intersect at point B, leading to a certain value for a_y/g. This value determines the steady-state point A on the handling curve. The horizontal distance between the intersection point B and the steady-state point A provides the steering angle value.

The handling diagram can be used to analyze the occurrence of steady-state solutions and the impact of these solutions on lateral acceleration and steady-state parameters δ, V, and R, when one of these parameters is changed. The following conclusions can be easily verified:

1. Increasing the steering angle for fixed speed V leads to a larger lateral acceleration (larger curvature). When the steering angle is increased beyond the value where the shifted speed curve intersects the handling curve, the stable steady-state solution vanishes. This means that the vehicle loses stability (there is saturation at the rear axle).
2. Increasing the steering angle for fixed curve radius R leads to a larger speed. Again, beyond a certain value of δ, the vehicle loses stability.
3. Increasing the speed V for a fixed curvature leads to similar results.
4. Increasing the speed for a fixed steering angle results in a larger lateral acceleration and in a larger curve radius (smaller curvature). This may lead to loss of stability for larger speeds, but only if the steering angle exceeds a certain value.

Depending on the axle characteristics, different handling curves and handling diagrams are possible, which results in different types and number of steady-state solutions (as shown for cases 2 and 3 in Figure 5.37). We treat three more situations in this section, all with a front axle steering angle of 4 [°].

1. Axle characteristics compared to case 2 in Section 5.5.1; understeer behavior in the linear range, and no intersection of these normalized axle characteristics. The speed is chosen as 70 [km/h].
2. Axle characteristics compared to case 3 in Section 5.5.1; oversteer behavior in the linear range, and there is no intersection of these normalized axle characteristics. Two speeds are considered, 40 and 90 [km/h].
3. Axle characteristics with oversteer behavior in the linear range; however here, these normalized axle characteristics intersect at a specific slip angle. Again, two speeds are considered, 40 and 90 [km/h].

Considering situation 1 first (depicted in Figure 5.46), the difference with Figure 5.44 is that the handling curve does not pass the a_y-axis.

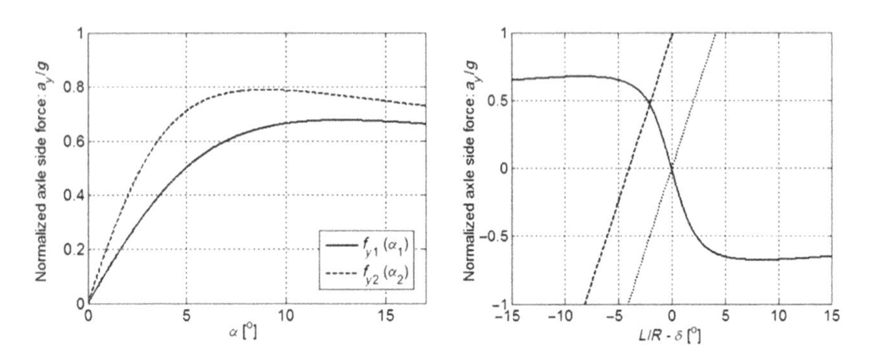

FIGURE 5.46 Axle characteristics and handling diagram for situation 1.

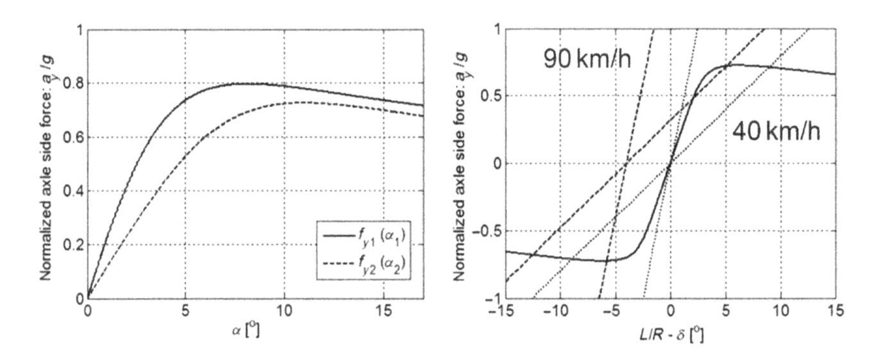

FIGURE 5.47 Axle characteristics and handling diagram for situation 2.

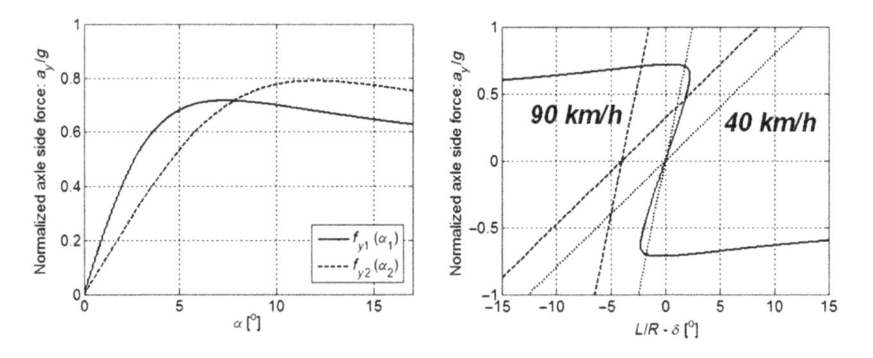

FIGURE 5.48 Axle characteristics and handling diagram for situation 3.

Consequently, only one steady-state solution is shown. Increasing the steering angle (or speed) leads to increasing $\alpha_1 - \alpha_2$. There is a maximum value for a_y where sliding at the front axle will occur, and where, for a larger steering angle, the lateral acceleration decreases.

The handling diagrams for situations 2 and 3 are shown in Figures 5.47 and 5.48, respectively. The steepest speed curves correspond to 90 [km/h]. The slope of the handling curves in Figures 5.47 and 5.48 in the linear range is now positive, meaning that the understeer gradient is negative. With no intersection of the axle characteristics, the handling curve for situation 2 is quite similar to the curve in Figure 5.46, but mirrors that curve. At the intersection of the speed curve with the handling curve, the slope of the speed curve is either smaller (for 40 [km/h]) or larger (for 90 [km/h]) than the slope of the handling curve.

The first case corresponds to stable oversteer, whereas the second case corresponds to unstable oversteer. Note that for a low speed, more than one steady-state solution may exist; where the two outer solutions correspond to a slope of the rear normalized axle characteristic that is positive but small, or negative. According to Figure 5.41, one would expect saddle points in the phase plane for these steady-state solutions.

The handling curve in Figure 5.48 is qualitatively similar to the curve in Figure 5.42, but again is mirrored with respect to this curve. The same conclusions can be drawn with regard to understeer and oversteer behavior for low and high speeds. For a large speed and small steering angle (smaller than indicated in the figure), three steady-state solutions may exist, with the two outer ones now corresponding to a slope of the front normalized axle characteristic that is positive but small or negative. According to Figure 5.41, one expects stable solutions with possible excessive understeer. Apparently, the stability at the intermediate steady-state solution in Figure 5.47 is transferred to the outer steady-state solutions with increasing speed.

To analyze this further, we determined the phase plane representation according to Guo (see Section 5.5.1), for the axle characteristics according to situation 3, and for speeds 40, 60, and 80 [km/h], see Figure 5.49, with a steering angle of 1 [°].

Observe that for a low speed, only one stable critical point exists, being a two-sided node (see Appendix 2). With increasing speed, this node moves to a stable focus. Increasing the speed further to 80 [km/h], three critical points arise: a saddle point, and two stable focus points. The steady-state yaw rate tends to decrease because the curve radius that corresponds to the highest intersection of speed curve and handling curve (shown in Figure 5.47) increases with speed. This can easily be concluded by following procedure (i), which has been included previously.

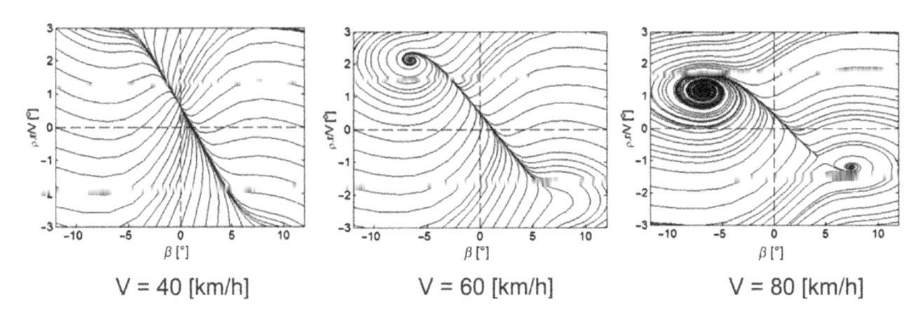

FIGURE 5.49 Phase plane representation for situation 3, for speeds 40, 60, and 80 [km/h], and for steering angle $\delta = 1$ [°].

5.5.4 The MMM Diagram

In the preceding sections, graphical assessment methods have been introduced to visualize and analyze vehicle state patterns, local and global stability, and the relationship between operating parameters under steady-state conditions. These methods apply for nonlinear axle characteristics. This means that axles may saturate, with the saturation limits depending on vehicle and tire parameters. With our focus on the vehicle states, these saturation limits have not been shown in the various "portraits" of vehicle handling and maneuvering performance discussed thus far.

Starting from the phase plane representations in Section 5.5.1, the following approach can be followed to correct that. Define output variables:

$$\text{lateral force coefficient}: C_{\text{F}} = \frac{F_y}{m \cdot g}$$

$$\text{yaw moment coefficient}: C_{\text{M}} = \frac{M_z}{m \cdot g \cdot L}$$

where F_y and M_z are the lateral force and yaw moment acting on the vehicle, respectively, resulting from the lateral tire forces, see Eq. (5.5). These coefficients were introduced by Milliken and Milliken in Ref. [26]. Let us use the axle characteristics corresponding to case 1 in Figure 5.36 and determine these coefficients for the combinations of Guo states (body slip angle and non-dimensional yaw rate) depicted in the phase plane representation in Figure 5.37 for a vehicle velocity of 70 [km/h] and a steering angle of 2 [°]. This means that we determine the phase plane representation in terms of C_{F} and C_{M} (Figure 5.50). As expected, the figure is bound due to the saturation limits of the axle characteristics.

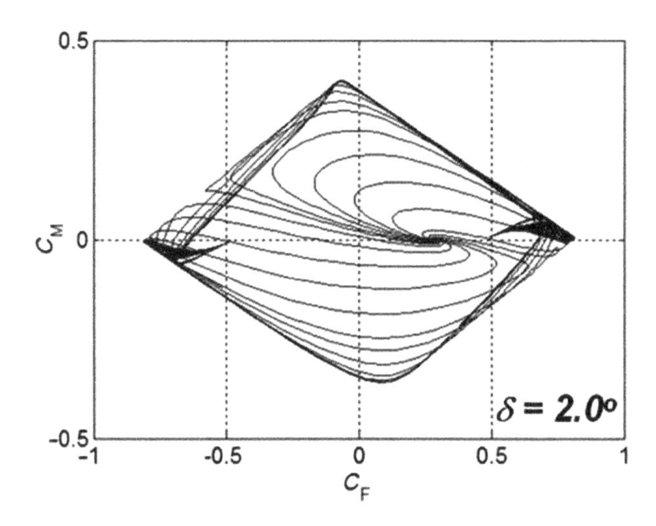

FIGURE 5.50 Phase plane representation of C_{F} versus C_{M} for the understeered vehicle cf. case 1 ($V = 70$ [km/h], $\delta = 2$ [°]).

From this figure, the following observations can be made:

1. Under steady-state conditions, no yaw acceleration exists and therefore $M_z = 0$. Consequently, the steady-state solutions are lying on the C_F-axis. This is clearly shown for the stable steady-state solution.
2. The position of the stable steady-state solution along the C_F-axis indicates the severity of the handling conditions. Approximately 40% of the side force potential has been used.
3. The other steady-state solutions (there are three steady-state conditions in this case, see Figures 5.47 and 5.44) correspond to extreme conditions and are therefore located at the edge of the diagram in Figure 5.50.
4. The vehicle lateral force F_y may act in two opposite directions. From the trajectories, it is clear that only the positive direction (for positive steering angle) is a candidate for the steady-state solution.

The C_M-axis corresponds to situations where a yaw moment, but no lateral force, is acting on the vehicle. Clearly, these situations are not steady state. When we increase the steering angle to large values, one does not expect the negative C_M value, with the vehicle yawing against the direction of the front axle steering, to be possible. We have determined the (CF, CM) phase plane for case 1, for steering angles 5 and 10 [°] (Figure 5.51). The steady state solution moves to larger values of CF, and the lower half of the diagram is reduced in size. In case of δ = 10 [°], no stable solution exists.

These (C_F, C_M) phase plane representations are related to the so-called MRA Moment Method diagrams, or MMM-diagrams for short, introduced by Milliken and Milliken [26]. These diagrams are a subset of the preceding phase plane representations, in which the yaw rate times speed is taken equal to the lateral acceleration. In this way, the yaw rate is expressed in terms of C_F, which reduces the dimension of the phase plane representation for fixed δ to 1, i.e., to a single curve for varying β. That

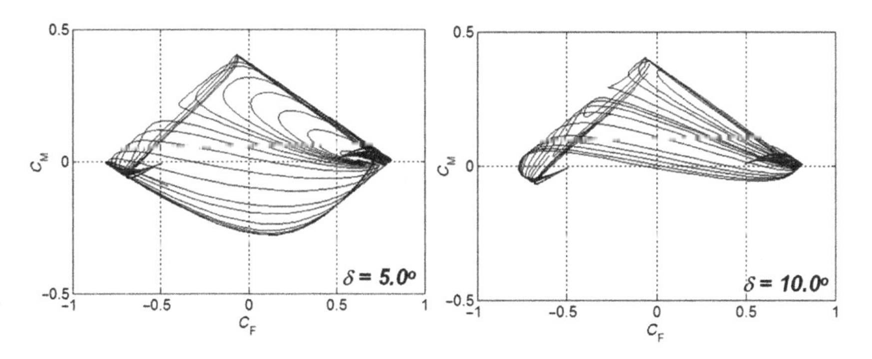

FIGURE 5.51 (C_F, C_M) phase plane representations for the understeered vehicle cf. case 1, for $\delta = 5$ and 10 [°] ($V = 70$ [km/h]).

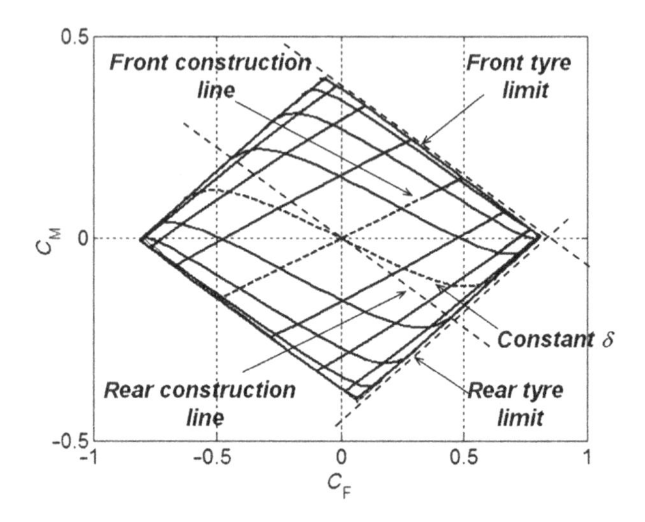

FIGURE 5.52 MMM-diagram for the understeered vehicle cf. case 1 (V = 70 [km/h]).

allows the variation of the steering angle, as well in the same plot, bringing this representation again back to a two-dimensional visualization. With β fixed and δ varying, we obtain other curves of the MMM diagram. The MMM diagram for the understeered vehicle cf. case 1 is shown in Figure 5.52. The overall size of the diagram is similar to Figure 5.50.

The δ curve through the origin corresponds to the behavior for $\delta = 0$. The intersection of other δ curves with the C_F-axis provides the steady-state solution for that steering angle. The outer boundaries of the diagram correspond to the front and rear tire saturation limits. With $\beta = b/R$ (see Figure 5.5) and $\delta \neq 0$, the only contribution to C_F and C_M comes from the front axle. Hence,

$$C_M = \frac{a}{L} \cdot \frac{F_{y1}}{m \cdot g} = \frac{a}{L} \cdot C_F$$

This line is denoted as the **front construction line** in Figure 5.52. In the same way, one can consider the situation with only the rear axle contributing to C_F and C_M, and denoted as the **rear construction line**. In that case, one finds

$$C_M = -\frac{b}{L} \cdot \frac{F_{y1}}{m \cdot g} = -\frac{b}{L} \cdot C_F$$

For *linear axle characteristics*, C_F and C_M can be expressed as

$$F_y = Y_\beta \cdot \beta + Y_r.r + C_{\alpha 1} \cdot \delta$$

$$M_z = N_\beta \cdot \beta + N_r \cdot r + a \cdot C_{\alpha 1} \cdot \delta$$

(see Eqs. (5.16a) and (5.16b)). We can eliminate the body slip angle β, to arrive at the following relationship between C_F and C_M:

$$C_M = \frac{a \cdot C_{\alpha 1} - b \cdot C_{\alpha 2}}{L \cdot (C_{\alpha 1} + C_{\alpha 2})} \cdot C_F + \frac{C_{\alpha 1} \cdot C_{\alpha 2}}{m \cdot g \cdot (C_{\alpha 1} + C_{\alpha 2})} \cdot \left(\delta - \frac{r \cdot L}{V} \right) \tag{5.68}$$

With yaw rate times speed replaced by the lateral acceleration $(g \cdot C_F)$ and introducing

$$C_0 = \frac{C_{\alpha 1} \cdot C_{\alpha 2}}{(C_{\alpha 1} + C_{\alpha 2}) \cdot m \cdot g}$$

and one finds

$$C_M = \left(M_s - C_0 \cdot \frac{g \cdot L}{V^2} \right) \cdot C_F + C_0 \cdot \delta \tag{5.69}$$

where M_s is the static margin defined in expression (5.42). Hence, for a fixed steering angle and linear tires, the relationship between C_F and C_M corresponds to lines in the MMM diagram with slope

$$\mathrm{SI} = \left(M_s - C_0 \cdot \frac{g \cdot L}{V^2} \right) = - \frac{C_{\alpha 1} \cdot C_{\alpha 2}}{m \cdot g \cdot (C_{\alpha 1} + C_{\alpha 2})} \cdot \left(\eta + \frac{g \cdot L}{V^2} \right) \tag{5.70}$$

This slope is referred to as the **stability index**, which is directly related to the undamped yaw eigenfrequency ω_0; see also Eq. (5.48):

$$\mathrm{SI} = - \frac{m \cdot a \cdot b}{(C_{\alpha 1} + C_{\alpha 2})} \cdot \omega_0^2$$

For an understeered vehicle, $\mathrm{SI} < 0$. For a stable oversteered vehicle, the stability index is negative as well, $\mathrm{SI} < 0$, whereas for the unstable oversteered vehicle it is $\mathrm{SI} > 0$. Varying speed between 0 and ∞ for the understeered vehicle, SI varies between $-\infty$ and $M_s < 0$. In Figure 5.21, we have depicted ω_0 versus speed.

From the preceding discussion, the MMM diagram is expected to be reduced in case that one or both of the axles have lower saturation limits. Consider Figure 5.53 where the friction at the front tire has been reduced. As seen from the MMM diagram, the width along the front construction line has been reduced, as expected.

We close this section with a discussion of the oversteered vehicle using axle characteristics, as shown in Figure 5.48. The discussion on handling and stability diagrams revealed that a stable steady-state solution exists for low speed (two-sided node), becoming unstable for increasing speed, and transferred to stable excessive understeer solutions when the speed is increased further. The (C_F, C_M) phase planes for speeds 40 and 120 [km/h] are shown in Figure 5.54.

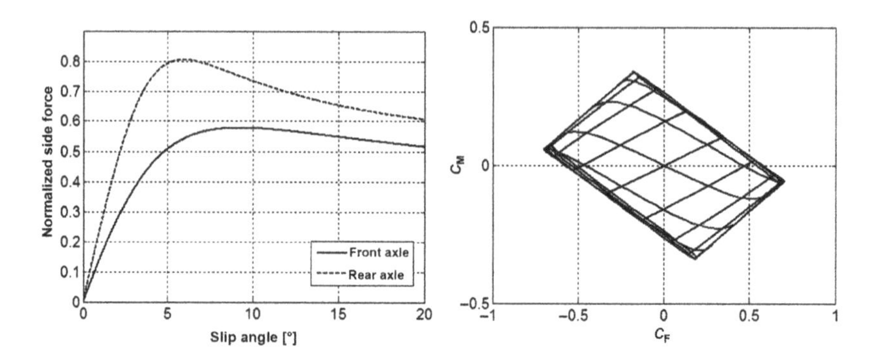

FIGURE 5.53 MMM diagram for an understeered vehicle with reduced friction at front axle.

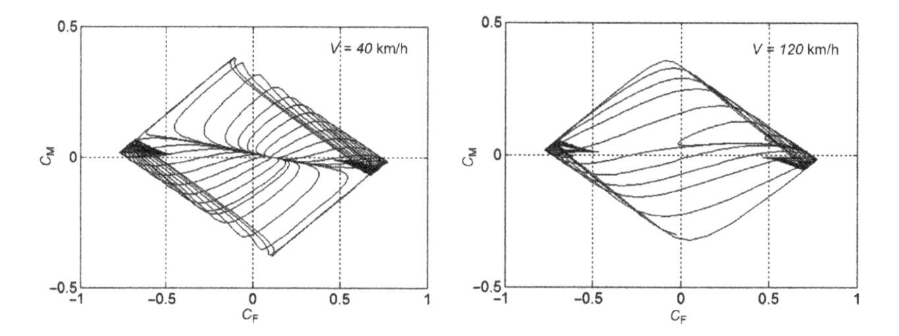

FIGURE 5.54 (C_F, C_M) phase plane representation for the oversteered vehicle cf. Figure 5.47 for $\delta = 1$ [°] and $V = 40$ and 120 [km/h].

The two-sided node is clearly shown in the left image of Figure 5.54. Notice that the steady-state point moves to the right for increasing speed, with the slope of the trajectories (for constant $\delta = 1$ [°]) changing from a negative sign to positive sign. This is consistent with the discussion on stable and unstable oversteer behavior. The corresponding MMM diagrams are shown in Figure 5.55.

One observes the same variation in stability index, with the solution for 120 [km/h] clearly unstable. As indicated earlier, the intersections of the constant δ curves with the C_F-axis provide the steady-state solutions. In the diagram for $V = 40$ [km/h], the difference between consecutive constant δ curves is 3 [°], meaning a maximum steering angle of less than 6 [°]. This confirms (in order of magnitude) the results in Section 5.5.3 (see also Figure 5.48).

5.5.5 The g-g Diagram

When a vehicle is under cornering conditions, load transfer occurs from the inner wheels to the outer wheels. When a vehicle is braking, load transfer occurs from rear wheels to front wheels. The effect of lateral and longitudinal

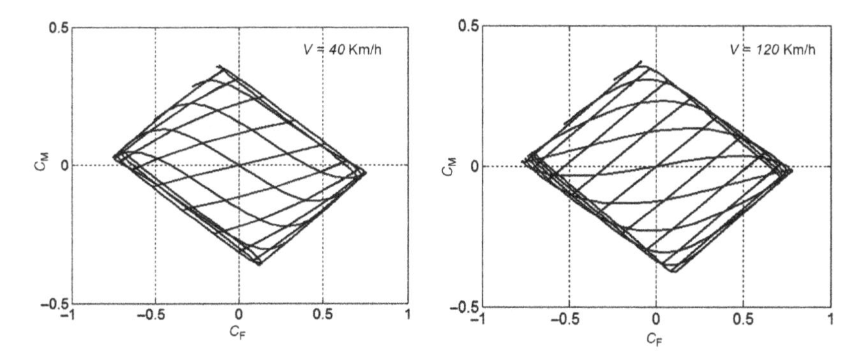

FIGURE 5.55 MMM-diagrams for the oversteered vehicle cf. Figure 5.47 for $\delta = 1$ [°] and $V = 40$ and 120 [km/h].

acceleration on wheel loads F_{zij} (with index i denoting front and rear, and index j denoting left and right) has been described in Section 5.2.2. The potential shear force per wheel is bounded by $\mu \cdot F_{zij}$ with road friction μ. The best performance of a vehicle is such that this potential is equally shared among the four wheels, which makes it important to consider this potential during arbitrary maneuvering conditions.

Considering shear forces at the tire–road interface as the only forces acting on the vehicle, the sum of wheel forces provides the total lateral and longitudinal forces acting on the vehicle, i.e., being the total cornering force $(m \cdot a_y)$ and the total driving force $(m \cdot a_x)$. This maximum potential in terms of acceleration in g was referred to by Milliken and Milliken as the g-g diagram [26].

Clearly, the vehicle's capability to accelerate, decelerate, and corner depends on the underlying separate friction circles per wheel. As discussed by Milliken and Milliken [26], these friction limits per wheel are not pure circle, but have traction limitations, which means that the maximum drive force is exceeded by the maximum brake force (in absolute sense). Suspension effects change the local wheel orientation, and therefore, the local tire forces. A vehicle is, in general, understeered, meaning that the rear axle saturation limits will not be reached under extreme (steady-state) cornering conditions.

In this section we will discuss a vehicle, completing a lane change. A lane change requires a driver; the vehicle-driver behavior during a lane change is covered in Section 6.4 in the next chapter. We use the driver preview time of 0.68 [s] and a driver lag time of 0.1 [s]. The vehicle is described using a two-track model with wheel loads determined using expressions (5.23a)–(5.23d), according to Klencke and Nielsen. The vehicle has an initial speed of 90 [km/h], and the driver tries to maintain this speed. At the same time, the driver will not accept lateral accelerations exceeding 0.7 [g], which means that the driver will decelerate the vehicle as soon as ay is expected to grow beyond this value. When the lateral acceleration drops below 0.7 [g] and the speed is still below 90 [km/h], the driver will accelerate again.

This acceleration and deceleration is assumed to depend on the deviation of the actual speed V_{act} from the intended speed V_{int}, with a maximum driving and braking torque per wheel M_{drive} and M_{brake}. We used the following approach for the brake torque T_{brake} in case the intended speed V_{int} is exceeded by the actual speed V_{act}:

$$T_{brake} = \theta_F \cdot M_{brake} \cdot (1 - e^{-|V_{act} - V_{int}| \cdot K_{brake}/M_{brake}}); \quad V_{act} > V_{int} \tag{5.71}$$

for the share θ_F of T_{brake} to the front axle (and share $1 - \theta_F$ to the rear axle), and some factor K_{brake}, describing the brake torque gain for small speed deviations. If the intended speed $V_{int} > V_{act}$, the brake torque is set to zero. The driving torque has been similarly defined. Brake torque is distributed over front and rear axles such that 70% is carried by the front axle. The vehicle is assumed rear driven.

The nonlinear normalized tire forces, loaded by half of the axle load (i.e., load transfer neglected) are shown in Figure 5.56. Combined slip characteristics have been accounted for.

The lane change maneuver resulted in time histories for the four tire loads shown in Figure 5.57.

Initially, the wheel load equals half of the axle load. When the maneuver begins, load is transferred to the right wheels, which are currently the outer wheels. Halfway into the second lane, the left wheels become the outer wheels.

Moving back to the original lane, a similar behavior is observed.

We selected four times for this maneuver, indicated in Figure 5.57. For each of these times, the friction potential per wheel is determined from the wheel load (for $\mu = 0.9$).

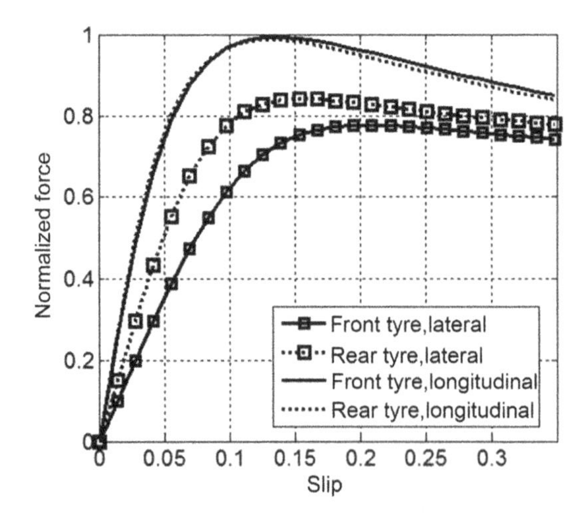

FIGURE 5.56 Tire characteristics used in lane change analysis.

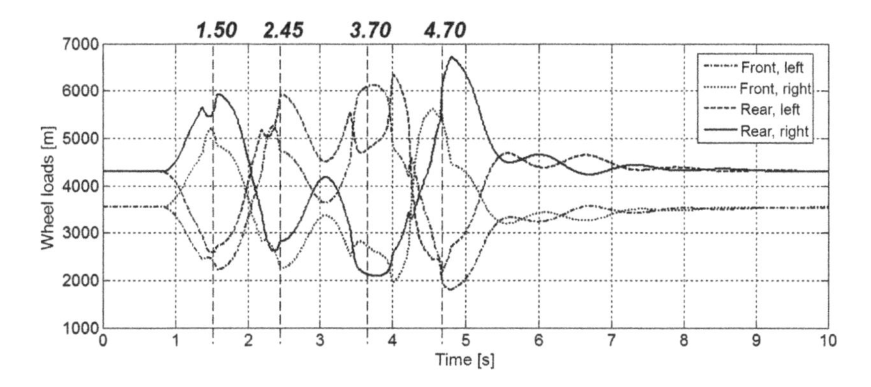

FIGURE 5.57 Wheel loads versus time for a lane change maneuver.

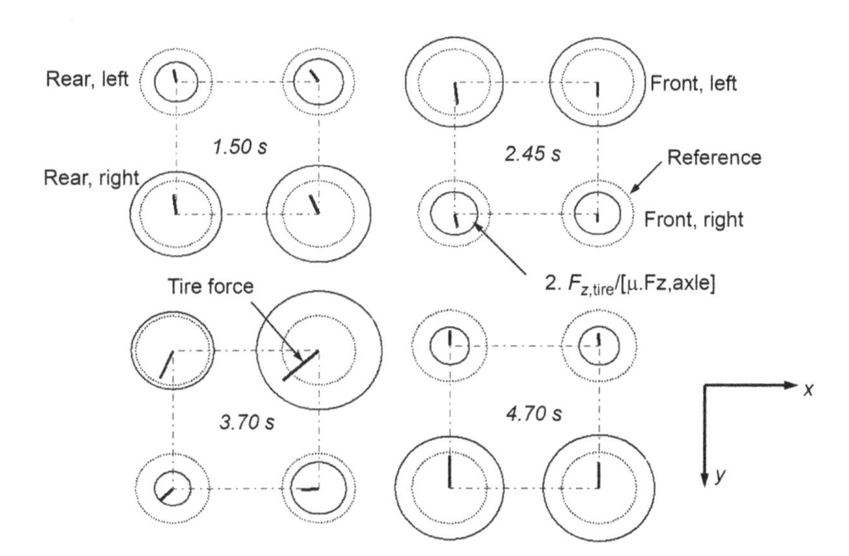

FIGURE 5.58 Wheel shear force diagrams for different times during a lane change.

This friction potential is shown in Figure 5.58 using a solid circle for each wheel with radius R_{ij} ($i = 1,2$ for front and rear axle, and $j = L$, R for left and right):

$$R_{ij} = \frac{\mu \cdot F_{zij}}{1/2F_{zi}}$$

for axle load F_{zi}. The reference value $R_{ij} = \mu$ is indicated with dotted circles. Tire forces are shown (divided by half of the axle load) as solid straight lines. Following Figure 5.58, the vehicle slows down at the first change of lanes to keep the lateral acceleration within reasonable bounds to prevent a loss of control. One observes that the inner wheels (left) are close to saturation.

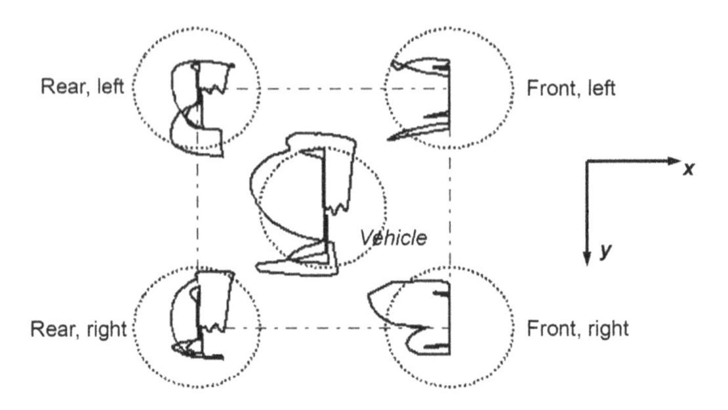

FIGURE 5.59 Wheel and vehicle shear forces during a lane change.

Entering the second lane is done with some slight acceleration at the rear wheels. Observe that the friction potential of the left wheels (now being the outer wheels) has been increased, whereas the opposite is shown (decrease of potential) for the right wheels.

Returning to the first lane requires further speed reduction, and Figure 5.58 shows significant braking at 3.70 sec. Consequently, the friction potential of the front wheels increases, relatively compared to the rear wheels. The front left wheel now has the highest potential for combined braking and cornering; this is where the largest wheel force occurs. Finally, at 4.70 sec. we have a situation where apparently the speed is close to the value where a lateral acceleration of 0.7 [g] is expected, and nearly pure cornering occurs. With the vehicle entering the first lane, the driver is expected to accelerate to the original 90 [km/h].

Figure 5.58 only shows four times during the lane change. An alternative visualization is found by plotting the variation of wheel forces in a similar plot as in Figure 5.58, but for the entire lane change. The friction potential circles will vary as well, but that will be omitted. This means that we show the nondimensional wheel forces (wheel forces, divided by half of the axle load). Likewise, we can show the total lateral and longitudinal forces ($m \cdot a_y$ and $m \cdot a_x$) acting on the vehicle and being the sum of the wheel forces in lateral and longitudinal direction. To arrive at the same scale, we have divided these forces by $m \cdot g$, which means that we showed the vehicle accelerations in g. This plot is referred to by Milliken and Milliken [26] as the g-g diagram. Results are shown in Figure 5.59 with the same reference circles as in Figure 5.58, i.e., having a radius μ. Observe that a drive torque only occurs at the rear wheels and for a negative side force. A negative side force occurs when the vehicle is just beginning the lane change and when it returns to the original lane at the end of the lane change maneuver. This last situation is when the more extreme maneuvering has passed and the intended speed of 90 km/h is restored.

Chapter | Six

The Vehicle–Driver Interface

Vehicle and driver are each interacting subsystems of the entire vehicle–driver system. The driver controls the vehicle's behavior by applying input signals, such as steering angle and throttle or brake pedal position. In doing so, the driver intends to follow the road or maintain a certain distance from the vehicle in front of him or her under relatively mild conditions. The driver may also be faced with unexpected situations where extreme maneuvering is required to avoid accidents.

The behavior of the vehicle in response to the driver's input assists the driver in predicting the vehicle's performance. It confirms that the driver is still in control, it informs about any deviations from an intended path or desired lead time, etc. Further, the vehicle response serves to predict and warn about danger ahead, and therefore affect the driver's perception and response at different levels. These levels can be considered with reference to the categories of human behavior and driving task hierarchy as distinguished by Donges [7] (depicted in Figure 6.1). At the left of this figure, the classic hierarchy in behavioral categories is shown with distinction between:

- *Knowledge-based behavior*, which corresponds to the response to unfamiliar situations.
- *Rule-based behavior*, which corresponds to associative response based on selection of the most appropriate alternative according to previous subjective experience.
- *Skill based behavior*, which can be regarded as an automatic unconscious reflex.

Comparing these classifications for the different driving task levels as shown in Figure 6.1 (transport mission), the vehicle–road contact is mainly relevant at the levels indicated as *guidance* and *stabilization*. The dynamic status of the vehicle involves changes in the input data to the driver, a major part of which is affected by the tires (steering feel, vibrations, noise, lateral motions, etc.). The driver responds partly at guidance level (such as

Essentials of Vehicle Dynamics.

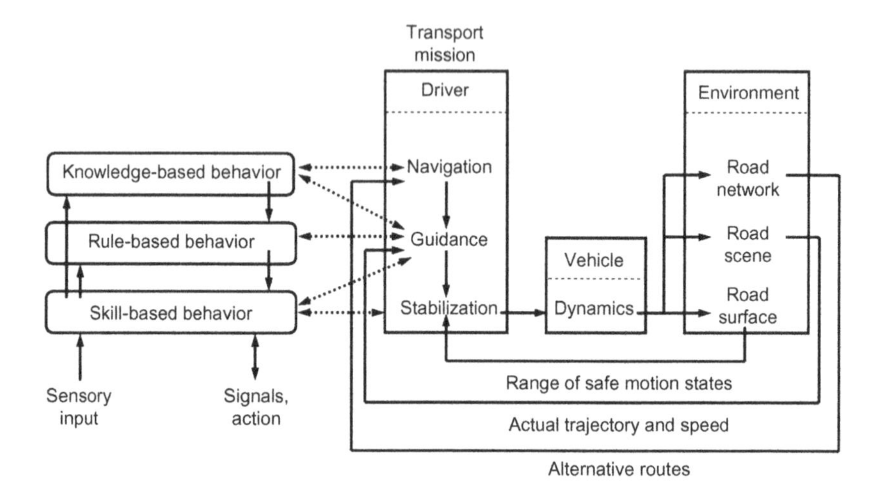

FIGURE 6.1 Human behavior and driving tasks.

corresponding to open-loop control), meaning that driver must choose a steering strategy to negotiate a curve, pass another vehicle, or maintain a set distance (a certain lead time) from the vehicle ahead.

The driver also responds partly at stabilization level (such as closed-loop control) when potentially critical conditions may be apparent. The distinction between these two levels depends on the driver and driver's experience with similar traffic situations. At the lowest level, information is obtained from the dynamics of the vehicle, yielding perceived data (such as friction level, tire–road contact, road unevenness, resulting cornering, and braking resistance) from which the driver must decide, consciously or unconsciously, about safe versus unsafe conditions and the necessary measures to overcome any unsafe circumstances. Anticipation of forthcoming situations will improve the driver's response and ability to avoid accidents.

Note that the driver is assisted at all levels of his driving task. Clearly, navigation is, in many cases, a matter of activating your navigating system. However, this support system is not yet capable of judging the conditions of the driver, such as the driver managing possible dense traffic conditions or driving in unfamiliar area. Both these examples can be serious problems for elderly drivers. Examples for guidance support available for these tasks are lane keeping, parking assistance, and adaptive cruise control. Stabilization supports include a collision avoidance system or electronic stability control (ESP). An outline of such driver support systems is given by the ATZ Fahrwerkhandbuch [16, Chapter 8]. As for the navigation support, these guidance and stabilization support systems do not account for driver's state, which means that the driver must use the same systems, regardless of his perceived driving skills under the circumstance at hand (mental workload).

Another schematic overview of the driver's reactions in emergency situations has been given by Braun and Ihme and reported by Käppler and Godthelp in [19] (see Figure 6.2). The three "partners" in any arbitrary traffic situation

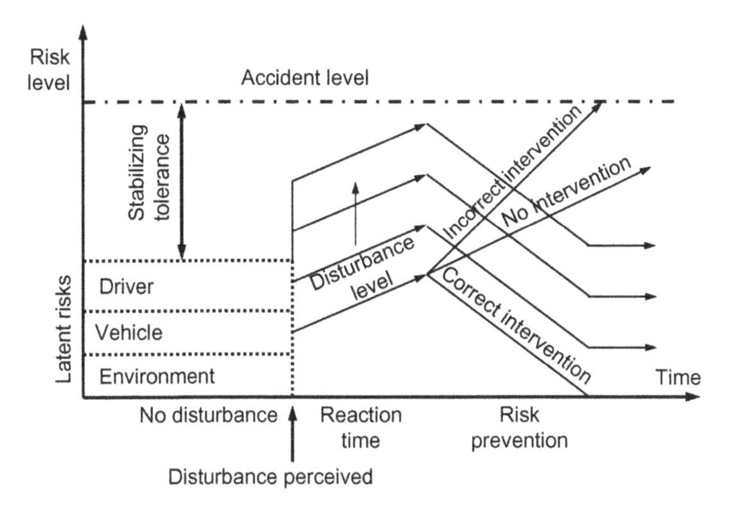

FIGURE 6.2 Driver response to potentially dangerous situations.

indicated in the right half of Figure 6.1 (i.e., driver, vehicle, and environment) are shown in Figure 6.2 as contributors to an experienced level of risk. Such "latent risks" could be affected by poor driving behavior (e.g., excessive speed), a vehicle deficiency (e.g., low tire pressure), or changes in the environment (e.g., slippery road, poor visibility, dense traffic). A sudden event may yield a sharp increase in risk level and, consequently, a reduced stabilizing tolerance. Following a reaction time, the driver may intervene correctly, may intervene incorrectly (e.g., braking on an icy surface), may not respond, or may respond too late (e.g., if the accident level has already been reached). In general, appropriate information largely based on tire performance helps the driver anticipate risky situations (i.e., reduce the reaction time), whereas the vehicle–driver system performance is crucial to overcoming and preventing emergencies.

The vehicle–driver system is schematically shown in Figure 6.3. The driver controls the vehicle through steering, acceleration, and braking. The vehicle responds, providing information to the driver in terms of path to be followed, orientation (i.e., yaw angle), lead time, and distance to the preceding vehicle. Vibrations and vehicle acoustic variations are feedback values used by the driver to assess the current and forthcoming vehicle status, with respect to the intended response. In addition, the driver receives information through the control devices. Low road friction will reduce the torque feedback on the steering wheel and activation of ABS is noticed through vibrations in the brake pedal. Further, the vehicle will experience external disturbances such as aerodynamic forces, road irregularities, and road friction variations. Consequently, the vehicle will not respond in the same way to the driver input in all cases. This impact of external disturbances and the limited ability of the driver to control the vehicle accurately means that the driver is constantly correcting his input to the vehicle. The driver responds to the

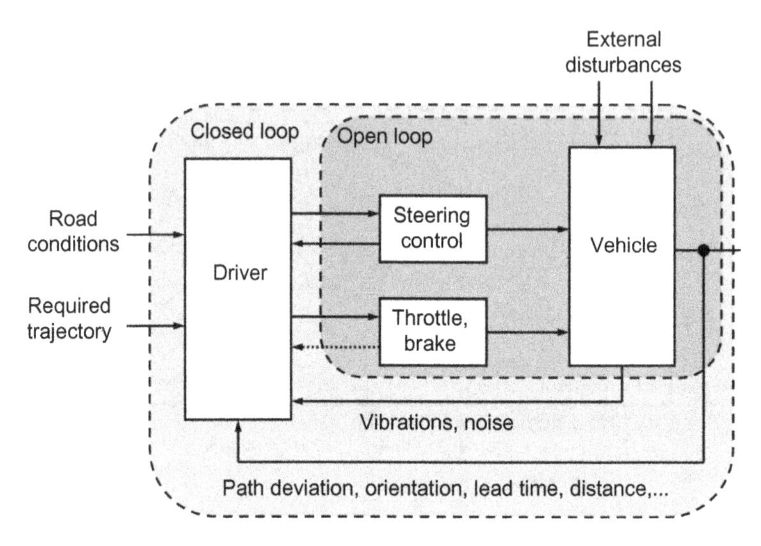

FIGURE 6.3 Vehicle–driver system.

vehicle behavior to fulfill a certain task, which is closed-loop behavior (discussed in Chapter 5), in contrast to the situation where one is considering vehicle response to driver input without driver feedback (denoted in Chapter 5 as open-loop behavior).

6.1 ASSESSMENT OF VEHICLE–DRIVER PERFORMANCE

From Figure 6.3, it is clear that a driver, as a subsystem of the entire vehicle–driver system, combines the following activities.

- Monitoring of vehicle performance and the environment by using all senses, which includes vestibular input (sense of balance, accelerations), proprioception (sense of motion and relative position of body parts and the effort applied to achieve such motion), haptic feedback (communication through touch), and visual observations.
- Actuation through the vehicle control inputs (steering wheel, throttle and brake pedal, and gear). Human limitations can be accounted for, such as reaction time (delay) and neuromuscular lag, reflecting the time required for muscle activity to occur following the required response.
- Processing, which is the transfer of monitoring input to actuation output. All the inputs (indicated in Figure 6.3) are used to make a decision about the control actions to be taken by the driver.

The ultimate goal is good performance of the vehicle–driver system. Good performance is defined as a predictable vehicle response, meaning that the effect of disturbances should be limited (immunity to external disturbances) and that the behavior of the vehicle is proportional to the driver input. For example, if steering with a fixed steering angle of 2° for a certain speed results in a lateral acceleration of 3 m/s^2, then a steering angle of 3° would be expected to result in a 50% higher acceleration, i.e., of 4.5 m/s^2. This proportional behavior holds under low acceleration and high road friction conditions, and the gains have been discussed in Section 5.3.2. It is clear that, under extreme conditions, this predictability is lost. Moreover, even if control systems in the vehicle were able to maintain such linear (predictable) behavior up to large accelerations, one would be faced with sudden transitions towards tire saturation and loss of control. Consequently, the driver does not receive sufficient prewarning for extreme situations, which misleads the driver's perception.

In addition to this predictability, good performance has been described in Chapter 5 in terms of time delays between command input and vehicle response (not too large), gains (not too large and not too small), body slip angle (not too large, but not close to zero either), roll response (preferably small), and a compromise between responsiveness and stability. A loss of stability means that small changes in the driver command input results in large variations in vehicle response. On the other hand, too high stability may lead to lack of responsiveness to driver command input and large delays, which should also be avoided. Finally, good performance includes comfort, which corresponds to the sensation of internal noise and vibrations experienced by the driver.

As discussed in Chapter 5, the assessment of such performance can be completed on an open- or closed-loop basis. In the first case, the driver is acting as a robot with a prescribed input to the vehicle. Hence, one is only considering the transfer properties (response to inputs) of the vehicle as an independent system. In the second case, the driver responds to vehicle behavior as indicated in Figure 6.3. However, one may question whether this is always an easy task for the driver. Under normal traffic conditions, a driver is expected to control his vehicle in a routine way. As traffic becomes denser, with large speed variations and perhaps with reduced visibility (fog, rain) and road friction, the driving task becomes more complex and the driver will experience a higher demand and require higher performance to control the vehicle to avoid safety risks. If the required driver performance approaches the driver's maximum ability to respond, the driver will have more difficulty dealing with the situation, resulting in an increased workload.

The concept of mental workload has extensively been studied by de Waard [56]. In addition to this, there exist other types of workload, such as physical workload and visual workload. In this book, we will mainly refer to mental workload, for which different definitions exist. De Waard formulates workload as follows:

> *Workload is the portion of the operator's limited capacity that is actually required to perform a particular task.*

This definition makes mental workload specific to each person, meaning to what extent one is willing or capable to use the available personal resources to fulfill the task demand. Demand describes the goal (as judged by the driver) in terms of task performance. The influence of this demand on the driver and his limited capacity corresponds to the workload. Figure 6.1 distinguishes between driving tasks at different hierarchical levels. Here, capacity may be insufficient to match the demand at each of these levels.

A distinction is made between the primary task (described in Ref. [56] as safe control of the vehicle within the traffic environment) and secondary tasks (which refers to all other tasks, such as checking the navigation system, making a phone call, or controlling the entertainment devices in the car). According to Ref. [56], factors influencing driver workload may be related to the driver's state (e.g., fatigue and alcohol use), to the driver's traits (experience and age), and to environmental factors (vehicle ergonomics, road and traffic conditions, poor visibility, or advanced driver support measures).

When the workload is high, the performance of the driver, and therefore of the vehicle–driver system, is expected to be low. With high task demand, the driver is expected to put more effort in his primary task to compensate. In this way, the driver may be able to maintain good closed-loop performance to avoid possible safety risks. This effort is task related. On the other hand, with very low demand, the driver may be lacking the reference to perform at a good level, which also results in an increased workload. A typical example is driving on an almost empty road, where a high level of attention is required for good driving performance. This effort is clearly state related.

De Waard [56] distinguishes four regions of primary task demand, shown in Figure 6.4, denoted as A, B, C, and D. Region A corresponds to normal–challenging traffic behavior with high performance. Increasing demand corresponds to a transition from region A to C through B. In region B, the driver capacity is insufficient to compensate for the increased workload and the vehicle–driver performance drops. In region C, workload is high and performance is low, meaning that the driver is overloaded, which results in high safety risks. In terms of Figure 6.2, this is the case where the stabilizing tolerance is completely lost. Lower demand may correspond to monotonous tasks

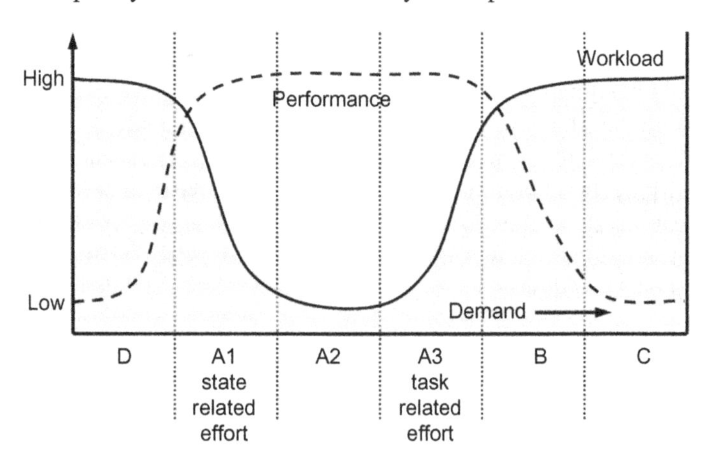

FIGURE 6.4 Regions with distinctive workload and performance, cf. De Waard [56].

which may result in a reduction of driver capacity (e.g., boredom) used for task performance, leading to a higher workload. The driver may turn attention to things other than primary task-related effects, and this may lead to increased stress reactions. This corresponds to region D.

The region A consists of three subregions: A1, A2, and A3. Region A1 is related to low density monotonous conditions. Region A2 corresponds to normal routine traffic behavior, in which the driver has no difficulty in adjusting to the traffic circumstances. Region A3 is related to more demanding traffic conditions.

Note that this model is simple in the sense that mental workload is displayed one-dimensionally. In practice, workload will have more dimensions, with different transitions for specific task demands.

Assessment of the vehicle–driver performance can be completed using subjective methodology strategies including primary task performance tests, rating scales, and open questions. The concept of rating scales means that test drivers use self-report measures that can account for both performance and workload. As indicated in Ref. [56], self-report measures can be used in combination with all the task demand areas except the A2 region. Considering Figure 6.4, we observe that distinction in performance requires test conditions to be linked to demand region $D \rightarrow A1$ or region $A3 \rightarrow B$. When we want to understand the impact of vehicle design or driver support modifications on mental workload, we need to design the test scenario in region A1 or region A3, depending whether the assessment refers to state- or task-related effort. We restrict ourselves to task-related effort, i.e., to region A3. One way to achieve such conditions is to add secondary tasks. When these tasks are sufficiently demanding, they affect the available capability of the driver to perform his primary driving tasks, which allows us to shift the workload conditions from A2 to A3. In addition, secondary task measures can be used to assess the workload.

Typical driver primary task measures are maximum speed, minimum errors, or deviations from optimal driver performance. More specifically, this may refer to lane departure, path deviations, or time to line crossing, which is the time required for any wheels of the vehicle to reach either edge of the driving lane given the vehicle original path [13]. It may also refer to distance gap or distance headway (the distance between two subsequent vehicles), the time headway (THW), which is the following time between two subsequent vehicles, and the time to contact (TTC), which is the time required for the following vehicle to make contact with the lead vehicle for unchanged speeds, i.e., if no corrective action is taken [44].

Typical driver secondary tasks are cognitive tasks (e.g., memory search, arithmetic tasks) or event response tasks (the driver responds to random visual or acoustic signals). Another way to establish a shift towards region A3 is to restrict the driver's primary tasks capacity. This can be done using an occlusion strategy. In that case, the driver is asked to perform a certain primary task with vision that is limited by glasses. It is left to the driver to judge the time (occlusion time) not required to look at the road for proper maneuvering, and to decide about restoring vision and visual feedback. Frequency of opening and closing, occlusion time, and the resulting primary task performance can be taken as measures of the workload performance.

When test design is such that workload is likely to be affected, it may be expected that the control behavior of the driver will change. Such control behavior includes steering and using the accelerator pedal. For example, when a driver is stressed, one may expect him to steer more frequently and with increased intensity. Measuring the steering wheel signal, more frequent steering is reflected in the so-called steering reversal rate (SRR), defined as the number of steering reversals per unit time (second, minute), exceeding a certain threshold value (gap size, for example, 2 degrees).

Savino [44] includes an extensive list of driving performance measures. Typical reversals with the intention to turn the vehicle are neglected; only the unintentional reversals are included. An alternative definition is the number of changes from negative (clockwise movement) to positive (counterclockwise) rotational velocity, with the positive rotational velocity exceeding a certain value (e.g., 3 [°/s]). Another definition is the number of events when the steering rate leaves a zero velocity dead band and returns to it, with a minimum change in steering angle. Note that SRR does not describe the magnitude of the steering corrections, only the frequency. Note also that a low SRR value may indicate a lack of attention in driving, which is not the same as a low workload. Observed changes in SRR indicate an effect on driver control behavior, but one should be careful in interpreting these changes in terms of driving performance and workload.

The SRR has been used in a study on a side stick steered vehicle [60] and in a study on the impact of changing steering interface characteristics (i.e., torsion bar stiffness) and vehicle handling performance (i.e., understeer gradient) on driver response [35]. Different maneuvers were used, including a single lane change and a U-turn. The discrimination of SRR appeared to be low, except for extreme variation in vehicle parameters.

An approach that includes the steering intensity is the power spectral density (PSD) of the steering wheel signal: PSD(δ; ω), where ω indicates the radial frequency. The PSD is further explained in Appendix 8. When the driver shifts to higher steering frequencies, the PSD will shift to increasing frequency as well, meaning that a specific radial frequency ω_s can be found such that the high frequency area HFA

$$\mathrm{HFA} = \frac{\int_{\omega_s}^{\infty} \mathrm{PSD}(\delta;\omega)\mathrm{d}\omega}{\int_{0}^{\omega_s} \mathrm{PSD}(\delta;\omega)\mathrm{d}\omega} \tag{6.1}$$

is expected to increase with higher workload. HFA depends on both steering frequency and steering intensity. It was shown in Ref. [35] that HFA was able to discriminate between variations in steering interface characteristics and vehicle understeer behavior.

For longitudinal following behavior, TTC and THW are defined as:

$$\mathrm{THW} = \frac{x_{\mathrm{rel}}}{V_F} \tag{6.2}$$

$$\mathrm{TTC} = -\frac{x_{\mathrm{rel}}}{V_{\mathrm{rel}}} = -\,\mathrm{THW} \cdot \frac{V_{\mathrm{F}}}{V_{\mathrm{rel}}} \tag{6.3}$$

Where x_{rel} and V_{rel} are the relative distance and relative velocity between lead car and following car, respectively, and V_{F} is the velocity of the following car. As discussed in Ref. [30], different phases can be distinguished in following scenarios. The first phase is the regulation phase with the driver of the following vehicle maintaining a headway, being sufficiently large and no perceived risk to hit the leading car. The second phase is the reaction phase, which is when the driver takes action to correct when the present or predicted distance is too small. In terms of THW and TTC, this means that the transition between the regulation and the reaction phases is a driver-dependent hyperbolic-shaped curve in the (THW, TTC) plane, as indicated in Figure 6.5.

It is shown in Ref. [30] that the decision of the driver to release the accelerator pedal and respond to the lead vehicle can be described by a simple curve (the decision line) in the plane, defined by the relative speed V_{rel} (replacing TTC) and the deviation of THW from the desired time headway, denoted as $\mathrm{THW}_{\mathrm{des}}$, with this curve being independent on $\mathrm{THW}_{\mathrm{des}}$. The shape of this curve (discussed further in Section 6.3) suggests characteristic driver parameters that are related to the driver's response as matched with a driver model, and which can be determined during actual following conditions. Changing the driver state is expected to change this curve, meaning that the curve-related characteristic driver parameters might be used to identify this driver state. In the same way, driver steering response may be described by a

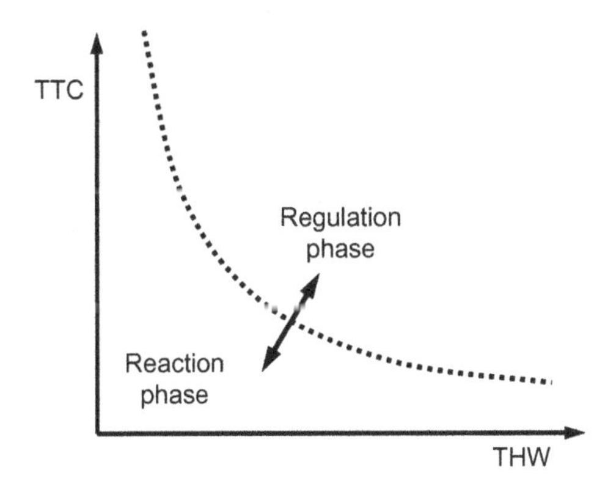

FIGURE 6.5 Driver response phase in terms of TTC and THW based on Ref. [30].

driver model with the magnitude of the relevant driver model parameters functioning as an indication of the driver state.

We close this section with comments on physiological measures, which are sensitive to driver mental workload variation. More specifically, as formulated by De Waard [56], physiological measures are shown to change with human activation level, or with information processing by the driver. This means a change of such measures may have various causes, not all related to the primary driving task, which may make it difficult to interpret these measures in terms of traffic and driving conditions. The "low workload" reference values are driver dependent. According to Ref. [56], physical efforts and emotional factors may have an impact on the resulting physiological measures. We mention some physiological measures here, where we refer to Ref. [56] for further information

The Inter-Beat-Interval

This is defined as the time between two subsequent heartbeats, in terms of electric impulses related to the heart contraction, obtained from an ECG (electrocardiogram). Increasing mental workload is expected to lead to a higher heart rate (beats per minute), and therefore a reduced mean inter-beat-interval (IBI). This measure is known to be sensitive to physical workload.

The Heart Rate Variability

This measure is defined as the ratio of the IBI standard deviation divided by the mean IBI. Heart rate variability (HRV) is often considered in the frequency domain, where a distinction is made between a low frequency band (0.02–0.06 Hz), the 0.10 Hz frequency band (0.07–0.14 Hz), and the high frequency band (0.15–0.5 Hz). The power over the mid frequency band is shown to change with mental workload and task demand and appears to be insensitive to physical workload, which makes it a better indicator for workload than the mean IBI.

Pupil Diameter and Endogenous Eye Blinks

Increasing task demand (e.g., visual demands, information processing) tends to result in an increase of pupil diameter, whereas eye blink latency related to stimulus occurrence and eye closure duration decrease. The problem is that ambient conditions (e.g., illumination, air quality) affect the eye response, which makes this type of measurement less suited for in-vehicle use.

Blood Pressure Variability

The HRV 0.10 Hz band is related to short-term blood pressure regulation. It is therefore expected that workload variations will affect blood pressure. Successful application of blood pressure variability (BPV) assessments in mental tasks is referred to in Ref. [56]. However, the technique (pressure monitoring using a cuff enclosing a finger) is less suited for in-vehicle use.

Skin Conduction Response

The conductance of the skin is determined using a small current, where it has been reported [56] that skin conduction response (SCR) is sensitive to information processing. As commented in Ref. [56], skin measurement results may be related to all undifferentiated kinds of emotional or physical stress. In addition, these results depend on many external factors (such as humidity, temperature, gender of the driver, time of the day, age, emotional arousal), which makes this measure less selective.

Facial Muscle Activity

Facial muscle behavior reflects human emotion and mental effort. Tools are available for visual recognition and interpretation of these emotions, which may be of interest to identify the driver's state.

An example of the application of the HRV frequency approach is reported by Monsma and Shrey [28], where subjective ratings for perceived mental workload from professional test drivers were compared with mean IBI and HRV measures. The drivers had to follow a curved path, indicated with cones, for different speeds and with different tire pressure conditions. Different cone spacing strategies were used, including different fixed distances and varying (unpredictable) spacing. In contrast to the HRV frequency bands, described above, a low frequency band between 0.04 and 0.15 Hz was chosen. The high frequency band was the same as given previously, i.e. between 0.15 and 0.5 Hz. In this way, low and high frequency powers can be determined. The ratio LF/HF was used as a measure for workload. Figure 6.6 shows the variation of mean IBI and the LF/HF ratio for a driver for varying

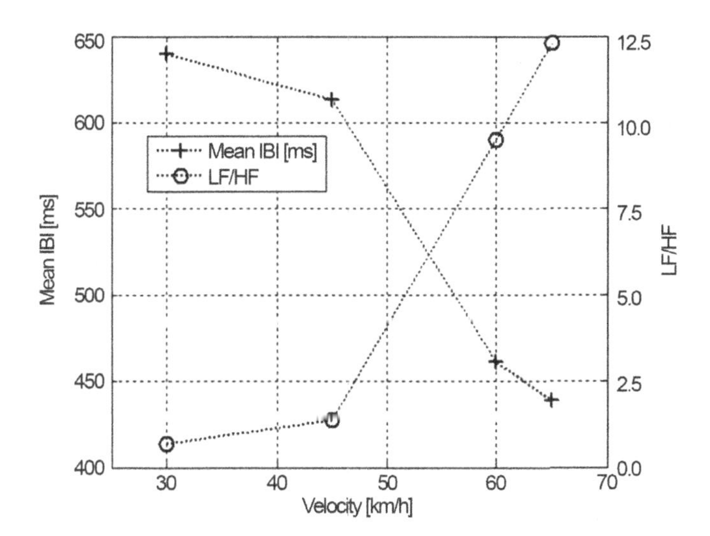

FIGURE 6.6 Mean IBI and LF/HF ratio for a professional driver following a handling curve for different speeds [57].

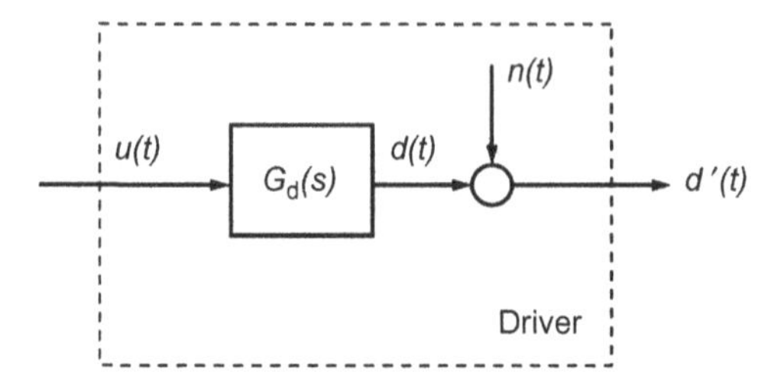

FIGURE 6.7 Human operator with quasi-linear transfer function G_d, where $d(t)$ is the linear response and $n(t)$ is the internal noise, accounting for nonlinear characteristics.

speed with fixed cone spacing, and normal tire pressures. Increasing the speed (using cruise control) will result in increasing mental workload, and the mean IBI and LF/HF ratio are expected to decrease and increase with speed, respectively. This is confirmed by Figure 6.6.

6.2 THE VEHICLE–DRIVER INTERFACE, A SYSTEM APPROACH

There are two reasons we want to find approximate simplified (i.e., linearized) descriptions of driver behavior. First, we want to analyze and predict the vehicle–driver closed-loop performance. Second, we want to understand the mechanism of human operation of the vehicle. The functionality, as well as the assessment, of the driver model parameters from realistic driving conditions can provide us with valuable information.

Linearized behavior means that the driver response to certain input $e(t)$ can be modeled using a linear transfer function $G_d(s)$ in the Laplace domain, which is equivalent to a linear differential equation in the time domain. (Appendix 4 discusses Bode plots and transfer functions further). This is illustrated in Figure 6.7, taken from Jagacinski and Flach [18]. The function $n(t)$ describes the internal noise, which is part of the driver behavior that does not follow the linear transfer assumption.

It is argued by Jagacinski and Flach [18] that quasi-linear driver behavior has a limited field of application. The driver perception has certain thresholds, but drivers are able to learn and adapt to situations. The more random (unpredictable) the input to the driver is, the better the quasi-linear assumptions hold.

6.2.1 Open-Loop and Closed-Loop Vehicle Behavior

According to Jagacinski and Flach [18], a human operator, controlling a plant (such as a vehicle) adapts behavior to the characteristics of this plant. Consider Figure 6.8, in which a driver is expected to follow an input signal $u(t)$ and we neglected the internal noise. The combined driver–vehicle (human operator–plant) response results in an output $x(t)$ that, due to external disturbances, will be different from $u(t)$.

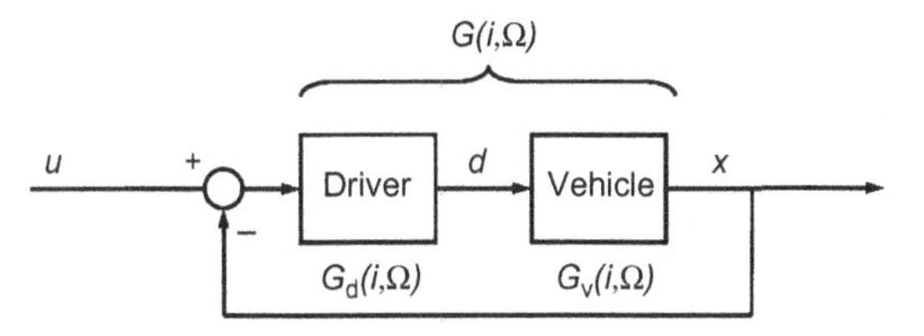

FIGURE 6.8 Closed-loop vehicle–driver system.

The transfer functions of the driver and the vehicle are denoted by $G_\mathrm{d}(i\Omega)$ and $G_\mathrm{v}(i\Omega)$, respectively, for radial frequency Ω (we replaced the Laplace variable s with $i \cdot \Omega$), with the combined transfer function $G(i\Omega)$, given by

$$G(i\Omega) = G_\mathrm{d}(i\Omega) \cdot G_\mathrm{v}(i\Omega) \tag{6.4}$$

Feedback of the error $x(t)-u(t)$ allows the human operator to compensate for this error. Suppose the vehicle acts as a simple gain $G_\mathrm{v}(i\Omega) = K$, meaning that the vehicle responds proportionally to the control input. It was demonstrated in Ref. [18] on the basis of a one-dimensional compensatory tracking study that the driver exhibits a first-order behavior (i.e., including a gain, lag, or integrator) and a delayed time response

$$G_\mathrm{d}(i\Omega) = K \cdot \frac{\mathrm{e}^{-i \cdot \Omega \cdot \tau_\mathrm{d}}}{i\Omega\tau_\mathrm{L} + 1} \quad \text{or} \quad G_\mathrm{d}(i\Omega) = K \cdot \frac{\mathrm{e}^{-i \cdot \Omega \cdot \tau_\mathrm{d}}}{i\Omega} \tag{6.5}$$

with a lag time τ_L, reaction (delay) time τ_d, and gain K. If the simple zero-order plant is replaced with a first-order plant behavior, the human operator appears to react as a simple gain with some reaction time delay according to Ref. [28], especially near the crossover frequency (the frequency where the gain $|G(i\Omega)| = 1$. Apparently, the operator adapts behavior to the plant to be controlled, such that the combined transfer function $G(i\Omega)$ behaves as a first-order system. This adaptive human behavior is consistent for a situation where the plant behavior becomes more complex, e.g., behaves as a second order acceleration control system. The human operator appears to respond with lead behavior, which means that the operator anticipates future error input by extrapolating the present trend in input error variation. In control terms, the operator acts as a differentiator for high frequencies, again in combination with a time delay.

The three situations described previously are illustrated in Table 6.1, assuming certain typical representative frequency response descriptions for the plant and the human operator (see Appendix 4).

The Bode diagrams for these transfer functions are shown in Figure 6.9 for certain parameter choices. The last column of plots includes the visualization of the transfer function $G(i\Omega)$ of the total vehicle–driver system,

TABLE 6.1 Typical Combinations of Plant (Vehicle) and Operator (Driver) Transfer Functions, Resulting in Combined First-Order Behavior

$G_v(i\Omega)$	$G_d(i\Omega)$
Zero-order M	Lag + time delay $\frac{K \cdot e^{-i \cdot \Omega \cdot \tau_d}}{i \cdot \Omega \cdot \tau_L + 1}$
First-order $\frac{M}{i \cdot \Omega \cdot \tau + 1}$	Gain + time delay $K \cdot e^{-i \cdot \Omega \cdot \tau_d}$
Second-order $\frac{M}{(i \cdot \Omega)^2}$	Lead + time delay $K \cdot (1 + i\Omega \cdot T_L) \cdot e^{-i \cdot \Omega \cdot \tau_d}$

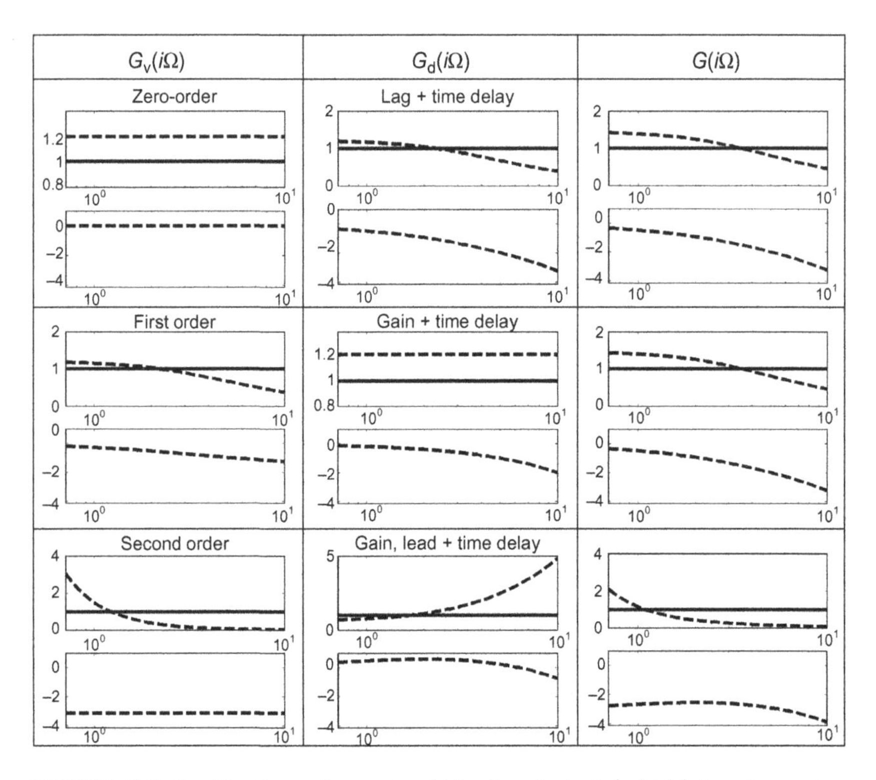

FIGURE 6.9 Combinations of operator (driver) and plant (vehicle) transfer performance, leading to a driver—vehicle transfer function, being approximately a gain in combination with an integrator and a delayed time response. Gain and phase (Bode diagrams, first the gain plot, then the phase plot) are shown against the radial frequency Ω. Phase is shown in radians.

confirming approximately a simple gain in combination with an integrator and a time delay for frequencies near the crossover frequency Ω_c:

$$G(i\Omega) = G_d(i \cdot \Omega) \cdot G_v(i \cdot \Omega) = K \cdot \frac{e^{-i \cdot \Omega \cdot \tau_d}}{i \cdot \Omega} \tag{6.6}$$

According to the Nyquist criterion (see Appendix 4), the vehicle–driver closed-loop system shown in Figure 6.8 is stable if the phase margin φ_m and the gain margin of the corresponding open-loop system are both positive. The phase margin is defined as the phase angle of the open-loop transfer function $G(i\Omega)$ on top of a phase of π, when this transfer function has unit magnitude, i.e., $|G(i\Omega)| = 1$. A positive phase margin ($|G(i \cdot \Omega)| = 1$, and therefore $K = \Omega = \Omega_c$) leads to

$$\varphi_m = \pi - \Omega \cdot \tau_d - \frac{\pi}{2} = \frac{\pi}{2} - \Omega \cdot \tau_d = \frac{\pi}{2} - K \cdot \tau_d > 0 \tag{6.7}$$

Consequently, closed-loop stability is lost if either the gain K or the delay time τ_d are large.

The behavior of $G(i\Omega)$, as shown in Figure 6.9, could also have been described by the combination of a gain, a time delay, and a simple lag:

$$G(i\Omega) = G_d(i \cdot \Omega) \cdot G_v(i \cdot \Omega) = K \cdot \frac{e^{-i \cdot \Omega \cdot \tau_d}}{i \cdot \Omega \cdot \tau_L + 1} \tag{6.8}$$

Near the crossover frequency, the qualitative difference between expressions (6.6) and (6.8) is small, especially for larger frequencies. The major difference is found in the low frequency range, where expression (6.8) is bounded and expression (6.6) is not. An unbounded transfer is unrealistic. The parameter τ_L in Eq. (6.8) expresses the fact that the driver is not able to keep track of high frequency error changes and responds *slower* than the error input. It takes time to reach a steady-state response in case of sudden error changes. This is indicated in Figure 6.10, where the first-order response to a unit step is shown. This is due to human neuromuscular restrictions. Appendix 4 and Ref. [54] offer a more extensive discussion of first- and second-order systems. The delay (or reaction) time shows that the driver is responding τ_d seconds *later* than the error input.

Suppose the overall open-loop transfer is given by Eq. (6.8). What does this mean for the stability of the closed-loop model of Figure 6.8? Again, applying the Nyquist criterion of positive phase margin, one finds:

$$\varphi_m = \pi - \Omega \cdot \tau_d - \arctan(\Omega \cdot \tau_L) > 0$$

$$|G(i\Omega)| = \frac{K}{\sqrt{1 + \tau_L^2 \cdot \Omega^2}} = 1$$

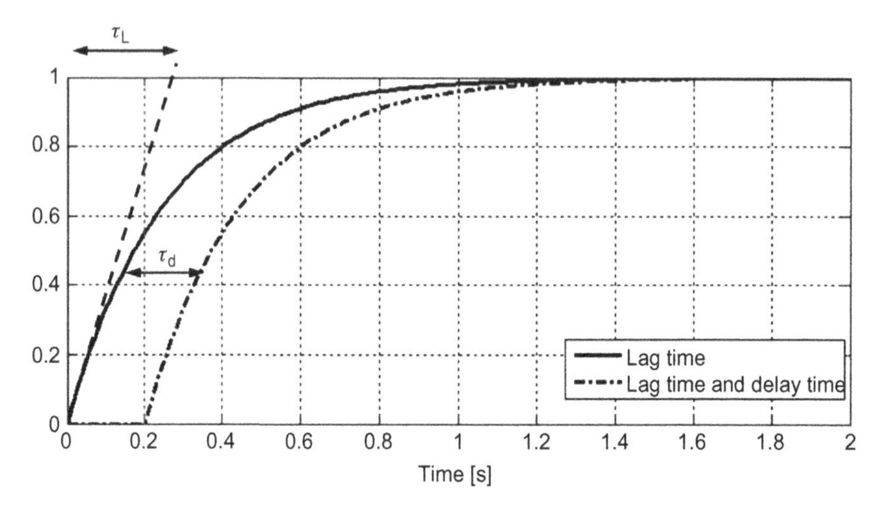

FIGURE 6.10 First-order time response for a unit step input at $t = 0$, with distinction between slower response (lag time) and later response (response time).

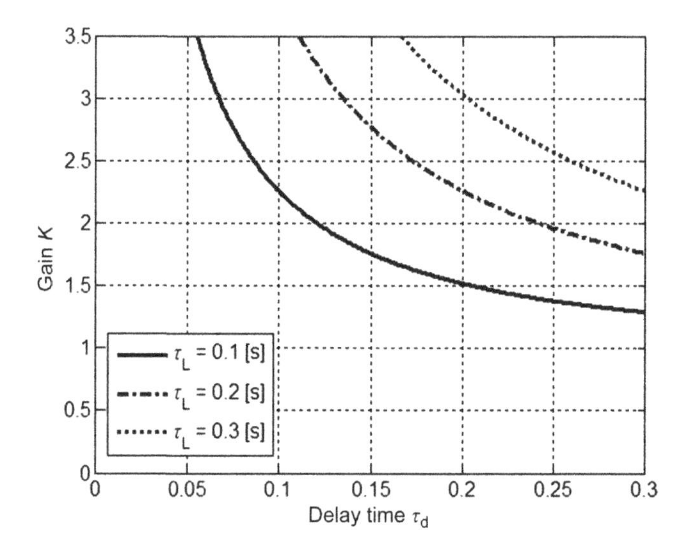

FIGURE 6.11 Stability depending on gain K and delay time τ_d for different lag time values τ_L.

leading to

$$\pi - S \cdot \frac{\tau_d}{\tau_L} - \arctan(S) > 0; \; S = \sqrt{K^2 - 1} \tag{6.9}$$

Hence, we arrive at a stability condition that depends on lag time, delay time, and gain. The stability boundary is shown in the (τ_d, K) plane in Figure 6.11 for different lag time values τ_L. For fixed lag time, this plot shows the same qualitative behavior with respect to stability as previously

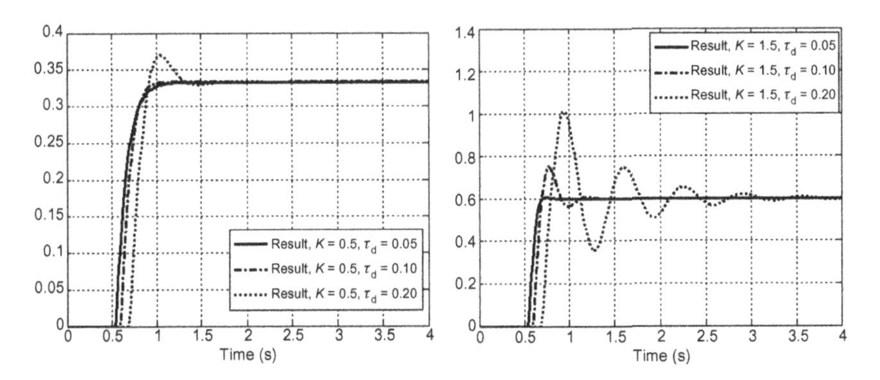

FIGURE 6.12 Closed-loop response for different gains and time delays.

observed. Large gain or large delay may lead to loss of stability. In other words, if a driver corrects his steering in response to path deviations, he will run into problems if the steering corrections are too large or if the driver needs too much time to respond. However, the driver will also have a difficult time if the path deviation is followed too precisely. More lag means more stable behavior, possibly at the cost of path accuracy.

We fixed the lag time at 0.2 [s], and determined the solution of the model as shown in Figure 6.8 for a step-input, with the transfer function G given by Eq. (6.8). Two separate cases were considered, $K=0.5$ and $K=1.5$, with the delay time taking values from 0.05 to 0.2 [s]. Results are shown in Figure 6.12.

Observe a good response for large gain and small delay time. For larger delay time, increasing the gain makes the response more oscillatory and less stable. For smaller gain, the closed-loop response becomes slower. Jagacinski and Flach [18] describe this behavior for low gain as "sluggish."

6.2.2 The McRuer Crossover Model

The general expression for a quasi-linear model of a human operator, with reference to McRuer et al. [24,25], is presented in Ref. [18] as follows:

$$G_{\mathrm{d}}(i\Omega) = K \cdot \left(\frac{\mathrm{e}^{-i\cdot\Omega\cdot\tau_{\mathrm{d}}}}{i\cdot\Omega\cdot\tau_{\mathrm{N}}+1}\right)\cdot\left(\frac{i\cdot\Omega\cdot T_{\mathrm{L}}+1}{i\cdot\Omega\cdot\tau_{\mathrm{L}}+1}\right) \tag{6.10}$$

with delay (reaction) time τ_{d}, lag times τ_{L} and τ_{N}, lead time T_{L}, and gain K. The τ_{N} lag time is related to the human's limited ability to follow quickly changing inputs due to neuromuscular inertial properties of the human system. The last factor in Eq. (6.10) is a combination of lead and lag, where lead describes the driver's ability to extrapolate on the expected control input in the near future. Typical values for the human operator (driver) related parameters are reported in Ref. [18] to be in the order of $\tau_{\mathrm{d}}\approx 0.1-0.25$ [s] (tracking studies) and $\tau_{\mathrm{N}}\approx 0.1-0.3$ [s], see also Ref. [1]. For longitudinal control, delay times are found to be higher, on the order of 0.5–1.0 [s], [30, 53]. The other model parameters, K, τ_L, and T_L, depend on the plant properties, as discussed in the preceding Section 6.2.1, and are used to tune the operator performance to the plant (i.e., vehicle) characteristics.

We discuss two specific driver behavior situations in the following sections:

1. Vehicle–driver longitudinal conditions are discussed in Section 6.3. In these conditions, the driver is following a lead vehicle and adjusting the accelerator and brake pedal position in response to the distance and velocity difference with this lead vehicle.
2. Vehicle–driver handling conditions, where the driver is adjusting the steering wheel angle to track a certain path to guide vehicle through a cornering situation, a lane change, etc.

In both situations, the expression (6.10) will be simplified and the impact of changing driver parameters will be discussed. It will be shown that the identified driver model parameters can be interpreted in terms of traffic safety, and may be used effectively as input for advanced driver support systems.

6.3 VEHICLE–DRIVER LONGITUDINAL PERFORMANCE

Consider the situation of a car following task, i.e., where one car is following a lead car. A schematic layout is shown in Figure 6.13. The position, speed, and acceleration of the following and leading cars are indicated with index F and L, respectively. Note that the positions of the front end of the following car and of the rear end of the leading car are used to define the relative distance x_{rel}.

The driver of the following car is observing errors in the desired headway distance, and the relative velocity, which both should be kept at a minimum. Assuming a desired time headway THW_{des} and considering a time delay τ_d, the following driver model is used:

$$P_a = K_1 \cdot \dot{x}_F(t-\tau_d) \cdot (THW(t-\tau_d) - THW_{des}) + K_2 \cdot V_{rel}(t-\tau_d) \quad (6.11)$$

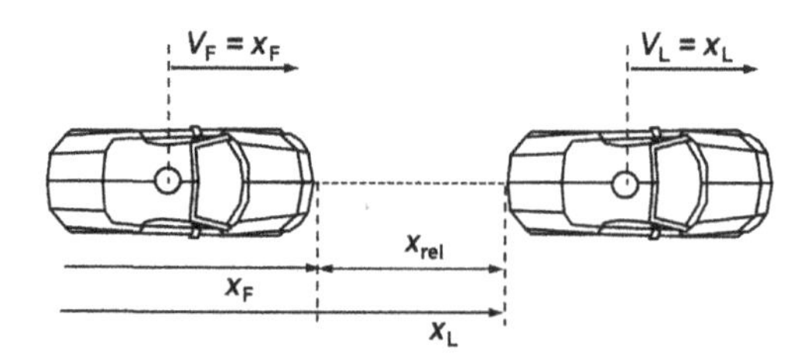

FIGURE 6.13 Schematic layout, following a lead car.

where P_a is the percentage of accelerator pedal depression. In the frequency domain, this model corresponds to the following two transfer functions:

$$G_{d,1}(i\Omega) = \frac{P_a}{THW - THW_{des}} = K_1 \cdot e^{-i \cdot \Omega \cdot \tau_d}$$

$$G_{d,2}(i\Omega) = \frac{P_a}{V_{rel}} = K_2 \cdot e^{-i \cdot \Omega \cdot \tau_d}$$

It is usually assumed that the desired headway distance is linear in the following vehicle velocity, and we take THW_{des} as a parameter in the driver model. The vehicle acceleration is taken as a gain G_a times the gas pedal stroke. A typical value of G_a is on the order of 0.025 [m/s^2/%] [41], which means that 50% pedal depression corresponds to 1.25 [m/s^2] longitudinal acceleration. In this situation, the variations in pedal stroke are assumed to be small, with the zero position (no acceleration) corresponding to a fixed small pedal depression. Releasing the pedal means a deceleration (engine braking) assumed to be on the same order as acceleration. One might choose different gains for acceleration and deceleration in case of extreme braking. One might also account for more accurate power train and engine performance. In our case of nonextreme following situations, we use the value for G_a as mentioned previously, i.e., 0.025 [m/s^2/%].

Typical values for the driver gains K_1 and K_2 appear to vary with the disturbance frequency of the leading vehicle. In Ref. [30], the dependency of these gains on the time headway was shown to be small. A larger THW_{des} tends to reduce the gains slightly. In Ref. [30], the lead vehicle speed was given a disturbance frequency, where values of 0.3, 0.5, and 0.7 Hz were chosen.

In Ref. [53], Urban examined an effective and accurate assessment method to determine the driver gains. His studies were completed using a driving simulator. He distinguished between normal driving conditions, conditions of limited visual input (modeled using occlusion), and conditions where the driver was performing a secondary task. These conditions are typical for driving in heavy rain, or with sun shining into the driver's eyes, and in case of a distraction (phone, radio, navigation, etc.), respectively.

In Ref. [41], values for these gains were determined for normal driving and deviated driving (the driver showed reduced attention to the leading vehicle's behavior). The gains were reduced in this case. The reaction time τ_d was neglected in Ref. [41]. Typical values for K_1 and K_2 are listed in Table 6.2.

We discuss the stability of Eq. (6.11) when the pedal depression is replaced by the longitudinal acceleration, divided by the gain G_a, with the delay approximated by

$$e^{-i \cdot \Omega \cdot \tau_d} \approx \frac{1}{1 + i \cdot \Omega \cdot \tau_d} \tag{6.12}$$

TABLE 6.2 Typical Values of the Driver Model Gains K_1, K_2 Introduced in Eq. (6.11), Based on Refs. [30], [41], and [53]

	[30] 0.3 Hz	[30] 0.5 Hz	[41] Normative	[53] Normal	[53] Occlusion	[53] Distraction
K_1 [%/m]	1.2	2.8	2.0	1.8	1.5	1.8
K_2 [s.%/m]	4.5	5.3	22.0	9.0	5.0	6.6

This leads to the system matrix A for Eq. (6.11), given as follows:

$$A = \begin{pmatrix} 0 & 1 & 0 \\ 0 & 0 & 1 \\ -\dfrac{G_a \cdot K_1}{\tau_d} & -\dfrac{G_a \cdot K_2}{\tau_d} & -\dfrac{1}{\tau_d} \end{pmatrix}$$

The stability boundary can be given by

$$\frac{K_1}{K_2} \cdot \tau_d = 1 \tag{6.13}$$

As a result, closed-loop stability is lost if either the delay time is large or the gain ratio K_1/K_2 is large, qualitatively confirming Figure 6.11.

6.3.1 Following a Single Vehicle

In this section, we discuss a vehicle following a lead vehicle that changes its speed with a certain frequency. The data used in the analysis are listed in Table 6.3.

These values are similar to the data used in Ref. [30]. The leading vehicle has a speed V_L that is a superposition of a constant part (90 km/h = 25 m/s), and three parts varying with 0.05, 0.1, and 0.3 Hz, respectively. The amplitudes of the parts with the lowest frequencies are chosen randomly and are not too large, whereas the amplitude of the part with 0.3 Hz is chosen as a fixed value of 2.0 [m/s]. For the first 10 [s], we multiplied the speed V_L with a smooth monotonously increasing function, growing from a value and slope of zero at initial time $t = 0$ [s] up to a value of 1 and again a slope equal to zero for $t = 10$ [s]. This allows a smooth transition from an undisturbed following situation to a situation of disturbed lead vehicle behavior.

The variations of the leading vehicle's speed V_L and the following vehicle's speed V_F are both shown in Figure 6.14. The headway distance (THW times V_F) and the desired headway distance (THW_{des} times V_F) are shown in Figure 6.15. Observe the low frequency behavior of the leading vehicle, with a

TABLE 6.3 Numerical Values for Model Parameters Used in Section 6.3.1

G_a [m/s^2/%]	K_1 [%/m]	K_2 [s.%/m]	THW_{des} [s]	τ_d [s]
0.025	1.8	4.0	1.0	0.6

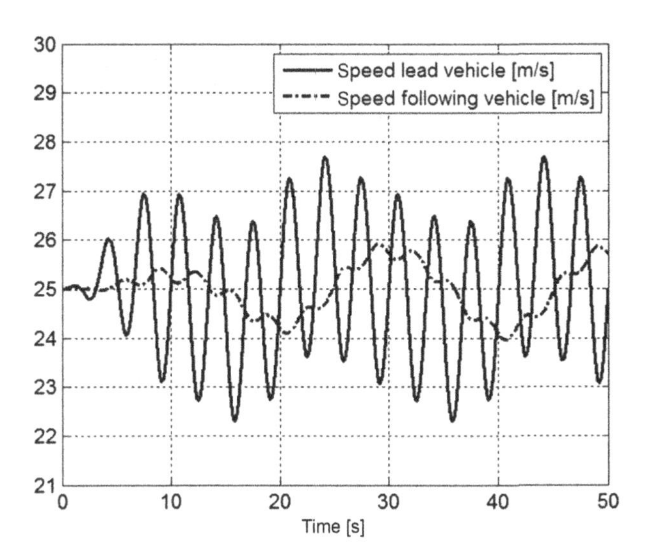

FIGURE 6.14 Velocity versus time for leading and following vehicle.

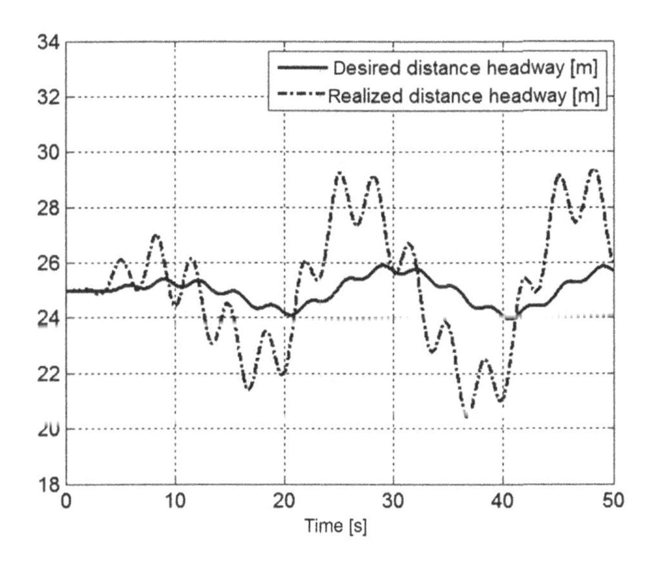

FIGURE 6.15 Realized and desired distance headway.

0.3 Hz frequency superimposed. The impact of the speed variation appears to be the variation in the following distance, varying between 22 and 29 [m], which shows a deviation from the desired headway distance of about 2–3 [m].

The driver is accelerating and decelerating to keep the speed difference small and the headway distance close to the desired distance. This means that both TTC and THW vary, with each of them being important for the transition of the regulation phase to and from the reaction phase, as indicated in Figure 6.5.

We plotted the THW deviation versus the relative speed in Figure 6.16. The same plots were used in Ref. [30] for different combinations of disturbing frequency and amplitude. Part of the curves is highlighted with additional markers. These situations correspond to a negative gas pedal depression, i.e., when the driver is in the reaction phase, aiming to slow down the vehicle.

The plot shows four quadrants in which the curves are drawn clockwise. The top two quadrants correspond to a negative relative speed, i.e., to a positive TTC, becoming smaller when moving toward larger values along the vertical axis. With a minimum relative distance of about 22 [m] and a maximum V_{rel} of 0.12 times V_F, the minimum TTC appears to be on the order of 7 [s], which is not very critical, but which already initiates some deceleration activity. In the two quadrants on the right-hand side, the actual time headway is less than the desired THW. Clearly, in the top-right quadrant, the driver is likely to take action to decelerate the vehicle because the TTC is decreasing and the THW is too small. Connecting the dots where the driver starts to decelerate (release of the accelerator) gives more or less a straight line (the dashed straight line in Figure 6.16) above which speed V_F is reduced. The position of this line is expected to depend on the disturbance frequency,

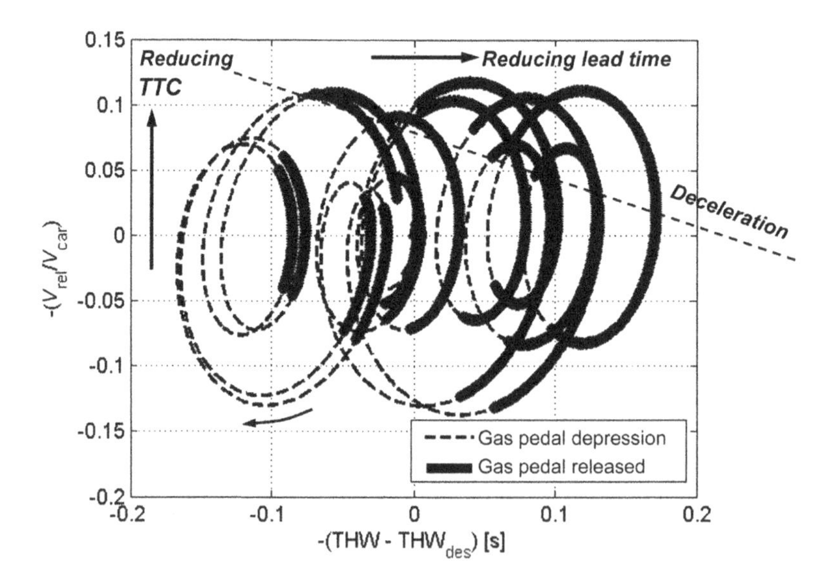

FIGURE 6.16 THW deviation versus relative vehicle speed.

where a maximum level 0.5 Hz was considered realistic in Ref. [30]. The driver is deciding to release the accelerator at a time of $\tau_d = 0.6$ sec before the actual accelerator l release. This means that a second line could be drawn in Figure 6.16, just below the dashed one, to indicate these decision points.

Mulder et al. [30] completed a Monte Carlo analysis by taking 100 lead car velocity disturbance profiles for different disturbance frequencies that lead to a linear band of decision points in the (V_{rel}/V_F, ΔTHW) plane. The combination of these bands gives a boundary in this plane, which can be described explicitly in terms of relative speed and realized THW, that can be used as a basis for an active driver deceleration support tool. This is further examined by Mulder et al. [30].

6.3.2 Driver Model and Driver State Identification

The driver parameters K_1, K_2, THW_{des}, and τ_d can be derived from observed speed differences, headway distances, and gas pedal depression (e.g., in Ref. [53], where a combination of Kalman filtering and Recursive Least Squares is used). For a discussion of these identification techniques, see Ref. [20]. It is to be expected that different driver states, such as workload, will correspond to different driver parameters. Moreover, driver parameters being changed might indicate a change in driver state, which is relevant for the way that driver support systems are being understood and used effectively.

Some of these relationships are shown in Table 6.2. When the driver is distracted by a secondary task, the gain K_2 tends to be reduced, meaning that the driver will respond less on speed differences. In case of reduced vision, both the headway deviation and the speed differences will have less impact on the driver's longitudinal vehicle control.

Mulder et al. [30] show that faster fluctuations of the lead vehicle speed changes the THW deviation and the relative speed for which the driver decides to release the gas pedal. A larger frequency results in a higher closing speed between both vehicles.

It was found in Ref. [41] that driving with less attention leads to reduced gains K_1 and K_2, confirming the results of Ref. [53].

6.4 VEHICLE–DRIVER HANDLING PERFORMANCE

Closed loop cornering behavior is defined as when a driver is controlling his steering wheel to follow a specific road layout. This control requires some reference, which could be a vehicle in front of the driver's vehicle, a far end-point, or one of the sides of the road. Using this reference, the driver knows from this visual feedback of the road and traffic conditions how to correct the steering so that crossing or hitting the side of the road is avoided. When changing lanes, the driver moves attention from the first to the second lane before actually arriving in the second lane. A well-known driver model is the path-tracking model, where the driver is assumed to follow a certain artificial path. We discuss this model, which fits with the McRuer general model (6.10), in some detail in Section 6.4.1 and discuss the corresponding closed-loop stability in Section 6.4.2, with reference to Section 6.2.1. The identification of the model parameters and their relationship with driver workload is discussed in Section 6.4.3.

6.4.1 Path-Tracking Driver Model

We consider the situation in which a vehicle aims to follow a specific path in the global (X, Y) plane, as indicated in Figure 6.17. As the vehicle follows its course, it will deviate from the intended path. The driver receives visual feedback of the expected path deviation, and is expected to change the steering angle to correct this deviation. In the single path-tracking model, the driver is assumed to observe the path deviation straight ahead of the vehicle, at a preview distance L_p from the vehicle's CoG. This deviation $D_p(t;L_p)$ consists of two parts. The first part is due to the initial deviation of the vehicle, denoted by $y(t)$. The second part is due to the difference in yaw angle orientation between the vehicle and the path. The orientation of the intended position along the path at distance L_p ahead of the vehicle, with respect to the path location near the vehicle (projected perpendicularly from the local longitudinal x direction), is denoted by the yaw angle ψ_p. Preview points ahead of the vehicle and along the path are denoted as A and T. The vehicle position is indicated as (X_V, Y_V).

According to the McRuer crossover model, one may expect a delay or lag in the driver steering response. In the frequency domain, this model can be described by

$$\frac{\delta(i\Omega)}{D_p(i\Omega)} = K_p \cdot \frac{e^{-i\cdot\Omega\cdot\tau_d}}{i\cdot\Omega\cdot\tau_L + 1} \tag{6.14}$$

where τ_L is assumed to include the neuromuscular lag. One might add a lead term for when the driver is able to anticipate forthcoming path deviations

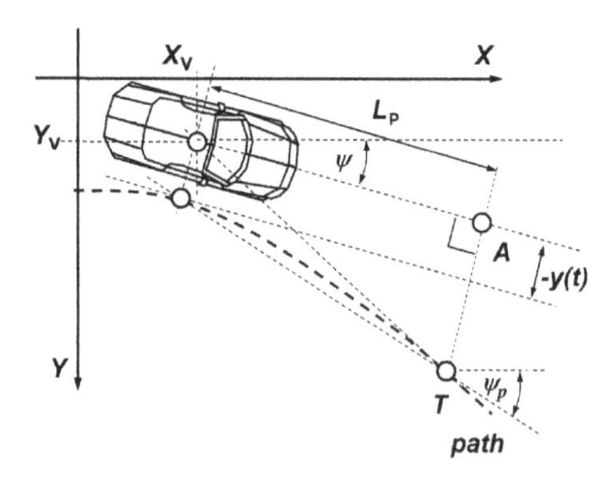

FIGURE 6.17 Vehicle following a path.

from the deviation change at distance L_p. When we approximate the delay term using lag, and assume the lag times to be small, we can neglect the delay term in Eq. (6.14). The lag time τ_L is then considered to cover both the actual lag and delay. In the time domain, this leads to the following differential equation:

$$\tau_L \cdot \dot{\delta}(t) + \delta(t) = K_p \cdot D_p(t; L_p) = K_p \cdot (L_p \cdot (\psi_p(t; L_p) - \psi(t)) - y(t)) \quad (6.15)$$

This model is used and explained in different textbooks, such as Refs. [1] and [11]. It has been examined extensively by Pauwelussen [38,39] and variations of this model are considered, such as the Salvucci model [42], where two preview lengths are used. Another extension of the model is when not only the path deviation, but also the deviation in yaw orientation, is considered. In this case, an additional term is added to the right-hand side of Eq. (6.15) that is proportional to the difference in yaw angle between the vehicle and the path at preview distances L_p. In this book, we restrict ourselves to path deviation and Eq. (6.15).

Let us start with a discussion on the driver model under stationary conditions, i.e., with the driver model reduced to

$$\delta = K_p \cdot (L_p \cdot (\psi_p(t; L_p) - \psi(t)) - y) \quad (6.16)$$

That means that the vehicle intends to follow a circular path with a curve radius R. Let us consider the situation that this vehicle is able to follow this path exactly, i.e., $y = 0$. This case was examined in Ref. [38], with the layout schematically shown in Figure 6.18. The preview length is usually on the order of 10–25 [m], which is much smaller than the curve radius. That means that higher-order terms in L_p/R can be neglected. Figure 6.18 shows relatively large L_p for reasons of clarity. The vehicle has a body slip angle β. The angle between the speed direction (perpendicular to the radial direction)

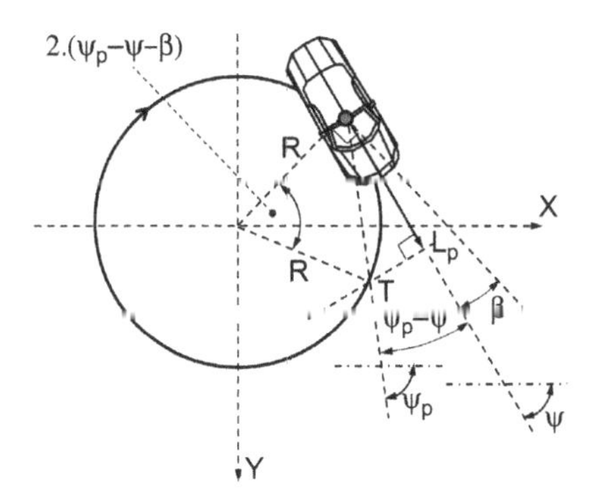

FIGURE 6.18 Steady-state path-tracking model.

and the line between the vehicle CoG and the point T on the circular path is given by $\psi_p - \psi - \beta$, which is half of the angle at the circle center in the triangle, given by this center, the vehicle CoG, and point T. From this, one may derive that the following expression is correct up to first-order in the angle ψ_p-ψ:

$$\psi_p - \psi - \beta = \arcsin\left(\frac{L_p}{2R}\right) \tag{6.17}$$

We refer to Section 5.3.2, where we derived the steering angle for a single-track vehicle model following a circular path

$$\delta = \frac{L}{R} + \eta \cdot \frac{V^2}{g \cdot R}$$

with speed V, acceleration of gravity g, vehicle wheelbase L, and understeer coefficient η. Substituting this expression and Eq. (6.17) into Eq. (6.16) for $y = 0$, and accounting for the steady-state gain expression for the body slip angle as derived in Section 5.3.2, we arrive at a relationship between the model parameters K_p (steering gain) and L_p (preview length)

$$L_p \cdot K_p \cdot \left[A_1 + \frac{L_p}{2}\right] \approx L_p \cdot K_p \cdot \left[A_1 + R \cdot \arcsin\left(\frac{L_p}{2 \cdot R}\right)\right] = A_2 \tag{6.18}$$

with

$$A_1 = g \cdot A_2 \cdot \frac{b - B \cdot V^2}{L \cdot g + \eta \cdot V^2}; A_2 = L + \eta \cdot \frac{V^2}{g}$$

We used the notation found in Section 5.3. The parameters B and η depend on the axle cornering stiffnesses as discussed in Chapter 5. Apparently, the parameters K_p and L_p cannot be chosen independently, but they satisfy a hyperbolic relationship

$$K_p = \frac{A_2}{L_p \cdot \left(A_1 + \frac{1}{2} L_p\right)} \tag{6.19}$$

Hence, the same steady-state vehicle performance is obtained for various combinations of K_p and L_p, where a large L_p corresponds to a small steering gain K_p and reducing the preview length should be compensated for with a larger steering gain. Considering Figure 6.18, this relationship is obvious. A larger preview length means a larger path deviation, which suggests a larger

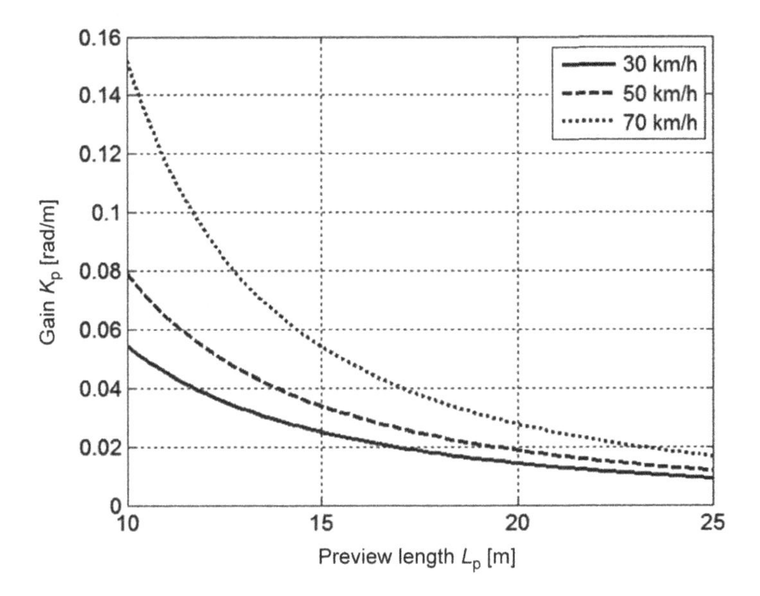

FIGURE 6.19 Preview length versus gain for various vehicle velocities.

steering angle. However, because the steering angle only depends on curve radius, speed, and vehicle properties, this is a contradiction. Consequently, the gain must be reduced in Eq. (6.16) to bring the path deviation and the steering angle to the original values. Note that (6.19) does not depend on the curve radius, which suggest that, if the vehicle is not too far from steady state conditions, the expression (6.19) holds for any arbitrary path, as long as the speed and the vehicle properties do not change. In Ref. [38], driver behavior was examined for oscillatory steering input for different frequencies resulting in lateral accelerations ranging from 0.02 to 0.5 [g]. The driver model was used to set a path and then track it. The driver parameters (L_p, K_p) were determined by matching the model steering output with the original steering input. It was observed that the resulting combinations (L_p, K_p) matched Eq. (6.19) quite well. Repeating the calculations in which the parameters were determined by matching the vehicle path instead of the steering angle resulted in the same results with maximum difference on the order of 2.5%.

We determined the (L_p, K_p) relationship for different speeds and different axle cornering stiffnesses, starting with the vehicle and axle cornering stiffness data included in Appendix 6. The results are shown in Figures 6.19 and 6.20, with reference speed 50 [km/h] used in Figure 6.20. One observes that a larger speed will lead to larger gains for the same preview length. In addition, a higher rear axle cornering stiffness or a lower front axle cornering stiffness will also lead to a larger steering gain. The latter corresponds to a

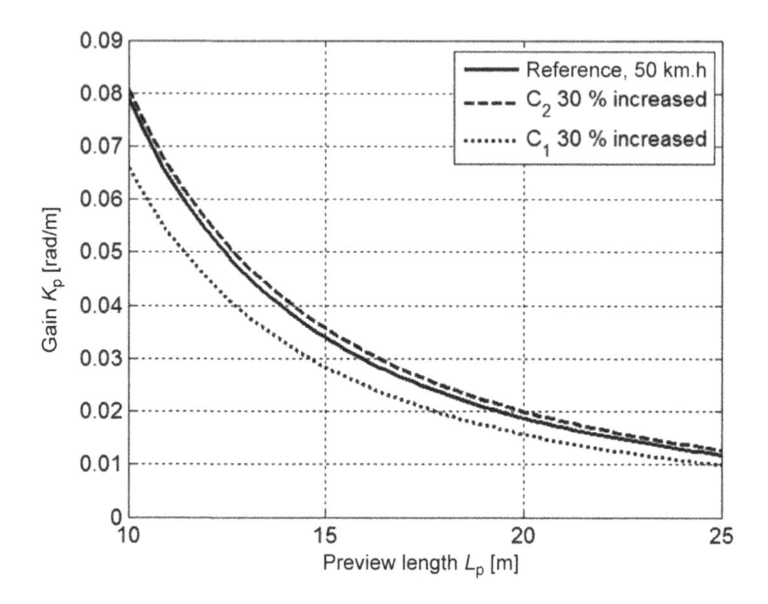

FIGURE 6.20 Preview length versus gain of various axle cornering stiffnesses.

higher understeer gradient, i.e., increased understeer where more steering would be expected. We summarize the results thus far:

1. Following a steady-state turn using a path-tracking driver model, the driver's steering gain and the preview length are related through a hyperbolic relationship.
2. The path dependency is small, meaning that this relationship (6.19) holds approximately for an arbitrary vehicle path.
3. The (L_p, K_p) relationship depends on the vehicle handling behavior (axle characteristics, understeer gradient, body slip angle gain, etc.) and vehicle speed.

Thus far, we considered the driver model in terms of steering gain and preview length. However, instead of the preview length, one may use the preview time, which is defined by

$$T_p = \frac{L_p}{V} \tag{6.20}$$

To maintain a similar relationship between preview time and steering gain, as given by Eq. (6.19), the steering gain is replaced by a corrected gain, which is defined as the product of the steering gain and vehicle speed:

$$K_{corrected} = K_p \cdot V \tag{6.21}$$

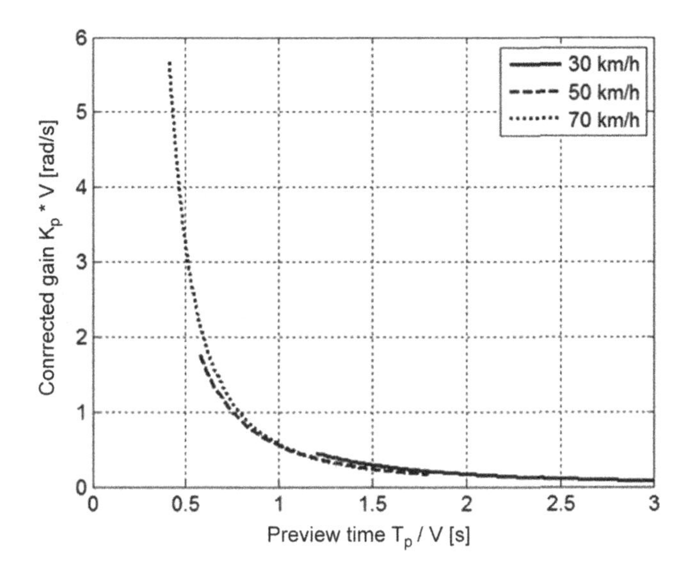

FIGURE 6.21 Driver preview time versus speed corrected steering gain.

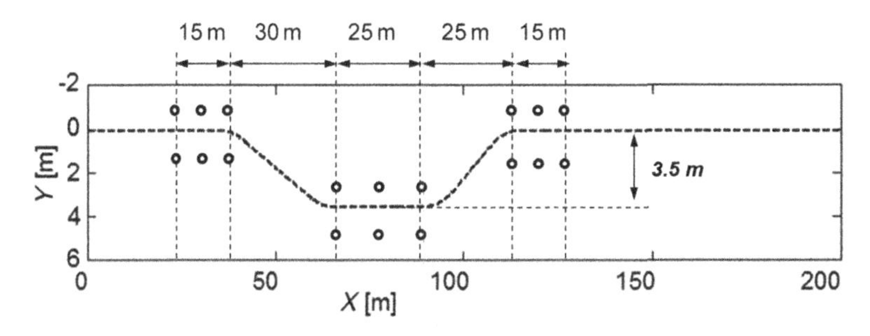

FIGURE 6.22 Geometry of the severe double lane change.

Again, we determined the impact of changing vehicle speed V, but here for the (T_p, $K_{corrected}$) relationship. The result is shown in Figure 6.21. This result is remarkable in the sense that this relationship seems to be almost invariant, with respect to the speed V. The invariance is not perfect, but does appear to be a good approximation. This is especially due to the replacement of the preview length by the preview time according to Eq. (6.20). We selected the same preview distance range, from 10 to 25 [m], as used in the previous analysis. Suppose one is interested in the identification of these driver parameters from realistic vehicle data, obtained under practical traffic conditions. Under those circumstances, one cannot expect a constant speed. Expressing the results graphically would therefore lead to a "cloud" of points in Figure 6.19, whereas these results would be closer to a curve when they are expressed as T_p versus $K_{corrected}$ and this would simplify the identification process.

The next step is to determine the closed-loop performance of a vehicle; for this, we selected the ISO 3888 severe lane change [62]. In this test, the driver is asked to change lanes through a restricted area marked by cones. This could be either a single lane change or a double lane change. Figure 6.22 illustrates a double lane change. The start of the lane change is at the left end in the figure. After 15 [m] between cones, the driver must move to the second lane. This means steering to the left while exiting the first lane and steering to the right to enter the second lane. Exiting the second lane and moving back to the first lane involves again steering to the right, followed by steering to the left. Note that we have taken the positive lateral position downward in Figure 6.22. The width of the vehicle excludes the side mirrors. The width of the lane between the cones is prescribed as follows: 1.1 times the width of the vehicle plus 0.25 [m] for the first section, 1.2 times the width of the vehicle plus 0.25 [m] for the second lane section, and 1.3 times the width of the vehicle plus 0.25 [m] for the final section of the first lane.

This test is a purely dynamic test, corresponding to a typical case in which a driver tries to avoid an obstacle or carries out an overtaking maneuver (in case of a single lane change). The double lane change test serves to verify excessive vehicle handling performance; for example, in relation to varied vehicle characteristics (tire properties) or driver support systems (active steering, wheel-by-wheel braking strategies).

Only a skilled driver should be involved in such tests and one should begin at a low speed and then gradually increase this speed while repeating the test. Assessment of vehicle performance can be completed based on:

- Peak values and lag (with respect to the steering input) for yaw rate, body slip angle response, and lateral acceleration, for different speeds.
- Subjective rating of vehicle handling performance by the test driver.
- The maximum speed applied without hitting any cones.

Instead of keeping the vehicle between the cones, we discuss the situation of a driver controlling the vehicle to keep the deviation from a lane change path as small as possible. This path (indicated in Figure 6.22) consists of straight and circular parts describing the transitions between these straight parts. We selected a curve radius of 40 [m] for these circular path transitions.

It means that the vehicle is assumed to follow a nearly straight path, with a small yaw angle ψ. The closed-loop problem can now be described in terms

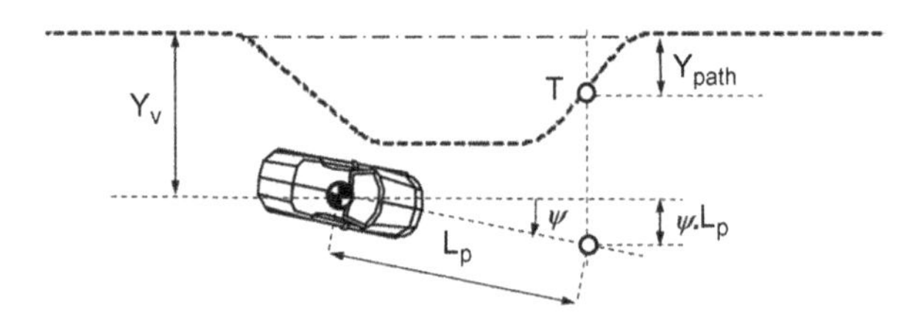

FIGURE 6.23 Schematic layout of river path-tracking model for nearly straight path.

of the global lateral vehicle position Y_V, as indicated in Figure 6.23. Abe [1] studied this case and we follow this approach.

Eq. (6.15) is now replaced by

$$\tau_L \cdot \dot{\delta}(t) + \delta(t) = K_p(D_p(t; L_p)) = K_p \cdot (Y_{path} - L_p \cdot \psi - Y_V) \tag{6.22}$$

where Y_{path} is taken at position $X_V + L_p$, with (X_V, Y_V) introduced earlier as the global vehicle CoG coordinates. The problem of a closed-loop path-tracking model is now a fifth-order system, with state vector $x = (\beta, r, \delta, \psi, Y_V)$. For linear axles, neglecting external forces and moments, and assuming ψ to be small (i.e., $\sin(\psi) = \psi$, $\cos(\psi) = 1$), this problem is described by the following matrix equation:

$$\underline{\dot{x}} = \begin{pmatrix} \frac{Y_\beta}{mV} & \frac{Y_r}{mV} - 1 & \frac{C_{\alpha 1}}{mV} & 0 & 0 \\ \frac{N_\beta}{J_z} & \frac{N_r}{J_z} & \frac{a.C_{\alpha 1}}{J_z} & 0 & 0 \\ 0 & 0 & -\frac{1}{\tau_L} & -\frac{K_p L_p}{\tau_L} & -\frac{K_p}{\tau_L} \\ 0 & 1 & 0 & 0 & 0 \\ V & 0 & 0 & V & 0 \end{pmatrix} \cdot \underline{x} + \begin{pmatrix} 0 \\ 0 \\ \frac{K_p}{\tau_L} \cdot Y_{path} \\ 0 \\ 0 \end{pmatrix} \tag{6.23}$$

using derivatives of stability Y_β, Y_r, N_β, and N_r introduced in Chapter 5, mass m, yaw moment of inertia J_z, and front axle slip stiffness $C_{\alpha 1}$.

We solved this set of equations for the double lane change using the vehicle data given in Appendix 6. The axle data were based on the Pacejka tire model, treated in Section 2.5.2, with parameters listed in Table 6.4. The understeer gradient η was determined from the axle cornering stiffness $C_\alpha = B \cdot C \cdot D$. Consequently, we modified the stiffness factor B of the front axle to change the understeer gradient η. All other parameters have been left unchanged. The lag time τ_L was chosen as 0.1 [s].

TABLE 6.4 Axle Pacejka Model Parameters Used in the Lane Change Analyses

	Strong Understeer		Limited Understeer	
Axle Parameter	**Front**	**Rear**	**Front**	**Rear**
C	1.3	1.3	1.3	1.3
B	6.3083	10.0546	0.5011	10.0545
D [N]	7107.3	8446.0	7107.3	8446.0
E	−0.0744	−1.3367	−0.0744	−1.3367
η	0.0482		0.0181	

If we denote the lateral position of the path at longitudinal position X_V by Y_p, then we can express the cumulative path error along the lane change by the root mean square value

$$\text{Error}_{\text{path}} = \sqrt{\frac{1}{t_2 - t_1} \cdot \int_{t_1}^{t_2} [Y_p(X_V(t)) - Y_V(t)]^2} \tag{6.24}$$

with times t_1 and t_2 corresponding to entry and exit of the lane change, according to the description in Figure 6.22.

Optimizing the lane change with respect to path error, we can determine the preview length as a function of vehicle speed $L_p(V)$, for which this error has a minimal value. Results are shown in Figure 6.24 for the two sets of parameters from Table 6.4. We depicted L_p versus speed V, as well as $L_p(V)$ versus gain $K_p(V)$ according to Eq. (6.19). Note that the right-hand plot does not describe the hyperbolic relationship (6.19) for constant speed. The pairs (L_p, K_p) in this plot are determined for different vehicle speeds.

An optimal value for L_p corresponds to a steering history, which will become more and more demanding for the driver with increasing speed. Further, it is not clear if the driver is actually capable of steering properly under such demanding conditions. Even if the driver is able to perform as required, it is expected that the driver will not act in that way and will accept increased path deviation. In other words, it is to be expected that some balance will be established between path error and the acceptable driver steering activity. This last aspect refers to steering effort, physical or mental workload, or other consequences of steering that may become too extreme. In this section, these effects are disregarded, and we focus only on path error as the dominant factor for the driver review length L_p.

From Figure 6.24, we conclude that increasing the vehicle velocity leads to a larger preview length and a smaller steering gain. Considering the difference in understeer, the results in Figure 6.24 show that a reduction in understeer only results in a change (increase) in preview length for high speed, i.e.,

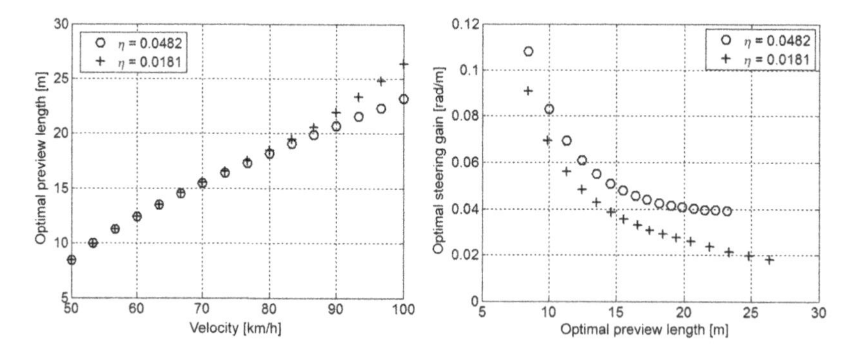

FIGURE 6.24 Optimal driver parameters for varying speed.

when the tires are expected to be loaded beyond the range in which they can be considered linear. A more responsive vehicle (smaller value of η) results in a lower value for the optimal steering gain, which is expected.

Next, we discuss the results of the lane change maneuver for two different speeds, $V = 60$ and 90 [km/h], and for the case of $\eta = 0.0482$. These two speeds correspond to $(L_p, K_p) = (12.45, 0.061)$ and $(20.70, 0.040)$, respectively, in the appropriate dimensions (cf. Figure 6.24). The lag time is again set at $\tau_L = 0.1$ [s]. Results are included in Figures 6.25 and 6.26, in which we have shown the vehicle path and the output variables steering angle, body slip angle, and lateral acceleration vs. time. Starting with 60 [km/h], one observes that the vehicle follows the path well, with a maximum path error of about 0.4 [m] just before the driver steers the vehicle back to the first lane. Steering angle and lateral acceleration are more or less in phase. There is some drifting (body slip).

Increasing the speed to 90 [km/h] leads to a significant path error increase, exceeding 0.9 [m]. In addition, drifting is more than doubled compared to the low speed case, which contributes to an increase in lateral acceleration (beyond 0.6 [g]). Also, observe the large lag between lateral acceleration and steering angle.

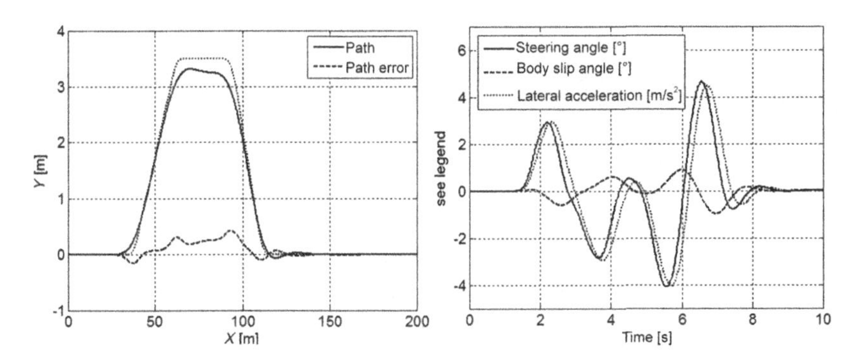

FIGURE 6.25 Vehicle path and vehicle behavior versus time for 60 km/h.

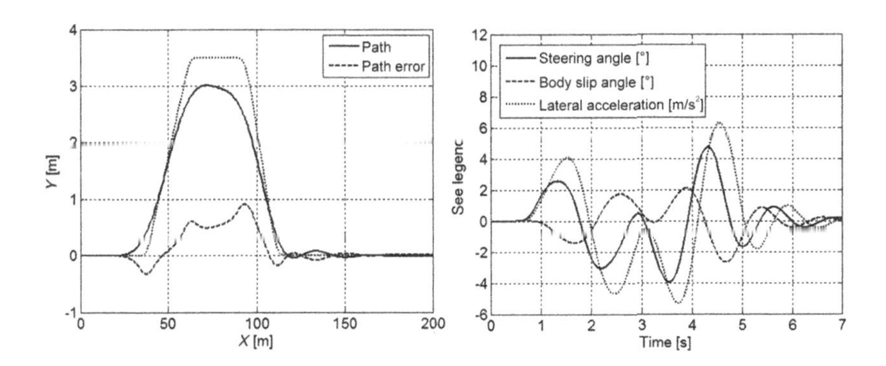

FIGURE 6.26 Vehicle path and vehicle behavior versus time for 90 km/h.

Clearly, the vehicle is no longer able to follow the steering input. Compare this with the frequency response plots in Chapter 5, where the yaw rate phase response was depicted for different speeds. It can be shown that the steering effort has increased for higher speed.

Thus far, we considered combinations of driver parameters (L_p, K_p) leading to a minimum path error, with a lag time equal to 0.1 [s]. Next we determine the vehicle path for the parameter sets listed in Table 6.5 if we move along the (L_p, K_p)-curve of Figure 6.19, or reduce the steering gain for the same preview length.

The reference case is chosen for a vehicle speed of 70 [km/h] (see Figure 6.19). The cases A, B, and C are indicated in the figure to the right in Table 6.5. Case D corresponds to an increase in the lag time. The results are shown in Figures 6.27–6.29. In Figure 6.27, one observes a later response due to the smaller preview length, which could be interpreted as a larger delay because it takes the driver longer to respond when approaching a change in the path.

One also observes oscillations in the path, indicating a reduced stability. Comparing this to Figure 6.11, which shows a loss of stability beyond a certain minimum level for the time delay, this makes sense. A larger preview

TABLE 6.5 Different Driver Parameter Sets to Determine the Vehicle Path

Case	L_p [m]	K_p [rad/m]	τ_L [s]
Reference	17	0.039	0.1
A	10	0.133	0.1
B	25	0.017	0.1
C	17	0.023	0.1
D	17	0.039	0.2

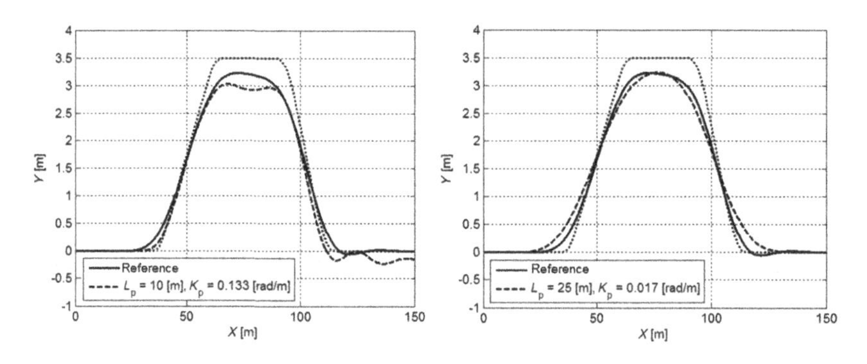

FIGURE 6.27 Vehicle path for low and high preview length, respectively, with the steering gain satisfying Eq. (6.19).

length shows an earlier response for both lane changes (moving to the second lane and back to the original lane), as well as a reduced overshoot. Reducing the gain and moving away from the ideal (L_p, K_p) curve shows a significant lag in Figure 6.28, as well as a reduced overshoot. It takes longer to build up a significant steering response, which also results in a larger integrated path deviation, cf. Eq. (6.24). Finally, an increased lag time leads to a more antisymmetric path and more overshoot (as shown in Figure 6.29).

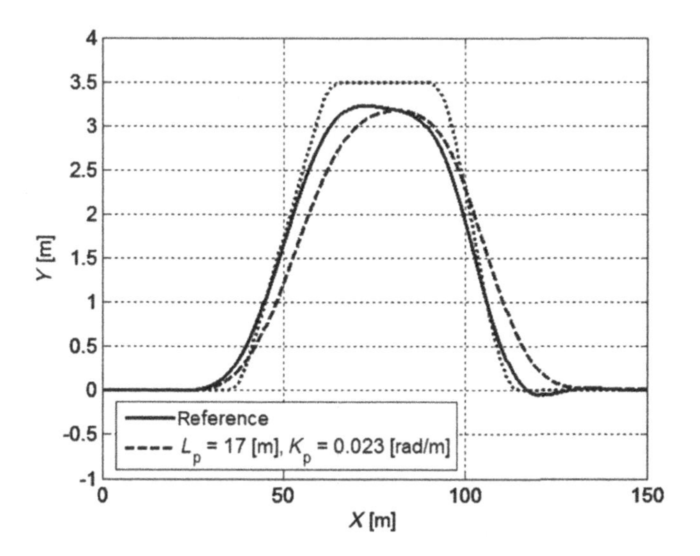

FIGURE 6.28 Vehicle path for reduced steering gain and unchanged preview length.

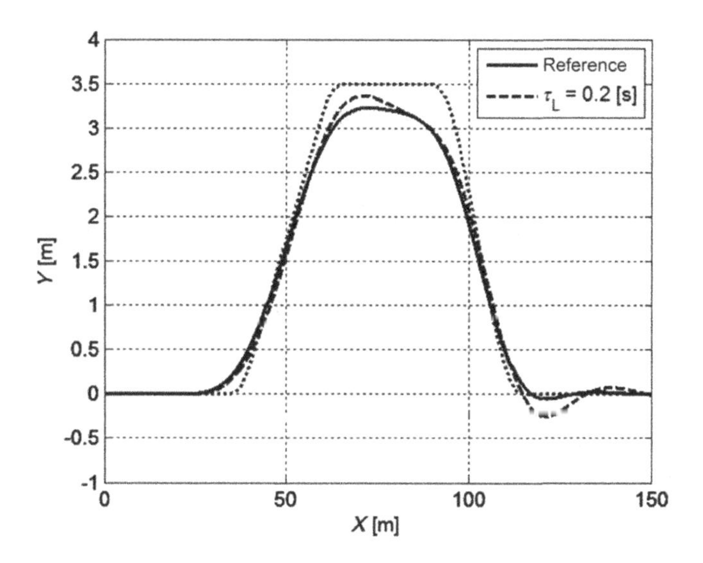

FIGURE 6.29 Vehicle path for increased delay time, $\tau_L = 0.2$ s.

6.4.2 Closed-Loop Handling Stability

The stability of the closed-loop problem (6.23) has been examined by Abe [1], assuming equal axle stiffness at front and rear. He arrived at a stability area in the (L_p, K_p) plane, described by

$$K_p \leq \frac{t_r + \tau_L}{t_r \cdot \tau_L} \cdot \frac{1}{L_p \cdot V} \cdot \left(1 - \frac{V \cdot (t_r + \tau_L)}{L_p}\right) \tag{6.25}$$

with

$$t_r = \frac{m \cdot V}{2 \cdot C_{axle}} \tag{6.26}$$

for axle cornering stiffness C_{axle}. Inequality (6.25) states that stability is lost if either the steering gain is too high or the preview length is too small. Another consequence of Eq. (6.25) is that, for a large preview length, the steering gain must be small to guarantee stability. A larger speed results in a smaller stability area, and the minimum preview length for which stability can be expected is the value where the right-hand side of the inequality (6.25) changes sign, i.e., when

$$L_{p,min} = V \cdot \left(\tau_L + \frac{m \cdot V}{2 \cdot C_{axle}}\right)$$

It was discussed in Ref. [38] that Eq. (6.25) might underestimate the critical gain values (at the boundary of the stability area). The stability of Eq. (6.23) is lost if the eigenvalues of the coefficient matrix have positive real parts. The corresponding characteristic equation is a fifth-order polynomial

$$a_5 \cdot \lambda^5 + a_4 \cdot \lambda^4 + a_3 \cdot \lambda^3 + a_2 \cdot \lambda^2 + a_1 \cdot \lambda + a_0 = 0 \tag{6.27}$$

with

$$a_0 = \frac{L \cdot C_{\alpha 1} \cdot C_{\alpha 2}}{m \cdot J_z \cdot \tau_L} \cdot K_p$$

$$a_1 = \frac{L \cdot C_{\alpha 1} \cdot C_{\alpha 2}}{m \cdot J_z \cdot V \cdot \tau_L} \cdot K_p \cdot (L_p + b)$$

$$a_2 = \frac{a \cdot C_{\alpha 1} \cdot K_p \cdot L_p}{J_z \cdot \tau_L} + \frac{C_{\alpha 1} \cdot K_p}{m \cdot \tau_L} + \frac{L^2 \cdot C_{\alpha 1} \cdot C_{\alpha 2}}{m \cdot J_z \cdot V^2 \cdot \tau_L} + \frac{N_\beta}{J_z \cdot \tau_L}$$

$$a_3 = \frac{L^2 \cdot C_{\alpha 1} \cdot C_{\alpha 2}}{m \cdot J_z \cdot V^2} + \frac{N_\beta}{J_z} - \frac{1}{\tau_L} \cdot \left(\frac{Y_\beta}{m \cdot V} + \frac{N_r}{J_z}\right)$$

$$a_4 = \frac{1}{\tau_L} - \frac{Y_\beta}{m \cdot V} - \frac{N_r}{J_z}$$

$$a_5 = 1$$

Note that the last three coefficients do not depend on the driver model parameters L_p and K_p. The coefficients a_0, a_1, and a_2 are proportional to the reciprocal lag time t_L^{-1}. Loss of stability occurs where the eigenvalues λ move into the right-hand part of the complex domain, i.e., when $\lambda = i \cdot \mu$. Substitution into Eq. (6.27), and splitting the equation into the real and imaginary parts, leads to two equations in μ^2:

$$\mu^4 - a_3 \cdot \mu^2 + a_1 = 0 \tag{6.28}$$

$$a_4 \cdot \mu^4 - a_2 \cdot \mu^2 + a_0 = 0 \tag{6.29}$$

If we multiply Eq. (6.29) by a_4 and subtract Eq. (6.29) from Eq. (6.28), we arrive at

$$\mu^2 = \frac{a_1 \cdot a_4 - a_0}{a_3 \cdot a_4 - a_2}$$

If we multiply the Eqs. (6.28) and (6.29) by a_0 and a_1, respectively, and subtract the resulting equations, we arrive at

$$\mu^2 = \frac{a_0 \cdot a_3 - a_1 \cdot a_2}{a_0 - a_1 \cdot a_4}$$

In order for both expressions for μ^2 to hold, the following equality must hold:

$$(a_0 - a_1 \cdot a_4)^2 + (a_0 \cdot a_3 - a_1 \cdot a_2) \cdot (a_3 \cdot a_4 - a_2) = 0 \tag{6.30}$$

which describes the boundary of the stability area in the (L_p, K_p) plane. One solution of Eq. (6.30) is found for $K_p = 0$, but it is not a relevant condition. The nontrivial stability boundary has been derived from Eq. (6.30) for two different speeds and two different lag times, using the same vehicle parameters as in Figures 6.19 and 6.20. The results are shown in Figure 6.30, with the area of stability lying beneath the curves.

Large gains or small preview lengths lead to a loss of stability, which confirms the stability results expressed in Figure 6.11, interpreting a preview length in terms of delay. Observe the nonlinear shape of the stability boundary, with the largest acceptable K_p range not far from the minimum preview length. From this figure, it is also clear that the stability is significantly reduced with increasing speed. Increasing speed means that the driver should either look further ahead of his vehicle or reduce his steering gain. Increasing the lag time also reduces the stability. To compensate for a larger τ_L, the driver should observe variations on the path earlier, i.e., have a larger value of the preview length.

When we combine the optimal path following (L_p, K_p) characteristics, expressed by Eq. (6.19), with the stability area for different speeds, one

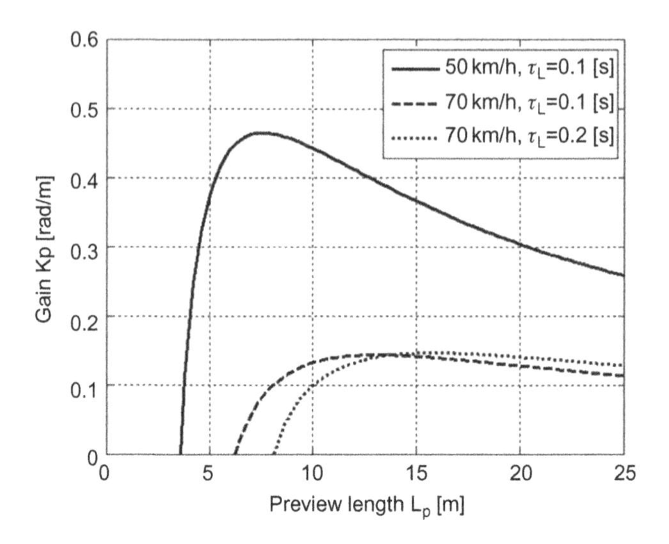

FIGURE 6.30 Stability areas for different speeds and lag times, in terms of driver gain K_p and preview length L_p.

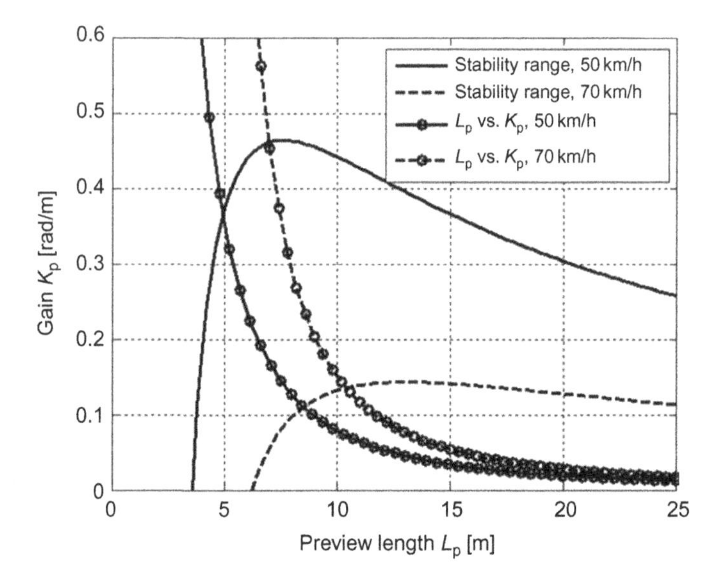

FIGURE 6.31 Stability areas and optimal path following characteristics for different speeds.

observes from Figure 6.31 that combinations of preview length and steering gain exist for which the closed-loop vehicle–driver system becomes unstable. The intersection of the optimal hyperbolic path following curve with the stability boundary determines the combination (minimum preview length, maximum gain) where stability is lost. Increasing the speed from 50 to 70 [km/h] means that (according to Figure 6.31) the minimum preview length is increased from approximately 5 to more than 10 [m], with the maximum steering gain reduced from about 0.37 to about 0.14 [rad/m].

The closed-loop stability boundary depends on the vehicle parameters. To obtain a better understanding of this dependency, we determined an explicit expression for the value $L_{p,min}$ that is the intersection of the stability boundary with the L_p-axis, i.e., for $K_p = 0$. It can be shown that, by dividing Eq. (6.30) by K_p and setting $K_p = 0$, this value can be expressed as follows:

$$L_{p,min} = -b + V \cdot \tau_L \cdot \left[1 - \frac{1}{\tau_L \cdot \omega_0^2} \cdot \left(\frac{Y_\beta}{m \cdot V} + \frac{N_r}{J}\right)\right] \tag{6.31}$$

with the undamped natural yaw frequency ω_0 introduced in Eq. (5.48) in Chapter 5 [under the simplification (5.10)], and given by

$$\omega_0^2 = \frac{C_{\alpha 1} \cdot C_{\alpha 2} \cdot L^2}{m^2 \cdot J \cdot V^2} + \frac{N_\beta}{J}$$

This frequency plays an important role in the open-loop stability, as discussed in Chapter 5. For an oversteered vehicle, ω_0 approaches zero for the speed close to the critical speed. Consequently, $L_{p,min}$ becomes unbounded, which means that closed-loop stability is also lost. Increasing $L_{p,min}$ corresponds to a reduced stability area and Eq. (6.31) can now be used for further analysis of the sensitivity of the closed stability, with respect to the vehicle parameters such as inertias and axle cornering stiffnesses. We reduced the front and rear axle cornering stiffnesses by 30% (see Figure 6.32). As expected, a

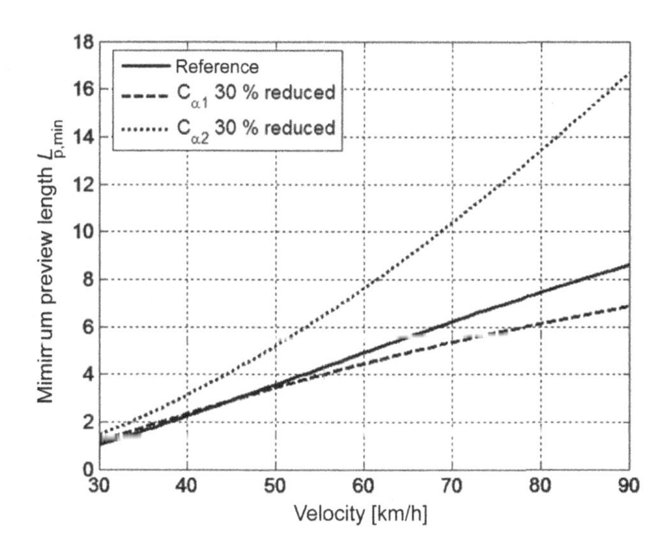

FIGURE 6.32 Minimum preview length $L_{p,min}$ versus vehicle versus vehicle speed for different axle cornering stiffnesses.

lower $C_{\alpha 1}$ leads to better stability beyond a certain speed, whereas a lower $C_{\alpha 2}$ has the opposite effect. These effects become stronger for larger speed.

6.4.3 Driver Model and Driver State Identification

In the preceding sections, the closed-loop vehicle driver behavior has been considered for fixed driver model parameters. The discussion on optimal path following model parameters suggests that such parameters can vary, as long as the preview length and the steering gain are related through a hyperbolic relationship. However, drivers may not follow this relationship precisely at the cost of a larger path deviation. We have seen that, even for this relationship of minimal path deviation to hold, the preview length should not be too small to avoid loss of closed-loop stability.

In this section, the driver model parameters are allowed to vary during normal driving, where we restrict ourselves to the parameters L_p and K_p, i.e., we choose τ_L to be constant.

We expect the model parameters to be, in some way, related to the driver state, and specifically to experience and workload. This means that on-line identification of L_p and K_p helps us to interpret the driver state and contribute to the improvement of driver support systems. Note that the lag time τ_L can only be identified under nonsteady-state conditions, which is one of the reasons we selected a fixed value for τ_L.

Starting from the path-tracking model (6.15), parameter identification requires information about the intended path, as indicated in Figure 6.17. This path is usually not available, and it cannot be derived from onboard data. The previewed path deviation in Eq. (6.15) consists of two parts: a path deviation due to differences in yaw orientation between vehicle and path and the present path deviation $y(t)$. The first part of the deviation is the dominant one in most situations with significant yaw motion, such as in cornering or changing lanes. However, these are precisely the cases when driver steering control is substantial and the driver model parameters are relevant.

Instead of basing driver behavior on the intended path, we use the realized path and neglect the local lateral displacement $y(t)$ in Eq. (6.15). This affects driver behavior, which has been examined by Pauwelussen and Patil [39]. They used a driver model with model parameters K_p and L_p that depend on the anticipated lateral acceleration in the forthcoming transitions in path curvature, with specified boundaries in, and desired value of driver preview time. A path was selected that consisted of several cornering conditions and a double lane change. The preview length was dependent upon the driver's speed, with vehicle acceleration and deceleration determined using desired speed and the same anticipated lateral acceleration. By using this model for a certain intended path, and then identifying the driver parameters in time from the realized vehicle path, the effect of neglecting the lateral displacement $y(t)$ was determined. Three different model settings (i.e., three different drivers) were applied. The first reference driver was assumed to vary the preview time between 0.5 and 2.0 [s], whereas the preview time for the second driver was bounded by 0.3 and 1.2 [s]. Hence, the second driver would respond later and the vehicle-driver system would possibly be less stable while approaching a maneuvering situation. The third driver was different from the first one in the sense that the lag time τ_L was reduced from 0.2 to 0.1 [s]. The conclusions in Ref. [39] were as follows:

- The delay time has only a limited effect on the identified preview length and steering gain.
- The global trend in model parameters was confirmed using the identified values but with a reduced variation. Apparently, using the real vehicle path has a filtering effect on the driver model parameters.
- The distinction of the three different drivers was maintained qualitatively, both in mean and in standard deviation (of the L_p variation along the path).

By neglecting the local lateral displacement $y(t)$, Eq. (6.15) reduces to

$$\tau_L \cdot \dot{\delta}(t) + \delta(t) = K_p \cdot D_p(t; L_p) = K_p \cdot L_p \cdot (\psi_p(t; L_p) - \psi(t)) \tag{6.32}$$

The parameters L_p, K_p are determined during practical driving conditions by matching the solution of Eq. (6.32) with the real steering angle $\delta_{test}(t)$ over finite time intervals (t_i, $t_i + \tau_{int}$) with interval length τ_{int}. This means that the identified values of L_p and K_p will give a minimum for the following functional:

$$\mathrm{Error}_\delta(L_p, K_p; \tau_L) = \sqrt{\int_{t_i}^{t_i+T_{int}} (\delta_{test}(t) - \delta(t))^2 \cdot \mathrm{d}t} \tag{6.33}$$

where $\delta(t)$ is the solution of Eq. (6.32):

$$\delta(t) = \frac{1}{\tau_L} \cdot \int_{t_i}^{t} \mathrm{e}^{-\frac{t-s}{\tau_L}} \cdot K_p \cdot D_p(s; L_p) \cdot \mathrm{d}s + \delta_{test}(t_i) \cdot \mathrm{e}^{-\frac{t-t_i}{\tau_L}} \tag{6.34}$$

The intervals may overlap. Choosing $t_i = i \cdot \Delta t$ for $i = 1, 2, 3, \ldots$, minimization of Eq. (6.33) leads to values L_p (t_i) and K_p (t_i). The accuracy depends on the choice of Δt and T_{int}, where we typically choose $\Delta t = 0.5$ [s] and $T_{int} = 3$ [s].

Following this approach, a problem arises with numerical efficiency. Minimization of Eq. (6.34) can be realized using Newton iteration, where the preview length varies and the path error must be determined from this preview length. This part of the analysis requires an additional iteration. A more efficient way to proceed is to first estimate the path deviation by a polynomial expression as follows:

$$D_p(t; L_p) = \sum_{j=0}^{N} a_j(t) \cdot L_p^j \tag{6.35}$$

This can easily be completed using interpolation from $N+1$ selected points on the path ahead of the vehicle. A value $N=3$ or 4 gives good results. Consider Figure 6.17 and note that the points T (on the path), A (preview point), and the vehicle CoG are on a circle that has the distance between the vehicle CoG and point T as the diameter. Hence, the position of point A can easily be derived from the vehicle position and orientation and the point T. Substituting Eq. (6.35) into Eq. (6.34) gives an explicit expression for the root mean square error (6.33) in terms of L_p and K_p, from which the minimum can be efficiently determined.

Tests were completed on a public road using an instrumented midsize test vehicle. These tests were completed by four different drivers: two experienced ones, and two inexperienced drivers (students). The vehicle parameters and the lag time τ_L are listed in Table 6.6.

All drivers drive around for about one hour, collecting data from cornering events, lane changes, roundabouts, etc. Such data included yaw rate, lateral acceleration, speed, GPS, and steering angle. Though the body slip angle was measured, we based our body slip angle data on a special observer from the match between model end test results for yaw rate and lateral acceleration. This has the advantage of being independent of body slip angle measurement data without losing accuracy in the assessment of this angle. One specific test part was a closed area at a nearby industrial and office area (see Figure 6.33),

TABLE 6.6 Vehicle Parameters and Lag Time τ_L

m [kg]	J_z [kg m^2]	$C_{\alpha 1}$ [N/rad]	$C_{\alpha 2}$ [N/rad]	a [m]	b [m]	τ_L [s]
1776	2750	88,800	124,000	1.508	1.252	0.2

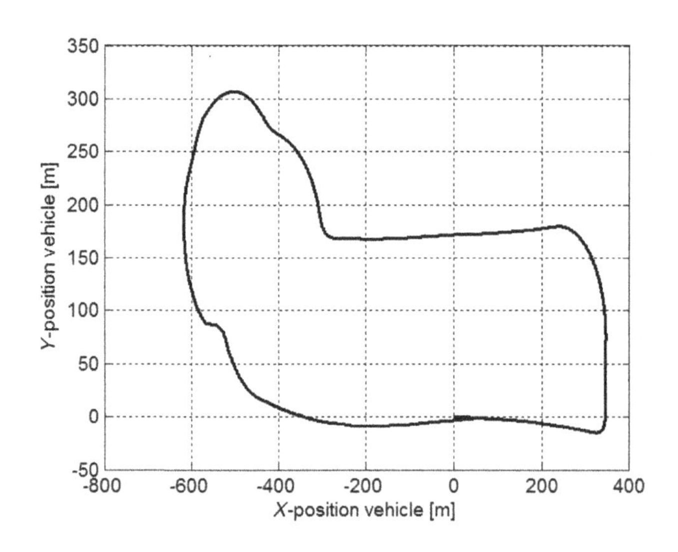

FIGURE 6.33 Vehicle path.

being driven clockwise and counterclockwise, for a total duration of about 2 min. Here we use the results from the counterclockwise test. Note that part of this path is straight, leading to larger preview length. We set a maximum to the preview length of 30 [m], which means that every L_p value larger than 30 [m] is set as 30 [m]. The maximum lateral acceleration on this closed path was 5.2 [m/s^2] for the experienced drivers and the speed varied between 8 and 18 [m/s]. For the inexperienced drivers, the maximum lateral acceleration was 4.5 [m/s^2] and the speed ranged between 6 and 14 [m/s].

The time history of the preview length in time for the experienced driver is shown in Figure 6.34. The preview length appears to vary between 5 and 17 [m], with a maximum value of 30 [m] for some (straight) parts of the path. This time history does not give much information on the actual driver behavior. Closer examination of the vehicle and driver results indicates reduction in preview length while approaching a certain transition in lateral acceleration, being restored to larger values when the driver is close to this transition (and beyond). Note that the preview length also depends on speed. Instead of this behavior in time, it is more interesting to plot the preview time T_p versus the steering gain K_p for both types of drivers. As we discussed previously (see Figures 6.19 and 6.21), this use of T_p is more robust with respect to speed than the optimal hyperbolic (L_p, K_p) relationship. Results are shown in Figure 6.35.

One recognizes the typical hyperbolic relationship in both plots, where the experienced driver is very consistent in behavior, whereas the inexperienced driver deviates further from this ideal behavior.

We also determined the frequency distributions for the preview time, preview length, and steering gain for both drivers. Results for the preview time are shown in Figure 6.36. The experienced driver clearly shows, in average, a higher preview time, and therefore a lower steering gain, compared to the

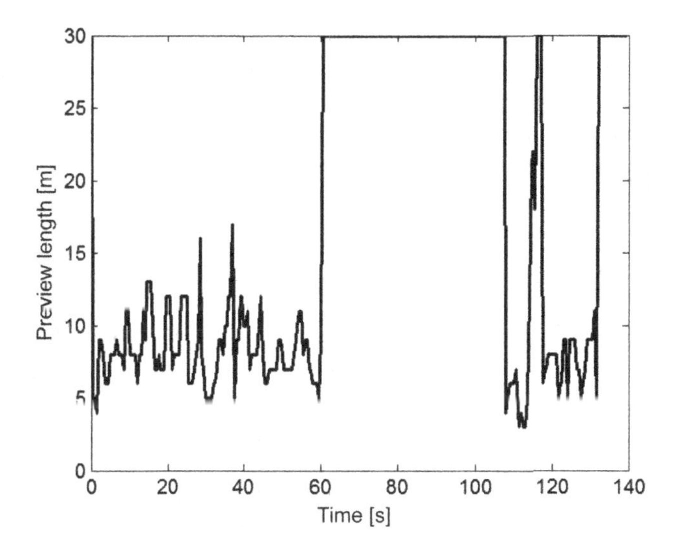

FIGURE 6.34 Preview length versus time for the experienced driver.

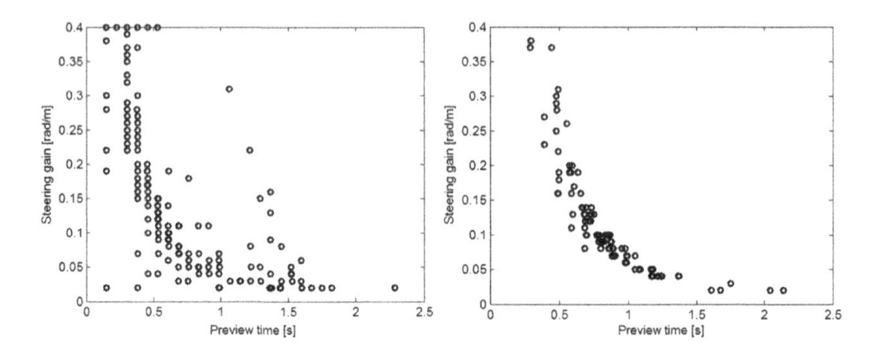

FIGURE 6.35 Preview time versus steering gain for the inexperienced driver (left) and the experienced driver (right).

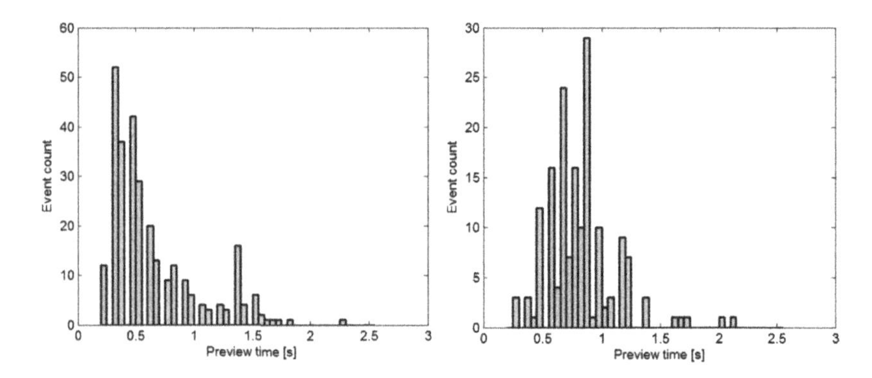

FIGURE 6.36 Frequency distributions for preview time for the inexperienced driver (left) and the experienced driver (right).

inexperienced driver. This suggests a higher workload for the inexperienced driver. This is further supported by SRR results, also presented in Ref. [39]. The largest values were obtained for the inexperienced driver.

When the inexperienced driver repeats the same path several times, the driver becomes more experienced on this specific path, and one would expect the frequency distribution to shift to larger values of T_p. This tendency was observed from the test results, which could be interpreted as a more feed-forward path-tracking performance.

We close this section with some remarks regarding the scientific value of these results. So far, some results are shown for only a limited number of drivers, and driving tests under practical traffic circumstances. Drawing general conclusions would therefore be incorrect. Nevertheless, it has been shown that driver model parameters can be derived under practical traffic conditions, that the results confirm earlier observations regarding the relationship between preview length and steering gain, and that distinction is obtained for different drivers.

Chapter | Seven

Exercises

7.1 EXERCISES FOR CHAPTER 2

Question 1

1.a Give *exact definition* of each of the following tire properties and input/output quantities:

- Effective rolling radius for a free rolling tire R_e
- Slip angle α
- Camber angle γ
- Practical longitudinal slip κ
- Longitudinal slip speed V_{sx}

1.b The rolling resistance force depends on many properties. Indicate in the next table with +, o, or − whether the rolling resistance force will increase, stay the same, or decrease. Justify your answer.

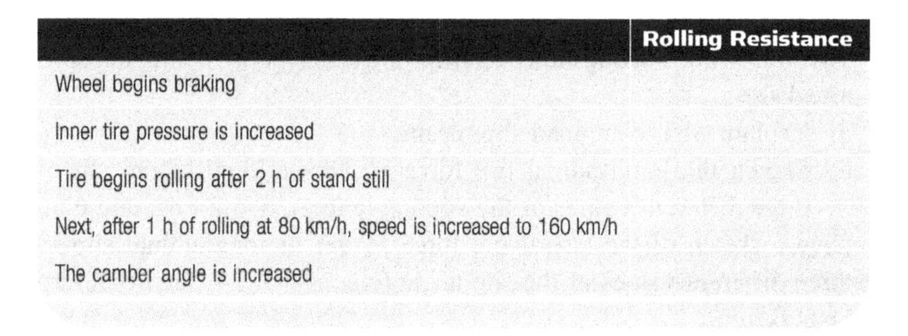

	Rolling Resistance
Wheel begins braking	
Inner tire pressure is increased	
Tire begins rolling after 2 h of stand still	
Next, after 1 h of rolling at 80 km/h, speed is increased to 160 km/h	
The camber angle is increased	

1.c We sketched a tire under free rolling conditions. We know that the peripheral speed of points at the circumference of the tire varies between ΩR and ΩR_l for unloaded and loaded radii R and R_l, respectively, and circumferential speed Ω. A third speed level is given by ΩR_e.

Essentials of Vehicle Dynamics.

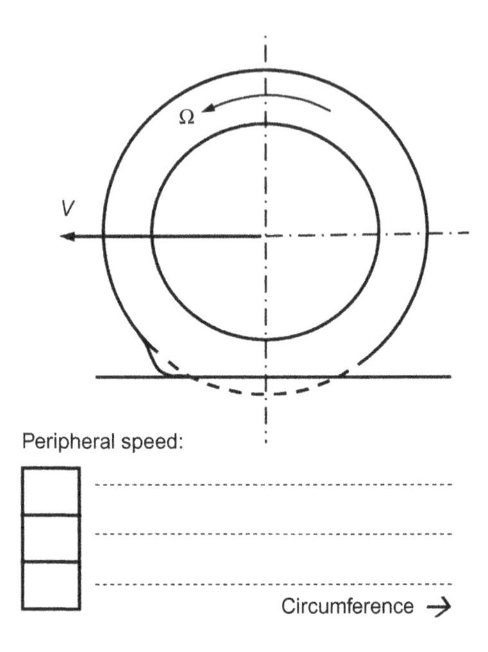

In the preceding figure, clearly indicate the behavior of the peripheral speed of the tire circumference as it approaches the contact area, passes through the contact area, and leaves the contact area.

Indicate in the boxes the peripheral speeds ΩR, ΩR_{l}, and ΩR_{e} to distinguish which values the peripheral speed is varying between.

1.d The following subquestions address the concept of a tire under combined slip.

  i. Explain what combined slip means.

  ii. Sketch the longitudinal tire force F_x versus longitudinal slip κ for three different values of the slip angle (e.g., $\alpha = 2°$, $4°$, and $6°$).

1.e Give a sketch of the lateral tire force F_y versus longitudinal slip κ, for three different values of the slip angle (e.g., $\alpha = 2°$, $4°$, and $6°$). Explain your sketch.

1.f Polar plot: Use the sketches from exercise 1.d and 1.e, plot lateral tire force F_y versus longitudinal tire force F_x for different (discrete) values of the slip angle.

1.g In the next figure, we plotted the aligning torque versus longitudinal tire force for varying longitudinal slip. Different lines correspond to different values of the slip angle α. Explain why this aligning torque changes sign if we move from a large brake force to a large driving force, as shown in this figure.

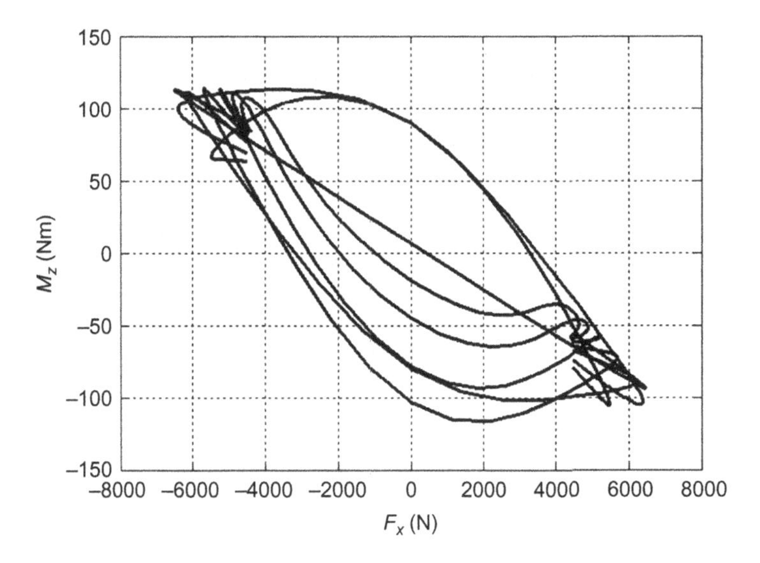

Question 2

2.a Tire design is a compromise of many output properties. Three of these properties (handling performance, rolling resistance, and comfort) are included in the following table, as well as three possible design/service adjustments to improve them.

	Handling Performance	Rolling Resistance	Comfort
Higher inner pressure			
Stiffer belt plies			
More hysteresis in the tread area rubber compound			

Indicate with a +, o, or − sign whether the specific tire performance aspect is positively influenced, not affected, or has a negative effect. Justify your answer.

2.b We have included a top-down view of a tire under a side force in which we consider, for the time being, pure lateral slip. We have also indicated the simple parabolic normal stress behavior in the contact area.

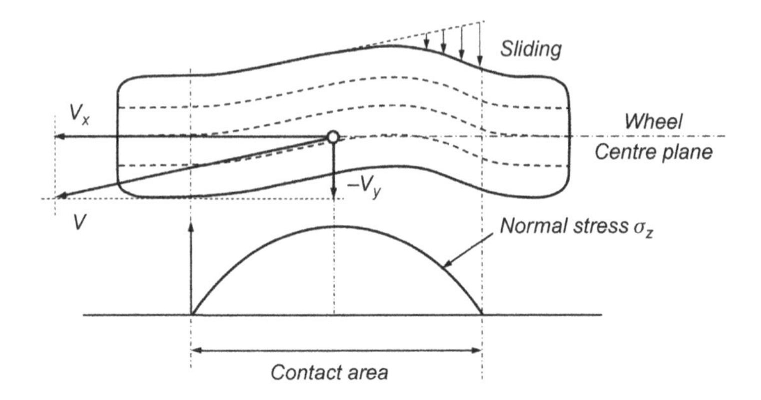

i. Please clearly indicate in the preceding top-down view figure:
  - Practical lateral slip α
  - Lateral force F_y
  - Pneumatic trail t_p
  - Aligning torque M_z

ii. Sketch the lateral shear stress pattern along the contact area in the same graph as the normal stress has been indicated. Explain your sketch.

2.c The Magic Formula tire model for the lateral force is described by the following mathematical expression:

$$F_y = D \cdot \sin[C \cdot \arctan\{B \cdot \alpha - E \cdot (B \cdot \alpha - \arctan(B\alpha)\}]$$

for slip α. We neglected the horizontal and vertical shifts and varied three of the four Magic Formula factors B, C, D, and E. Mark (X) in the following table to indicate which curve corresponds to the variation of which factor. *Each column should only contain one mark.*

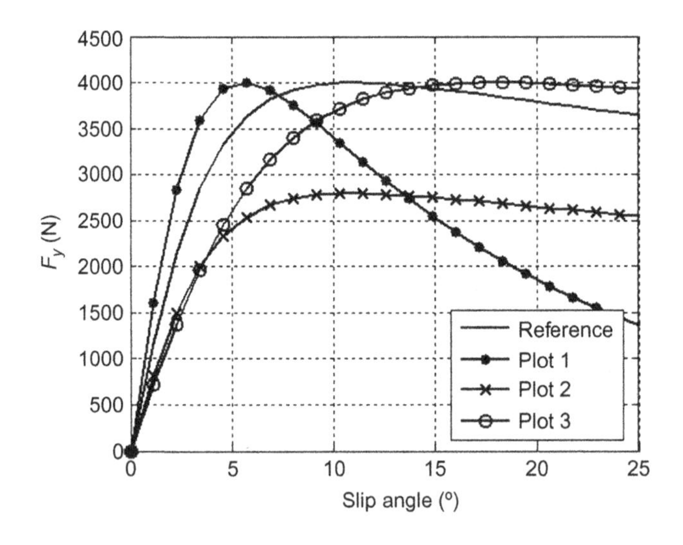

	Plot 1	Plot 2	Plot 3
Change of B			
Change of C			
Change of D			
Change of E			

Question 3

3.a The Magic Formula (Pacejka) tire model for the longitudinal force is described by the following mathematical expression:

$$F_y = D \cdot \sin[C \cdot \arctan\{B \cdot \kappa - E \cdot (B \cdot \kappa - \arctan(B\kappa)\}]$$

for pure longitudinal (brake/drive) slip κ. Give the *exact* definition of practical longitudinal slip.

3.b Give the values for the practical longitudinal slip for a locked wheel or excessive spinning wheel in case of braking and driving, respectively.

3.c The behavior of longitudinal slip versus tire brake force (pure slip) is shown in the next figure.

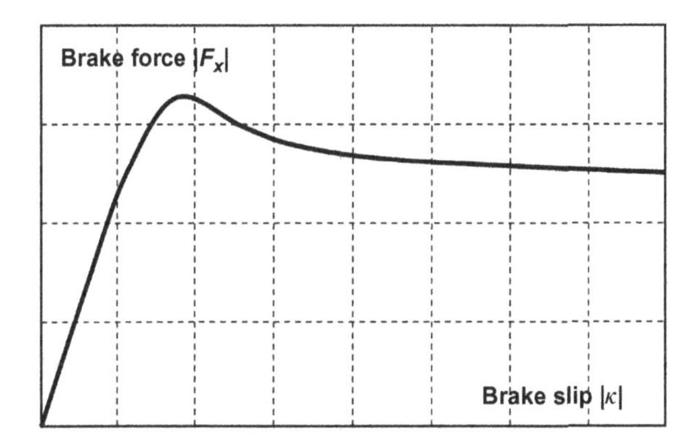

Indicate in this figure:

- The longitudinal force versus slip for road friction reduced by 50%
- The longitudinal force versus slip in case of combined slip (braking combined with side slip)
- *The lateral force* versus longitudinal slip for a fixed slip angle (braking while cornering).

7.2 EXERCISES FOR CHAPTER 3

Question 1

1.a Consider the accelerating single wheel vehicle, shown next.

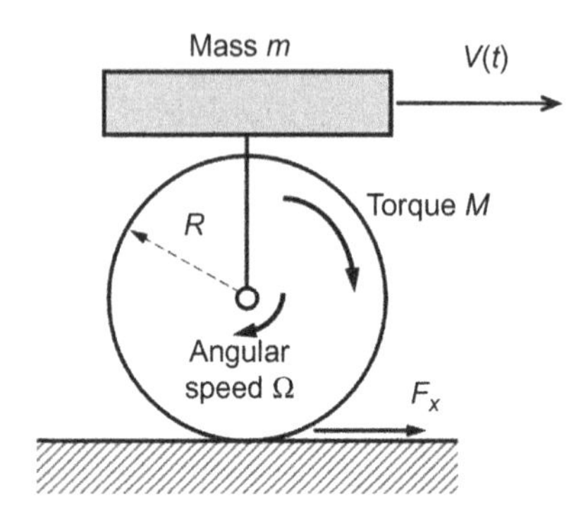

Use practical slip quantities, i.e., the stationary longitudinal slip

$$\kappa = \frac{\Omega \cdot R - V}{V}$$

The following symbols for parameters are used:

- Relaxation length σ_κ
- Total mass vehicle m
- Radius tire R
- Rotational wheel inertia J (proportional to R^3)
- Longitudinal slip stiffness tire C_κ

Use the following data:

- Relaxation length $\sigma_\kappa = 0.6$ m
- Longitudinal slip stiffness tire $C_\kappa = 75{,}000$ [N/rad] (linear tires)

Give the value of the longitudinal spring stiffness C_x of the tire in N/m.

1.b For a fast change in driving torque M, the slip in the contact area κ' will not be the same as the slip κ at the axle level. Give the differential equation in κ', describing the transient behavior of the tire in longitudinal direction.

1.c Assume that the speed V is varying very slowly, which means that we can take the speed V constant compared to tire slip and wheel speed. Then, the problem of a single wheel driven by a drive torque M is described by two equations, with tire slip κ' and rotational speed Ω. The equation in Ω is given by

$$J \cdot \dot{\Omega} = M - R \cdot F_x$$

Express this equation in terms of κ' and Ω assuming linear tire behavior.

1.d We have now derived two first-order equations. They can be transferred to a second-order equation in κ' or Ω. Derive this equation. Assume M constant.

1.e The system of the accelerating wheel has a rotational undamped eigenfrequency ω and a damping ratio ζ. Which of the following statements are true? Justify your answer.

- the eigenfrequency ω increases with C_κ
- the eigenfrequency ω increases with R
- the eigenfrequency ω increases with σ_κ
- the eigenfrequency ω increases with V
- the damping ratio ζ increases with V
- the damping ratio ζ increases with σ_κ

Question 2

2.a We consider a vehicle under cornering conditions, moving with constant speed V (only lateral slip), under transient tire behavior and assume linear tire behavior. The z-axis is pointing downward.

The following data are given:

Cog to front axle	: $a = 1.1$ [m]
Cog to rear axle	: $b = 1.5$ [m]
Mass vehicle	: 1600 [kg]
Normalized *axle* cornering stiffness, front	: 18
Normalized *axle* cornering stiffness, rear	: 15
Lateral spring stiffness for each *tire*	: $2 \cdot 10^5$ [N/m]

Is this vehicle understeered or oversteered? Justify your answer.

2.b Calculate the relaxation lengths σ_1 and σ_2 for the front and rear *axles*.

2.c With transient tire behavior, the slip angles between tires and road are unknown vehicle states, following from the slip angles at front and rear axles. Show that the front and rear axle slip angles satisfy the following equations:

$$\frac{\sigma_1}{V} \cdot \dot{\alpha}_1 + \alpha_1 = \delta - \beta - \frac{a \cdot r}{V}; \quad \frac{\sigma_2}{V} \cdot \dot{\alpha}_2 + \alpha_2 = -\beta + \frac{b \cdot r}{V}$$

with vehicle speed V, steering angle δ, body slip angle β, and yaw rate r.

2.d

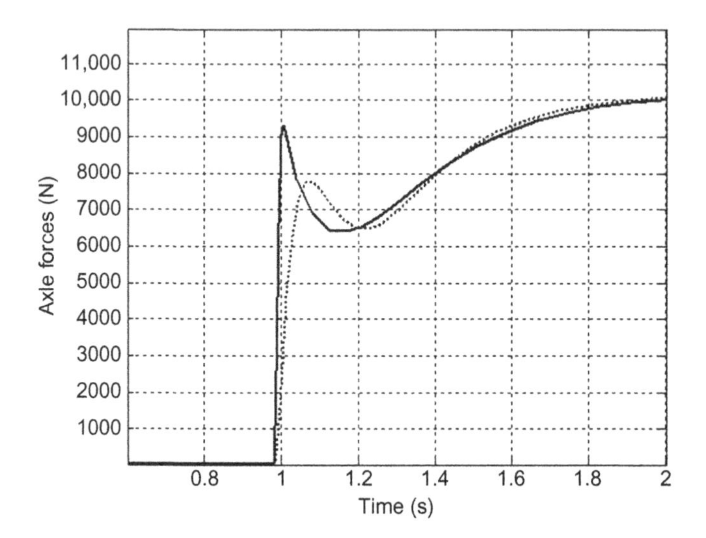

We have carried out a step steer analysis for a vehicle model with linear tires, with and without accounting for transient effects. Results for the front axle side force are shown in the preceding figure.

- Mark which curve has transient effects and which curve does not. Justify your answer.
- What will happen with the transient curve when we reduce the vehicle speed?

7.3 EXERCISES FOR CHAPTER 4

Question 1

1.a Explain what Ackermann steering means.

1.b See the top-down (planar) sketch of the low speed cornering situation for a tractor–semitrailer combination.

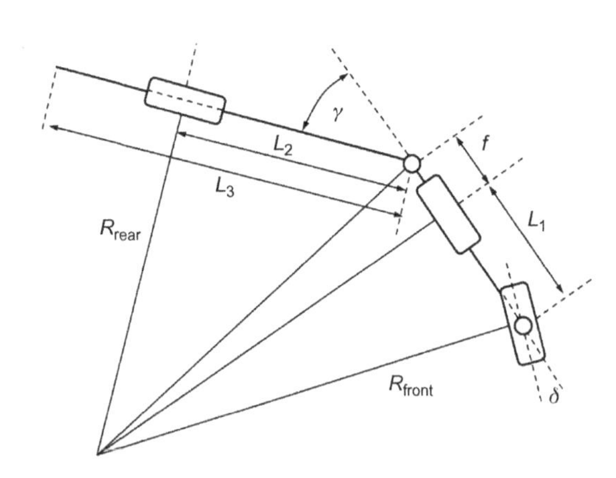

The following data are given:

$L_1 = 3.70$ [m]
$f = -0.60$ [m]
$L_2 = 7.5$ [m]
$L_3 = 11$ [m], distance from kingpin up to end of the semitrailer
$\delta = 4°$

Calculate the articulation angle γ.

1.c The track width of tractor and trailer is equal to 2 [m]. Give the value of the swept path (difference between inner and outer radius).

1.d What would happen to the swept path if you:

i. Choose the kingpin behind the rear tractor axle

ii. Increase the tractor wheel base L_1.

1.e To minimize the off-tracking, a rear axle steering angle δ_2 is applied such that $R_{rear} = R_1$ (see the following figure). Determine the ratio δ_2/δ_1 (the trailer axle gain).

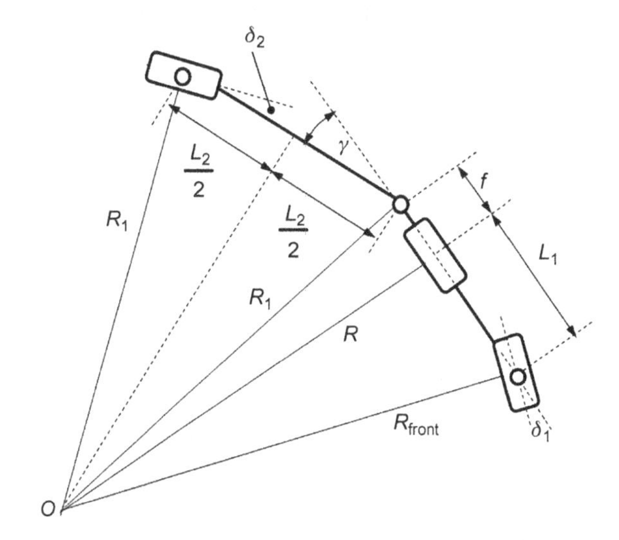

7.4 EXERCISES FOR CHAPTER 5

Question 1

1.a We consider a vehicle under a lateral force F_y (e.g., one initiated by a crosswind), as shown in the next figure.

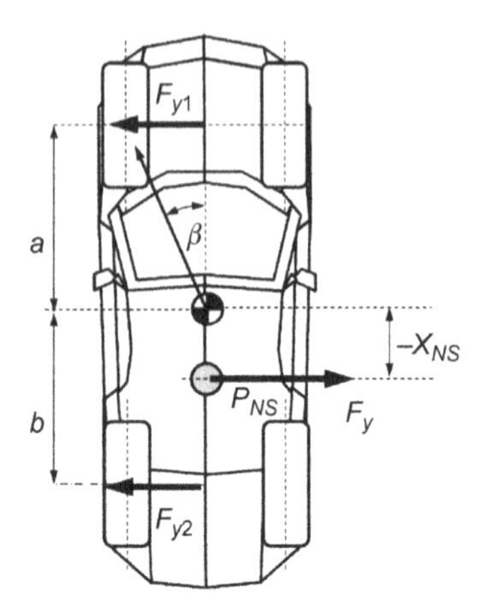

P_{NS} indicates the *neutral steer point*. The z-axis is taken upward.

i. The lateral force acts in the neutral steer point. How does this affect the resulting yaw behavior of the vehicle?

ii. The lateral force is counteracted by two reactions forces at the front and rear axles. Suppose linear tires, with cornering stiffnesses C_1 and C_2 for front and rear axle, respectively. Derive an expression of the neutral steer point in terms of parameters a and b and the cornering stiffnesses.

1.b We show two vehicles next, again under a lateral force F_y, now acting in the CoG, and resulting in a yaw response for vehicles A and B, as indicated. Which vehicle is understeered and which vehicle is oversteered? Justify your answer.

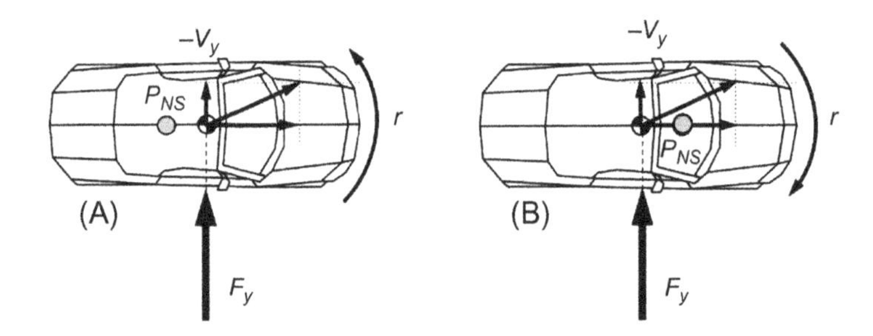

1.c An oversteered vehicle is driving on a straight flat road slope. At a certain point, the road is changed from flat to a transverse slope (see figure). The driver is keeping the steering angle fixed. Which of the following cases is true? Justify your answer.

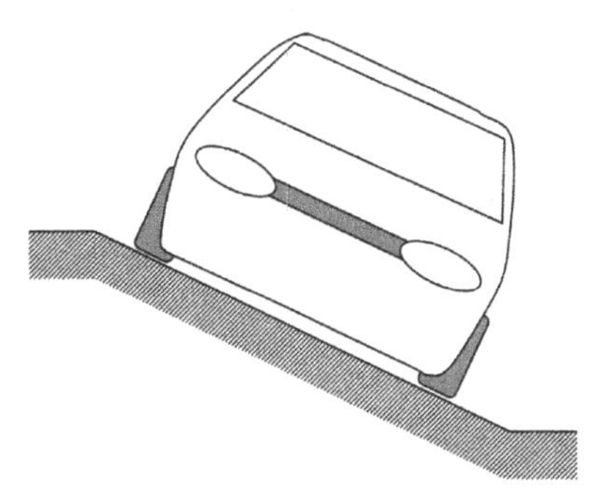

i. The vehicle will move up the slope.
ii. The vehicle will move down the slope.
iii. The vehicle will remain where it is.

1.d The relationship between lateral acceleration a_y and front axle steering angle δ under steady-state conditions on a flat road is given as follows:

$$\delta = \frac{L}{R} + \eta \cdot \frac{a_y}{g}$$

i. Give the meaning of L, R, and η.
ii. Derive an expression for the steady-state yaw rate gain from this expression.

1.e Suppose $\eta < 0$. One may sketch the steady-state yaw rate gain versus vehicle speed. In the next graph, we show three curves. Which one is the correct one? Justify your answer.

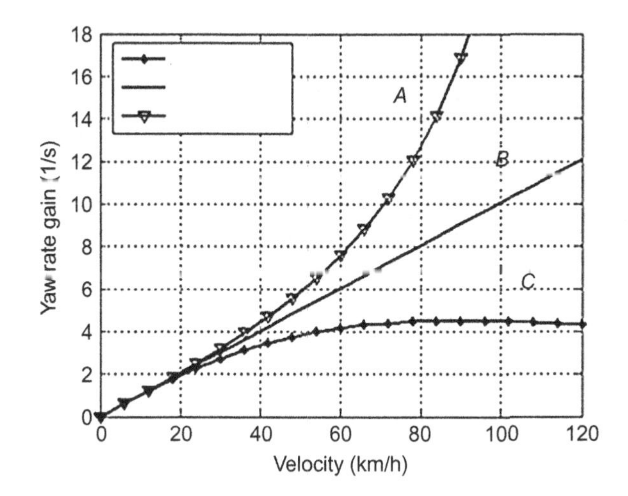

Question 2

2.a We have included four definitions of understeer behavior. Select the correct definition. Justify your answer.

i. A vehicle is understeered if the axle steering angle must be increased for increasing vehicle forward speed to negotiate the same curve.

ii. A vehicle is understeered if the rear axle slip angle exceeds the front axle slip angle under steady-state conditions: $\alpha_1 < \alpha_2$ (or, in more general terms, $|\alpha_1| < |\alpha_2|$).

iii. A vehicle is understeered if the understeer gradient $\eta > 0$.

iv. A vehicle is understeered if the rear axle normalized axle cornering stiffness is exceeded by the front axle normalized axle cornering stiffness.

2.b Provide the exact definition of the understeer gradient.

2.c A vehicle is driving a circular path with constant speed. At some moment, the driver decides to accelerate but keeps the steering wheel angle fixed. The vehicle tends to move out of the curve. Is the vehicle understeered or oversteered? Justify your answer.

2.d The driver decides to increase the roll stiffness at the front axle. Will the response as described under 2.c increase (i.e., moving more out of the curve) or decrease? Justify your answer.

Question 3

3.a In this exercise, the dimensions along the x- and y-axis are always in degrees ([°]). The figure shows the *energy phase plane* for a speed of 70 [km/h] and a front axle steering angle of 2 [°].

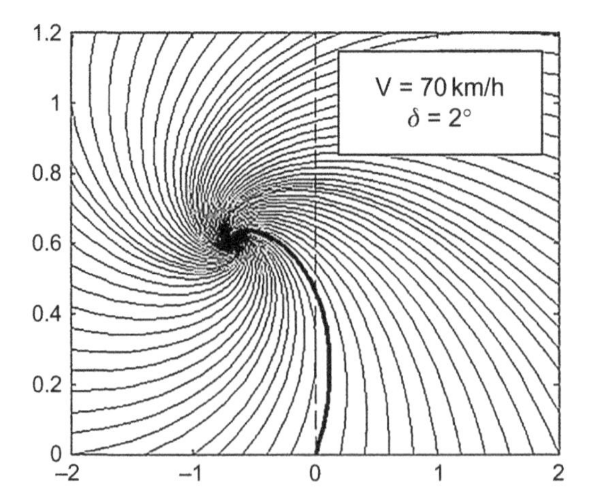

Explain what phase plane means.

3.b The *energy phase plane* is based on the notion that the ratio of cornering energy T_c and longitudinal kinetic energy T_k corresponds to the distance of a solution point to the origin.

- **i.** Define the x and y variables in terms of yaw rate and body slip angle.
- **ii.** Explain the drifting and yawing phase based on the solution curve in the above figure, passing through the origin.

3.c We have depicted three energy phase plane representations.

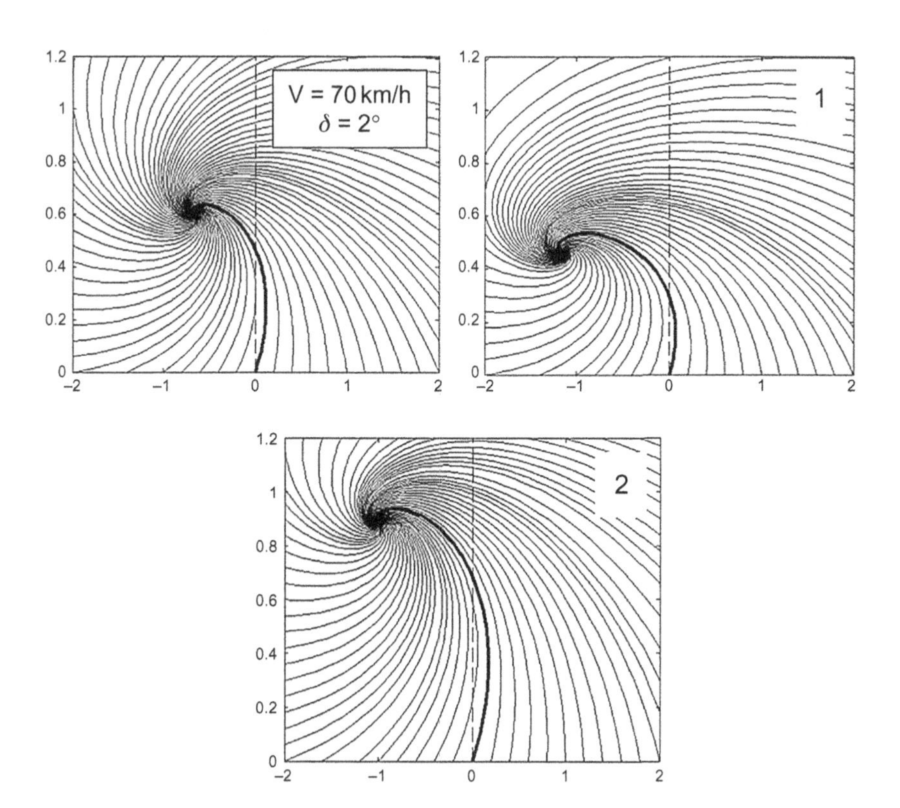

We have increased the speed in one figure and increased the steering angle in the other.

- **i.** Which figure shows the increased speed and which shows the increased steering angle? Justify your answer.
- **ii.** The steady-state solution in the middle figure corresponds to a radius of $R = 200$ [m]. Estimate the radius of gyration.

3.d The next figure shows the energy phase plane with the lines of constant angle α_2 (rear axle slip angle) and $\alpha_1 - \delta$ (difference of front axle slip angle and steering angle). The energy phase plane is divided in four areas, 1, …, 4.

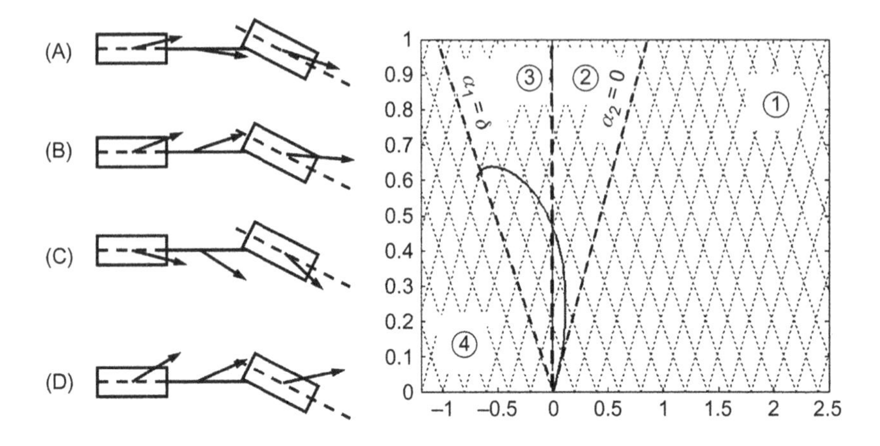

Which situation (A,. . .,D), indicated for the bicycle one-track vehicle model, corresponds to what area in the phase plane, and why?

3.e Next, we show three energy phase plane representations and three plots of *normalized axle characteristics*. Which axle characteristics plot corresponds to what phase plane and why?

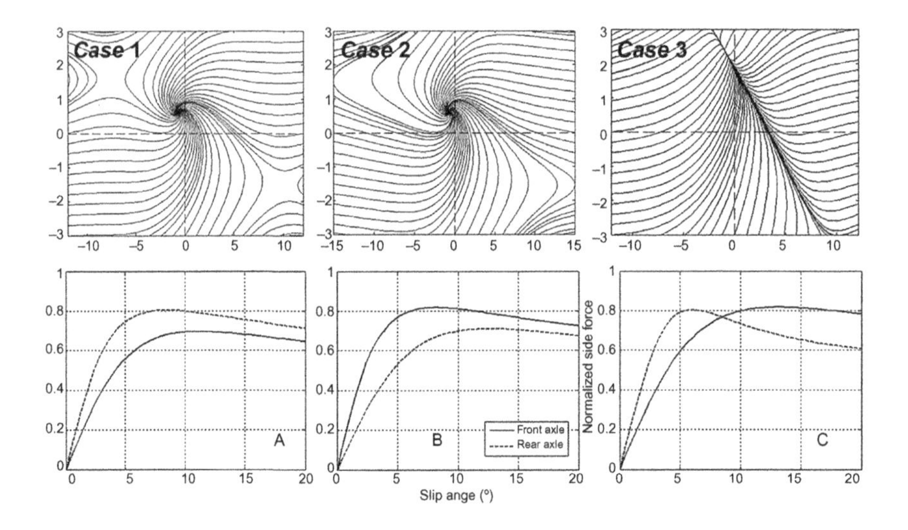

Question 4

4.a The next figure shows a handling diagram for a passenger car (z-axis downward) with a wheelbase of 2.5 [m].

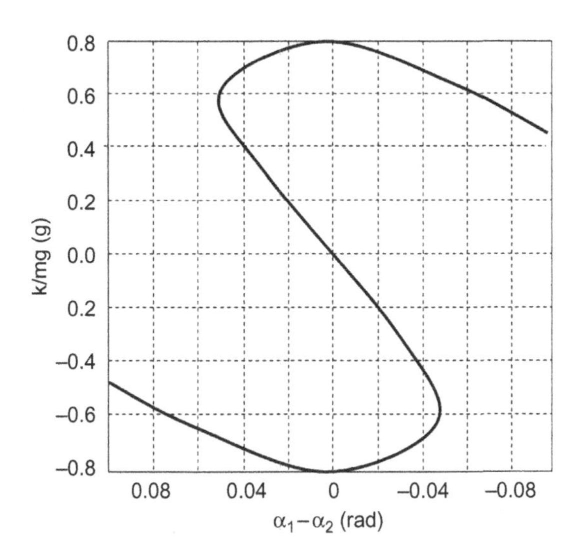

Sketch the *normalized* axle characteristics that lead to this type of handling diagram.

4.b Estimate the understeer gradient.

4.c Determine the steering angle *graphically* for the car moving in a steady-state circle with radius 50 [m] for two different speeds:

$V = 36$ km/h

$V = 54$ km/h

Take the acceleration of gravity as $g = 10$ m/s^2.

4.d Assume a front axle steering angle of 0.06 [rad]. Determine *graphically* the curve radius for a speed of 54 [km/h].

4.e Below, four phase (energy) planes are shown. Two of them are connected to the preceding handling diagram for $V = 36$ and 54 [km/h]. Which one corresponds to which speeds? Justify your answer.

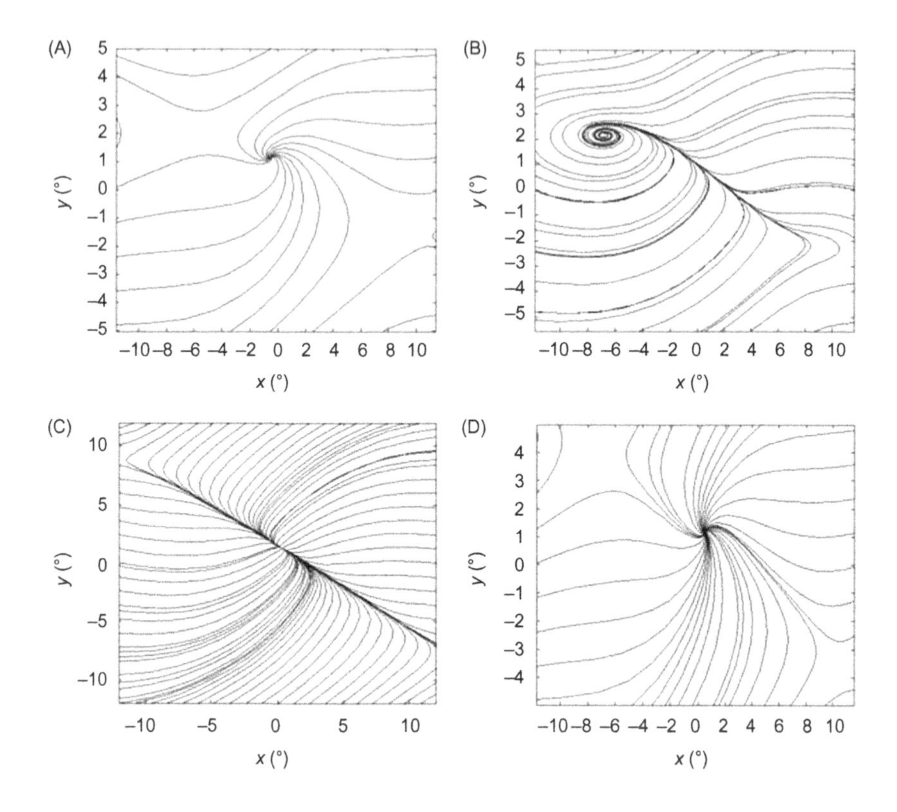

7.5 EXERCISES FOR CHAPTER 6

Question 1

1.a To model human behavior, McRuer performed a tracking experiment, in which systems with different behavior (zero-order, first-order, and second-order) should be controlled by the human. By measuring input and output of the human controller, Bode diagrams were made of the human behavior for these different systems. What was McRuer's main conclusion from these experiments?

1.b For this experiment, the Bode diagram of a first-order plant and of the human + plant is given next. Draw the gain curve and the phase delay curve in the Bode diagram for the human. Dimension of phase angle is in radians.

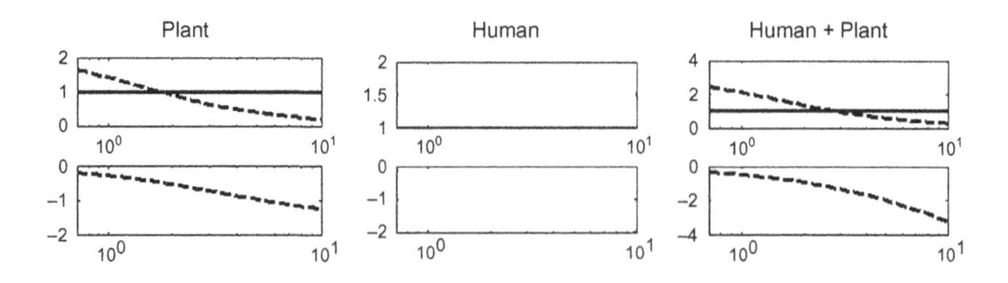

1.c Give an expression for this human transfer function $G(i \cdot \Omega)$.

1.d The Bode plot of the human + plant is given again next.

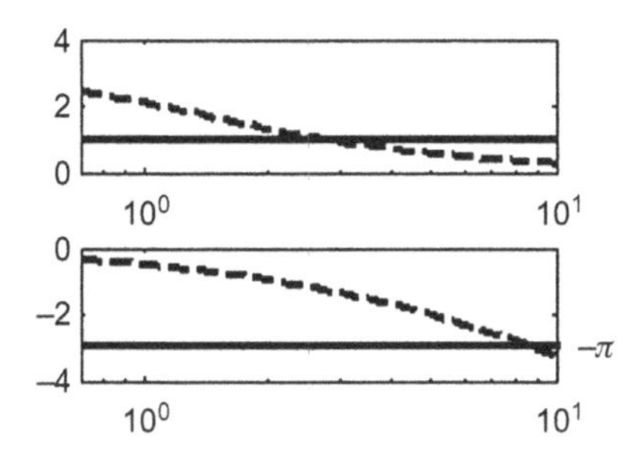

The closed-loop behavior of this first-order plant + human controller is stable. Explain how you can see this from the Bode diagram for the human + plant.

Question 2

2.a The McRuer crossover model is given next. Explain all five parameters of this model.

$$G_d(i\Omega) = K \cdot \left(\frac{e^{-i \cdot \Omega \cdot \tau_d}}{i \cdot \Omega \cdot \tau_N + 1} \right) \cdot \left(\frac{i \cdot \Omega \cdot T_L + 1}{i \cdot \Omega \cdot \tau_L + 1} \right)$$

2.b Next, the simplified version of the crossover model in the s-domain is given as a driver model with input $u(s)$ and output $d(s)$ for following a given path:

$$G(s) = \frac{d(s)}{u(s)} = K \cdot \frac{1}{1 + \tau \cdot s}$$

Given that:

- The input $u(s)$ corresponds with the error $\Delta\psi(t)$, defined as the error between the actual value of the vehicle yaw angle and the yaw angle corresponding to the required path.
- The output $d(s)$ corresponds to the steering wheel angle $\delta(t)$

Provide the time domain differential equation of this driver model

2.c The resulting vehicle−driver system behavior is quite bad for following a given path. Explain why.

2.d Suggest (and justify) an extension of this driver model so that path tracking behavior will improve.

Appendix 1: State Space Format

We have seen in Section 5.2 that the simplest way to describe handling of a vehicle with linear tire characteristics, under action of external force F_{ye}, moment M_{ze}, and/or steering angle input δ, is by using the following equations:

$$m \cdot V \cdot (\dot{\beta} + r) = Y_\beta \cdot \beta + Y_r \cdot r + C_{\alpha 1} \cdot \delta + F_{ye} \tag{A1.1}$$

$$J_z \cdot \dot{r} = N_\beta \cdot \beta + N_r \cdot r + a \cdot C_{\alpha 1} \cdot \delta + M_{ze} \tag{A1.2}$$

where the derivatives of stability Y_β, Y_r, N_β, and N_r are defined by (see also Section 5.2):

$$Y_\beta = -(C_{\alpha 1} + C_{\alpha 2}) \tag{A1.3}$$

$$Y_r = -\frac{a \cdot C_{\alpha 1} - b \cdot C_{\alpha 2}}{V} \tag{A1.4}$$

$$N_\beta = -(a \cdot C_{\alpha 1} - b \cdot C_{\alpha 2}) \tag{A1.5}$$

$$N_r = -\frac{a^2 \cdot C_{\alpha 1} + b^2 \cdot C_{\alpha 2}}{V} \tag{A1.6}$$

where we neglect the contributions of the aligning torque. The variables and parameters in Eqs. (A1.1)...(A1.6) are the yaw rate r, body slip angle β, the vehicle forward speed V, the mass m, yaw inertia J_z, the front and rear axle cornering stiffnesses $C_{\alpha 1}$ and $C_{\alpha 2}$, and the distances a and b between front axle and vehicle's CoG, and rear axle and vehicle's CoG, respectively. The preceding system can be considered in a general generic form:

$$\dot{\underline{x}} = A \cdot \underline{x} + B \cdot \underline{u} \tag{A1.7}$$

where $\underline{x}$ and $\underline{u}$ are the state and input vectors, respectively,

$$\underline{x} = \begin{pmatrix} \beta \\ r \end{pmatrix}; \quad \underline{u} = \begin{pmatrix} \delta \\ F_{ye} \\ M_{ze} \end{pmatrix} \tag{A1.8}$$

and where system matrix A and input matrix B are given by

$$A = \begin{pmatrix} \frac{Y_\beta}{m \cdot V} & \frac{Y_r}{m \cdot V} - 1 \\ \frac{N_\beta}{J_z} & \frac{N_r}{J_z} \end{pmatrix}; \quad B = \begin{pmatrix} \frac{C_{\alpha 1}}{m \cdot V} & \frac{1}{m \cdot V} & 0 \\ \frac{a \cdot C_{\alpha 1}}{J_z} & 0 & \frac{1}{J_z} \end{pmatrix} \tag{A1.9}$$

Suppose, we are interested in obtaining output in terms of the lateral speed v_y $(=\beta \cdot V)$ and lateral acceleration a_y. Then, we can write the output vector $\underline{y}(t)$ in generic form as follows:

$$\underline{y} = \begin{pmatrix} v_y \\ a_y \end{pmatrix} = C \cdot \underline{x} + D \cdot \underline{u} \tag{A1.10}$$

with the matrices C and D given by

$$C = \begin{pmatrix} V & 0 \\ \frac{Y_\beta}{m} & \frac{Y_r}{m} \end{pmatrix}; \quad D = \begin{pmatrix} 0 & 0 & 0 \\ \frac{C_{\alpha 1}}{m} & \frac{1}{m} & 0 \end{pmatrix} \tag{A1.11}$$

The formulation (A1.7) and (A1.10) is known as the *state space format.* It describes the general system of n first-order equations in state vector $\underline{x}$. The input is given by the m-dimensional vector $\underline{u}$. The dimension of the output vector $\underline{y}$ is denoted as k. Consequently, the system matrix A is an $n \times n$ matrix, B is an $n \times m$ matrix, C is a $k \times n$ matrix, and D is a $k \times m$ matrix.

The system in state space format may be described using a block diagram, as shown in Figure A1.1.

The symbol "$1/s$" indicates integration and is in correspondence with the s-domain properties. Note that integration in the time domain is created by dividing by s in the s-domain (obtained after using the Laplace transformation).

Various tools exist in Matlab–Simulink® to treat state space models. First, one must define the state space model in terms of the matrices $A, \ldots, D$ as follows:

$$\text{SYS} = \text{SS}(A, B, C, D);$$

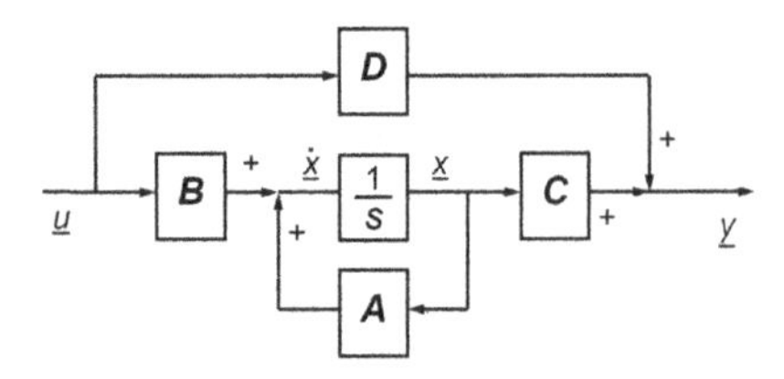

FIGURE A1.1 Block diagram for state space model.

Next, one must define the initial conditions $X0$ for the state vector X (a matrix with a number of rows similar to the number of time steps, and with a column for each separate scalar state). In the same way, the input vector U is a matrix with rows corresponding to time stamp, and the same number of columns as the number of entries of the vector $\underline{u}(t)$. The time stamp is stored in an array T.

For a predefined input U, the solution of the state space problem is obtained by a single statement:

$$[Y, T, X] = \text{LSIM}(\text{SYS}, U, T, X0)$$

The columns of X and Y are the states and outputs of the problem, respectively.

As an illustration, we determined the step response for the vehicle data specified in Appendix 6. The response in lateral acceleration is shown in Figure A1.2.

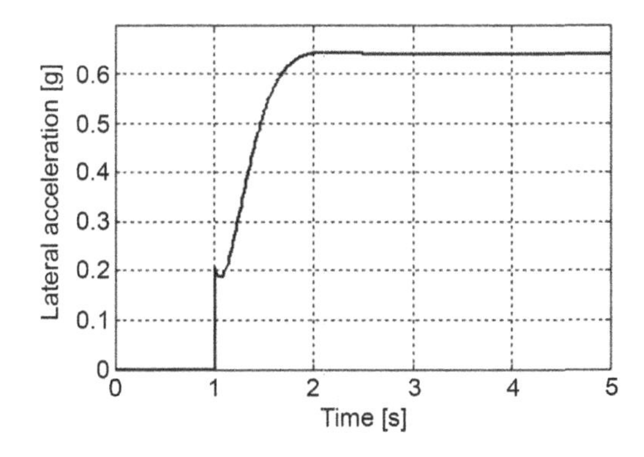

FIGURE A1.2 Output (lateral acceleration) of the state space model (Eqs. (A1.7) and (A1.10)).

Appendix 2: System Dynamics

A2.1 GENERAL APPROACH IN *N* DIMENSIONS

Consider the n-dimensional autonomous ordinary differential equation

$$\dot{\underline{x}} = \underline{F}(\underline{x}; \underline{u}) \quad t > 0 \tag{A2.1}$$

where $\underline{x}$ is an n-dimensional function in time and $\underline{u}$ is an m-dimensional time-dependent input, i.e.,

$$\underline{x}:[o, \infty) \to IR^n \quad \underline{u}:[o, \infty) \to IR^m$$

Likewise, the right-hand side function $\underline{F} = (F_1, F_2, \ldots, F_n)^T$ will also be n-dimensional and it may be nonlinear. For example, the bicycle model in Chapter 5 (with axle behavior described by the Pacejka model) is a nonlinear differential equation with $n = 2$ and $m = 1$ (assuming the steering angle to be the only input).

In many situations, one first examines the steady-state solutions for Eq. (A2.1), i.e., with the input being constant in time:

$$\underline{u}(t) = \underline{u}_s \quad t > 0 \tag{A2.2}$$

That means that we search for solutions that are constant in time, satisfying the equation

$$\underline{F}(\underline{x}_s; \underline{u}_s) = \underline{0} \tag{A2.3}$$

These solutions are referred to as the *equilibrium points*, *critical points*, or *singular points* of Eq. (A2.1). The next question one may pose is whether solutions of Eq. (A2.1) for constant input $\underline{u}$, starting close to $\underline{x}_s$ at $t = 0$, stay close to $\underline{x}_s$ or increase in time. This is the problem of stability for steady-state solutions.

A2.1.1 Definition: Stability

A steady-state solution $\underline{x}_s$ of Eq. (A2.1) with constant input ($\underline{u}_s$) is *stable* if for every $\varepsilon_1 > 0$, a value $\varepsilon_2 > 0$ exists such that

$$|| \underline{x}(t=0) - \underline{x}_s || < \varepsilon_2$$

implies

$$|| \underline{x}(t) - \underline{x}_s || < \varepsilon_1 \quad t > 0$$

In general, the behavior of the solution may be even better, in the sense that deviation from the equilibrium solution becomes arbitrarily small with increasing time, which is referred to as asymptotic stability.

A2.1.2 Definition: Asymptotic Stability

A steady-state solution $\underline{x}_s$ of Eq. (A2.1) with constant input ($\underline{u}_s$) is *asymptotically stable* if a value $\varepsilon_2 > 0$ exists such that

$$|| \underline{x}(t=0) - \underline{x}_s || < \varepsilon_2$$

implies

$$|| \underline{x}(t) - \underline{x}_s || \to 0 \quad t \to \infty$$

Most readers will be familiar with the fact that the size of a vector, indicated with $|| \cdot ||$, is usually determined as the square root of the sum of squares of the separate entries:

$$||\underline{x}|| = \sqrt{\sum_{i=1:n} x_i^2} \quad \text{for} \quad \underline{x} = (x_1 \quad x_2 \quad \ldots \quad x_n)^T$$

The practical relevance of the preceding definitions can be explained as follows. Consider a system that is behaving stationary because of an input $\underline{u}_s$, for example, a vehicle with a chosen steering angle. The vehicle should follow a perfect circle. However, the world is not perfect, and the vehicle path may suffer from wind, road disturbances, inclinations, etc. Consequently, the circle will not be perfect either; there will be a small deviation in the vehicle state. One expects the vehicle to be forgiving, in the sense that these small deviations will not lead to excessive vehicle behavior, without the driver interfering. This means we expect the vehicle to be stable with respect to yaw and drift disturbance. We define

$$\underline{d}_x \equiv \underline{x}(t) - \underline{x}_s$$

with $\underline{x}(t)$ being a solution of Eq. (A2.1) for $\underline{u} \equiv \underline{u}_s$. Using the previous equations, this function is found to satisfy the following vector equation:

$$\dot{\underline{d}}_x = \underline{F}(\underline{x}(t); \underline{u}_s) - \underline{F}(\underline{x}_s; \underline{u}_s) \quad t > 0 \tag{A2.4}$$

Assuming $\underline{F}$ to be differentiable, and assuming $\underline{x}$ to be close to $\underline{x}_s$, this equation can be approximated by

$$\dot{\underline{d}}_x = D_x \underline{F}(\underline{x}_s; \underline{u}_s) \cdot \underline{d}_x \quad t > 0 \tag{A2.5}$$

where $D_x \underline{F}$ is the Jacobian matrix of $\underline{F}$, with entries

$$(D_x \underline{F})_{ij} = \frac{\partial F_i}{\partial x_j} \tag{A2.6}$$

This Jacobian is an $n \times n$ matrix, which is constant in time. It means that the behavior of the solution of Eq. (A2.1) can be described locally near an equilibrium (critical) point by a linear n-dimensional differential equation

$$\dot{\underline{d}} = A \cdot \underline{d} \quad t > 0 \tag{A2.7}$$

where we denoted the Jacobian with A, and removed the index x in the vector function $\underline{d}(t)$. Solutions for Eq. (A2.7) can be expressed as a superposition of exponential functions

$$\underline{d}(t) = \sum_{i=1:n} \underline{a}_i \cdot e^{\lambda_i t} \quad t > 0 \tag{A2.8}$$

for constant eigenvectors $\underline{a}_i$ and where $\lambda_1, \lambda_2, \ldots, \lambda_n$ are the eigenvalues of the (Jacobian) matrix A, i.e., for which the determinant of $A - \lambda_i \cdot I$ vanishes:

$$|A - \lambda_i \cdot I| = 0 \quad i = 1, 2, \ldots, n \tag{A2.9}$$

where I is the unit matrix with the diagonal terms equal to one and all other entries as zero. The stability of the steady-state solution $\underline{x}_s$ depends on these eigenvalues. If $\mathrm{Re}(\lambda) > 0$ for at least one of these eigenvalues, then $\underline{d}(t)$ will become infinitely large. Consequently, there will be no stability. On the other hand, if $\mathrm{Re}(\lambda) \leq 0$ for all $i = 1, 2, \ldots, n$, then $\underline{x}_s$ is stable according to the definition given previously. If $\mathrm{Re}(\lambda) < 0$ for all $i = 1, 2, \ldots, n$, then $\underline{d} \to \underline{0}$ and $\underline{x}_s$ is asymptotically stable.

A2.2 SYSTEM DYNAMICS IN TWO DIMENSIONS

Let us consider the situation for $n = 2$. This allows us to draw solution curves in two dimensions, i.e., plot x_1 versus x_2, or d_1 versus d_2, if we consider only the deviations from the equilibrium solution. This plot of solution curves is called the *phase plane* and the solution curves are called *trajectories*. Starting from Eq. (A2.7) means that we consider solution curves around an equilibrium solution that are equal to $\underline{d} = 0$, which leads to the same local behavior as for Eq. (A2.1) near $\underline{x} = \underline{x}_s$. The preceding discussion explains the local behavior near the critical points, in terms of the eigenvalues of the system matrix A (Jacobian of the right-hand side in case of a nonlinear set of equations). We now have, at most, two eigenvalues (λ_1 and λ_2), which means that we can identify different situations for this local behavior, depending whether the eigenvalues are real or nonreal, and whether the real part of the eigenvalues is positive, zero, or negative. The situation of one of the eigenvalues being equal to zero corresponds to the matrix A being singular.

i. *λ_1 and λ_2 are both real, $\lambda_1 \cdot \lambda_2 < 0$*

One of the eigenvalues is positive, and the other is negative. Local solutions are described by

$$\underline{d}(t) = \underline{a}_1 \cdot e^{\lambda_1 \cdot t} + \underline{a}_2 \cdot e^{\lambda_2 \cdot t} \tag{A2.10}$$

Part of this solution tries to approach the origin 0, whereas the other part of the solution tries to move away from it. If $\lambda_1 < 0$, then all solutions in the (d_1, d_2) plane along the orientation given by $\underline{a}_1$ are moving toward 0. This is the only orientation for which this occurs. All other orientations will have some share in the second part of $\underline{d}(t)$ and, after having first approached 0, will change direction and move away again from 0. The only orientation for solution curves to move away immediately is given by $\underline{a}_2$. Plotting the solution curves results in the image shown in Figure A2.1. The dashed lines correspond to the vectors $\underline{a}_1$ and $\underline{a}_2$, where the intersection is the critical point. Solution curves (the solid

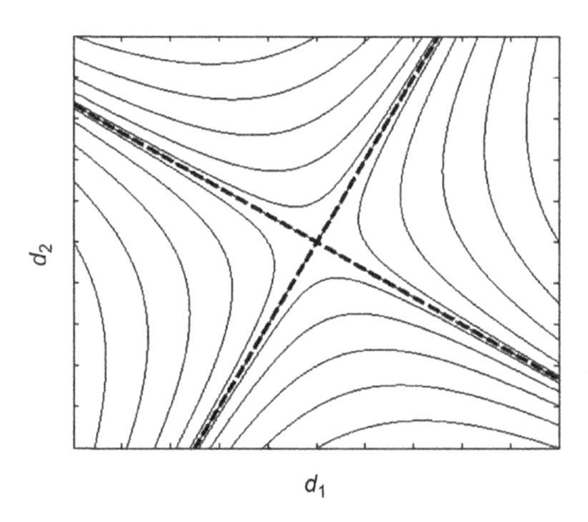

FIGURE A2.1 Local behavior near critical point case (i) (saddle point).

lines) approach this point but never reach it, bending off away from it. A critical point with this local behavior is referred to as a *saddle point*.

ii. *λ_1 and λ_2 are both real, $\lambda_1 \cdot \lambda_2 > 0$, $\lambda_1 \neq \lambda_2$*

Both eigenvalues are positive or negative. This means that all trajectories move away from the critical point (a *source* of trajectories) or approach it (a *sink* of trajectories), respectively. Suppose all trajectories are approaching the steady-state point, i.e., the eigenvalues are negative. Each term in Eq. (A2.10) tends to zero. The term for the eigenvalue with the smallest absolute value dominates the behavior near 0. It still has a significant value away from zero when the other term is already quite small. We selected the coefficients $\underline{a}_i$, $i = 1, 2$, as in case (i). Plotting the solution curves then results in the image shown in Figure A2.2. The two dashed lines again correspond to the eigenvectors $\underline{a}_1$ and $\underline{a}_2$. Most curves enter along the line corresponding to the eigenvalue with the smallest absolute value. A critical point with this local behavior is referred to as a *two-sided node*.

iii. *λ_1 and λ_2 are both real, $\lambda_1 \cdot \lambda_2 > 0$, $\lambda_1 = \lambda_2$*

There are two possibilities here: the eigenvalue can have a multiplicity of 1 or 2. A multiplicity of 2 means that all vectors in IR^2 are eigenvectors. The only possible way for this to occur is when matrix A is a diagonal matrix where the diagonal terms are identical. In that case, the arbitrary solution of Eq. (A2.7) is given by

$$\underline{d}(t) = \underline{a} \cdot e^{\lambda_1 \cdot t}$$

for any vector $\underline{a}$. This implies that the vector $\underline{d}(t)$ moves along straight lines toward the critical point 0 (if $\lambda_1 < 0$) or away from 0 (if $\lambda_1 > 0$). Consequently, the local behavior is star-shaped, as indicated in

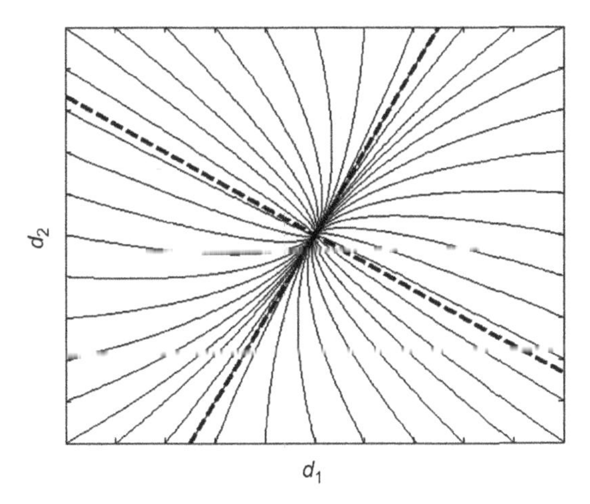

FIGURE A2.2 Local behavior near critical point, case (ii) (two-sided node).

Figure A2.3, and the critical point is called a *star*. For a multiplicity of 1, meaning that the set of eigenvectors is one-dimensional, the general solution of Eq. (A2.7) can be written as

$$\underline{d}(t) = [C_1 \cdot \underline{a}_1 + C_2 \cdot \underline{a}_2] \cdot e^{\lambda_1 \cdot t} + C_2 \cdot \underline{a}_1 \cdot t \cdot e^{\lambda_1 \cdot t}$$

for arbitrary coefficients C_1 and C_2, eigenvector $\underline{a}_1$, and some vector $\underline{a}_2$ depending on $\underline{a}_1$. Now, all trajectories move along a single line toward or away from the critical point, given by the orientation of eigenvector $\underline{a}_1$. This is called a *one-sided node* (Figure A2.4).

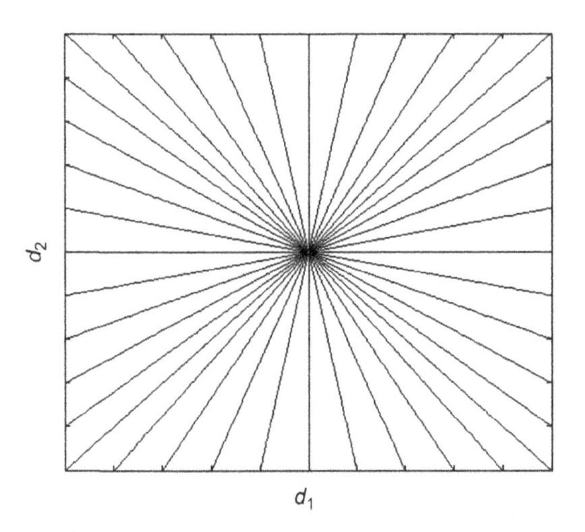

FIGURE A2.3 Local behavior near critical point, case (iii) (star).

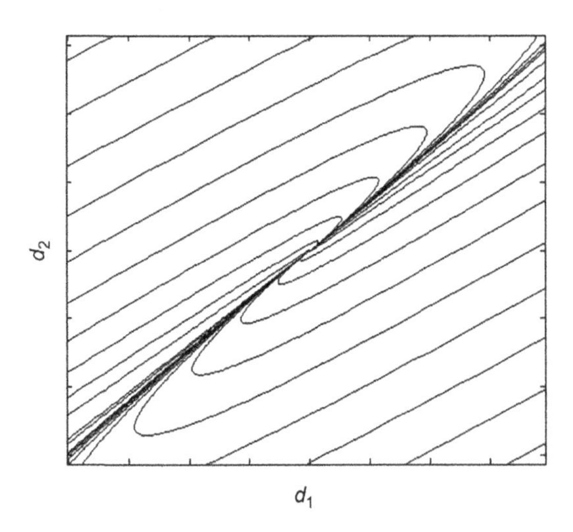

FIGURE A2.4 Local behavior near critical point, case (iii) (one-sided node).

iv. *λ_1 and λ_2 are both nonreal, with Re $\lambda_i \neq 0$, i = 1, 2.*

The eigenvalues are complex and conjugate each other (same real value, opposite imaginary value). We obtain solutions of Eq. (A2.7) of the form

$$\underline{d}(t) = e^{\mu_1 \cdot t} \cdot [\underline{a}_1 \cdot \cos(\mu_2 \cdot t) + \underline{a}_2 \cdot \sin(\mu_2 \cdot t)] \tag{A2.11}$$

with the eigenvalues λ_i, $i = 1, 2$, expressed in real numbers μ_1, μ_2:

$$\lambda = \mu_1 \pm i \cdot \mu_2$$

and for some vectors $\underline{a}_1$ and $\underline{a}_2$. The imaginary part μ_2 corresponds to the radial eigenfrequency. Solution curves (trajectories) are spiraling in (sink) and out (source) around the critical point for $\mu_1 < 0$ and $\mu_1 > 0$, respectively (Figure A2.5). A critical point with this local behavior is referred to as a *focus* or a *spiral point.*

v. *λ_1 and λ_2 are both nonreal, with Re $\lambda_i = 0$, i = 1, 2.*

This is the final case, with the eigenvalues being purely imaginary. The expression (A2.11) now becomes

$$\underline{d}(t) = \underline{a}_1 \cdot \cos(\mu_2 \cdot t) + \underline{a}_2 \cdot \sin(\mu_2 \cdot t) \tag{A2.12}$$

Hence, there is no decay to zero or unbounded growth of the solution $\underline{d}(t)$. The critical point is neither sink nor source and trajectories move around this point as ellipses, which is indicated in Figure A2.6. A critical point with this local behavior is referred to as a *center.*

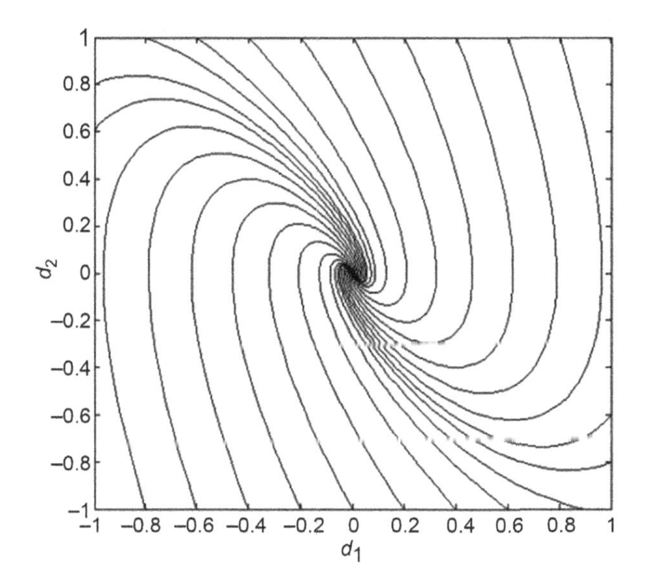

FIGURE A2.5 Local behavior near critical point, case (iv) (focus).

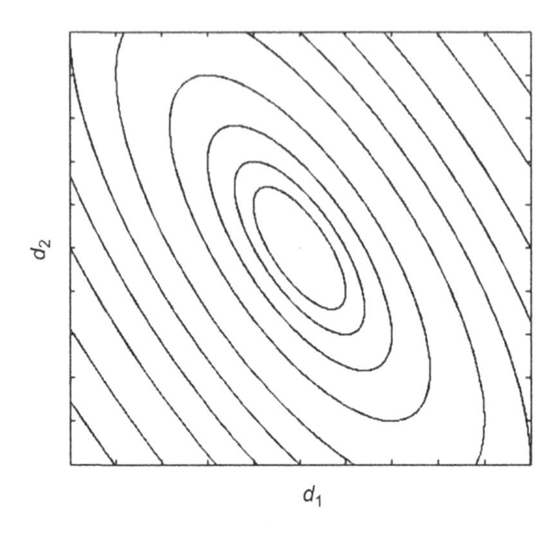

FIGURE A2.6 Local behavior near critical point, case (v) (center).

A2.2.1 Concluding Remark

Considering the different types of critical points, the saddle point, two-sided node, and focus are the most common. The situations of Figures A2.3, A2.4, and A2.6 are all rare, but are discussed here for the sake of completeness of the theory. In the discussion of global stability of the vehicle bicycle model in Chapter 5, these types of critical points are not of interest.

A2.3 SECOND-ORDER SYSTEM IN STANDARD FORM

A set of two first-order linear equations, obtained from expression (A2.7) in two dimensions, can be rewritten as one single second-order system. Begin with the general form of Eq. (A2.7), written as two coupled linear equations:

$$\dot{x} = a_{11} \cdot x + a_{12} \cdot y$$
$$\dot{y} = a_{21} \cdot x + a_{22} \cdot y$$

Differentiating the first equation in time, and using both equations for substitutions, one arrives at the second-order equation in x:

$$\ddot{x} - (a_{11} + a_{22}) \cdot \dot{x} + (a_{11} \cdot a_{22} - a_{12} \cdot a_{21}) \cdot x = 0$$

In terms of the trace (sum of diagonal terms) and determinant of the matrix A, this equation can also be written as

$$\ddot{x} - tr(A) \cdot \dot{x} + |A| \cdot x = 0$$

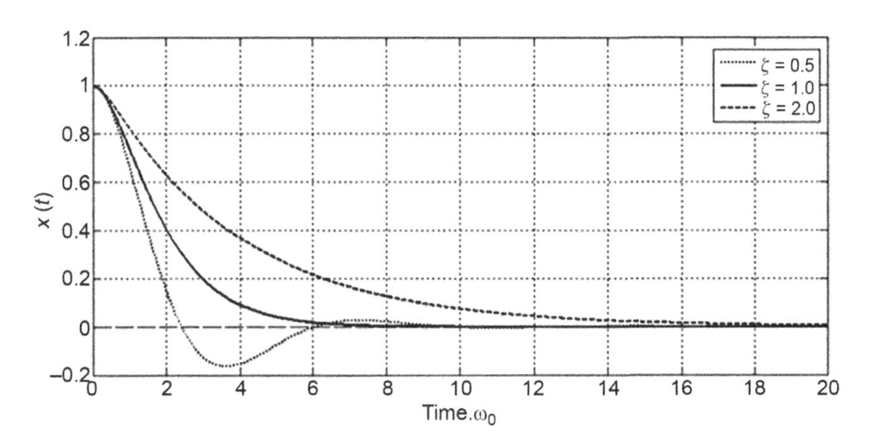

FIGURE A2.7 Second-order response to initial step for different damping values.

A second-order linear differential equation can be written in standard form

$$\ddot{x} + 2 \cdot \zeta \cdot \omega_0 \cdot \dot{x} + \omega_0^2 \cdot x = 0 \tag{A2.13}$$

with undamped natural frequency $\omega_0 > 0$ and damping ratio ζ (see Ref. [54]). The general solution of Eq. (A2.13) can be expressed as

$$x = C_1 \cdot e^{\lambda_1 \cdot t} + C_2 \cdot e^{\lambda_2 \cdot t}$$

for coefficients C_1, C_2 and with

$$\lambda_{12} = -\zeta \cdot \omega_0 \pm \omega_0 \cdot \sqrt{\zeta^2 - 1} \tag{A2.14}$$

Clearly, if $\zeta > 1$, all solutions will decay monotonously to zero, which corresponds to case (ii) of Section A2.2. If $0 < \zeta < 1$, solutions will decay to zero, but in an oscillatory way, which corresponds to case (iv) of Section A2.2. The damped radial eigenfrequency (ω) is then given by

$$\omega = \omega_0 \cdot \sqrt{1 - \zeta^2} \tag{A2.15}$$

We plotted the solution of expression (A2.13) with initial value $x(0) = 1$ and zero initial slope in Figure A2.7.

Appendix 3: Root Locus Plot

When dealing with differential equations, such as those described in state space format in Appendix 1,

$$\underline{\dot{x}} = A \cdot \underline{x} + B \cdot \underline{u} \quad \underline{x}: [0, \infty) \rightarrow IR^n \quad \underline{u}: [o, \infty) \rightarrow IR^m,$$

the eigenvalues λ_i, $i = 1, 2, \ldots, n$ of the system matrix A play an important role. Each eigenvalue, being a root of the characteristic equation,

$$|A - \lambda \cdot I| = 0$$

corresponds to an eigenvector (eigenmode) that is a nontrivial state vector of the form

$$\underline{x} = \underline{a} \cdot e^{\lambda_i . t} \quad t > 0$$

satisfying the differential equation for zero input $\underline{u} = 0$. In general, the eigenvalues are complex numbers $\lambda_i = \mu_{i1} + i.\ \mu_{i2}$ with their real part indicating the damping properties of the system at hand for that specific eigenmode, and their imaginary part corresponding to the radial eigenfrequency

$$\underline{x} = e^{\mu_{i1} . t} \cdot (\underline{a}_1 \cdot \cos((\mu_{i2} \cdot t) + \underline{a}_2 \cdot \sin(\mu_{i2} \cdot t))) \quad t > 0$$

When discussing handling analyses, one is interested in yaw and roll frequencies. When discussing comfort analysis, the eigenmodes of the vehicle suspension in heave and pitch are the main topics of investigation. Motorcycle handling strongly depends on eigenfrequencies of front assembly (wobble) and the total frame (weave). Eigenvalues, and therefore eigenmodes, also depend on vehicle parameters.

Hence, to improve the dynamic vehicle performance, one must find the relationship between eigenvalues and these parameters, considering vehicle speed as one of the most important parameters. The variation of the eigenvalues when one of these parameters is changed, when plotted in the complex domain, is called a *root locus*. The combination of these plots for all eigenvalues is called the *root locus plot*. We saw in Appendix 2 that stability of the steady-state solution requires that all of the eigenvalues must be in the left-hand half of the complex domain, i.e., in negative real part.

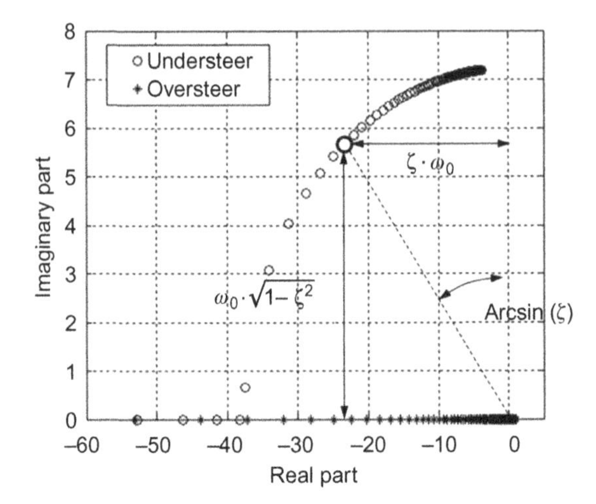

FIGURE A3.1 Root locus plot of one-track vehicle model.

Let us consider the characteristic equation for the single-track vehicle handling problem (3.53). Here, we varied the vehicle velocity and plotted one of the eigenvalues for both understeer and oversteer situation in Figure A3.1. Observe that, for the understeered vehicle, the eigenvalue becomes nonreal at a certain speed, as expected from the discussion in Section 5.4. The eigenvalue for an oversteered vehicle remains real and passes the imaginary axis at a certain speed (the critical speed), resulting in a loss of stability.

For nonreal eigenvalues (understeered vehicle), Eq. (3.53) can be related to a second-order differential equation in standard form (cf. Eq. (A2.13) in Appendix 2):

$$\ddot{x} + 2 \cdot \zeta \cdot \omega_0 \cdot \dot{x} + \omega_0^2 \cdot x = 0$$

with complex eigenvalues

$$\lambda_{12} = -\zeta \cdot \omega_0 \pm i \cdot \omega_0 \cdot \sqrt{1 - \zeta^2}$$

for $\zeta < 1$. We selected an eigenvalue for a specific speed and indicated the real and imaginary parts in Figure A3.1. From this, the following conclusions are given:

(i) The imaginary part (if it exists) corresponds to the damped radial eigenfrequency, which is connected to the specific eigenmode.

(ii) The real part of the eigenvalue describes the decay to zero. Hence, the further the eigenvalue is positioned to the left, the faster any disturbance to steady-state behavior will disappear.

(iii) The distance between origin and eigenvalue in the complex domain is equal to the undamped radial eigenfrequency.

(iv) The angle between the imaginary axis and the eigenvalue as a vector in the complex domain describes the damping ratio (ζ).

When more modes exist, this graphical interpretation is a strong tool for interpreting vehicle performance and identifying significant vehicle parameters.

Appendix 4: Bode Diagram

It is assumed that the transfer function in the s-domain for the solution of a linear differential equation can be written as follows:

$$G(s) = \frac{K}{s^n} \cdot \frac{S_1 \cdot S_2 \cdots S_k \cdot Q_1 \cdot Q_2 \cdots Q_l}{S_{k+1}.S_{k+2}\ldots.Q_{l+1} \cdot Q_{l+1}\ldots} \tag{A4.1}$$

for some n, l, $k \geq 0$, where S_i and Q_j are linear and quadratic expressions in s:

$$S_i = s \cdot \tau_i + 1 \tag{A4.2}$$

$$Q_j = \left(\frac{s}{\omega_j}\right)^2 + 2 \cdot \zeta_j \cdot \frac{s}{\omega_j} + 1 \tag{A4.3}$$

The parameters τ_i, ω_j, and ζ_j are all assumed positive. The functions (A4.2) and (A4.3) determine the poles (eigenvalues, eigenfrequencies) and zeroes of the transfer function [$G(s)$]. Here, we follow the treatment of Bode plots by van de Vegte [54]. Hence, for an input $u(s)$, a solution $x(s)$ in the s-domain can be found in the form

$$x(s) = G(s) \cdot u(s) \tag{A4.4}$$

which is obtained by Laplace transformation of the differential equation.

Considering a complex input in the time domain

$$u(t) = A \cdot e^{i \cdot \Omega \cdot t} = A \cdot [\cos(\Omega \cdot t) + i \cdot \sin(\Omega \cdot t)] \tag{A4.5}$$

a solution $x(t)$ in the time domain is found of the form

$$x(t) = M(\Omega) \cdot A \cdot e^{i \cdot \Omega \cdot t} \cdot e^{i \cdot \varphi(\Omega)} = M(\Omega) \cdot A \cdot [\cos(\Omega \cdot t + \varphi) + i \cdot \sin(\Omega \cdot t + \varphi)] \tag{A4.6}$$

Hence, an oscillatory (e.g., sinusoidal) input results in a forced response being oscillatory (sinusoidal) as well, with the amplitude being increased with a factor M and the phase shifted with an angle φ, that both depend on the forced input frequency Ω. Comparing the last two expressions (A4.5) and (A4.6) with Eq. (A4.4), one may conclude that

- The multiplication factor $M(\Omega)$ corresponds to the magnitude of the transfer function $G(s)$, where s is replaced by $i \cdot \Omega$

$$M(\Omega) = |G(i \cdot \Omega)|$$

This magnitude is typically expressed in terms of decibels (dB)

$$M_{\mathrm{dB}} = 20 \cdot {}^{10}\log M$$

- The phase angle φ between output and input corresponds to the argument of the *frequency transfer function* $G(i \cdot \Omega)$.

A set of Bode diagrams shows M versus Ω (or M_{dB} versus Ω) and φ versus Ω. For very small Ω, M will correspond to the steady-state gain (M_0). For large Ω, one expects the system is unable to follow the input anymore, which leads to small values of M and large negative values of φ (large phase shift).

We introduced the *bandwidth* Ω_{bw} and the *equivalent time* T_{eq} in Section 5.1. These properties were defined as the range of frequencies for which the magnitude exceeds $M_0/\sqrt{2}$, and $2 \cdot \pi/\Omega_{\mathrm{eq}}$ for which Ω_{eq} corresponds to a phase lag of 45°, respectively (see also Figure 3.7). This bandwidth corresponds to a reduction of 10. ${}^{10}\log(2) = 3.0103$ in M_{dB}, hence, in a reduction of approximately 3 dB.

Where $G(i\Omega)$ is a product of factors with magnitude M_j and phase φ_j:

$$G(i \cdot \Omega) = \prod_j M_j \cdot e^{i \cdot \varphi_j}$$

it follows that

$$M_{\mathrm{dB}} = 20 \cdot \sum_j {}^{10}\log M_j \quad \varphi = \sum_j \varphi_j$$

Hence, one composes the Bode diagrams from the magnitude and phase of the basic elements of Eq. (A4.1), including gain, integrators, differentiators, first- and second-order lead (numerator in Eq. (A4.1)), and lag (denominator in Eq. (A4.1)).

i. *A simple gain K*

For a simple gain K, the magnitude and phase are found to be independent of Ω:

$$M_{\mathrm{dB}} = 20 \cdot {}^{10}\log K \quad \varphi = 0$$

ii. *An integrator*

In case of an integrator, the transfer function G is given by:

$$G(i \cdot \Omega) = \left(\frac{1}{i \cdot \Omega}\right)^n$$

for some $n = 1, 2, 3, \ldots$. The magnitude and phase are found to be

$$M_{\mathrm{dB}} = 20 \cdot {}^{10}\log|i.\Omega|^{-n} = -20 \cdot n \cdot {}^{10}\log\Omega \quad \varphi = -n \cdot \frac{\pi}{2}$$

On a log scale (for Ω), the magnitude versus frequency is a straight line, passing through $(\Omega, M_{\mathrm{dB}}) = (1, 0)$ with slope $-20 \cdot n$ dB/decade. By selecting $n < 0$, we obtain a differentiator that leading to plots being the mirror image of an integrator with respect to the 0 dB and 0° lines in the magnitude and phase plots.

iii. *A simple lag*

In case of a simple lag, the frequency transfer function is given by:

$$G(i \cdot \Omega) = \frac{1}{i \cdot \Omega \cdot \tau + 1}$$

For low frequency (Ω), one finds

$$M_{\mathrm{dB}} \to 0 \quad \varphi \to 0 \quad \text{if } \Omega \downarrow 0$$

For large frequency, one finds

$$M_{\mathrm{dB}} \to -20 \cdot {}^{10}\log\Omega - 20 \cdot {}^{10}\log\tau \quad \varphi \to -\frac{\pi}{2} \quad \text{if } \Omega \to \infty$$

Both approximations are straight lines in the magnitude plot on a log scale, intersecting at $\Omega \cdot \tau = 1$. Bode plots for a first-order lag are shown in Figure A4.1.

We make a remark here on a delay, described in the frequency domain as

$$G(i \cdot \Omega) = e^{-i\Omega \cdot \tau_R}$$

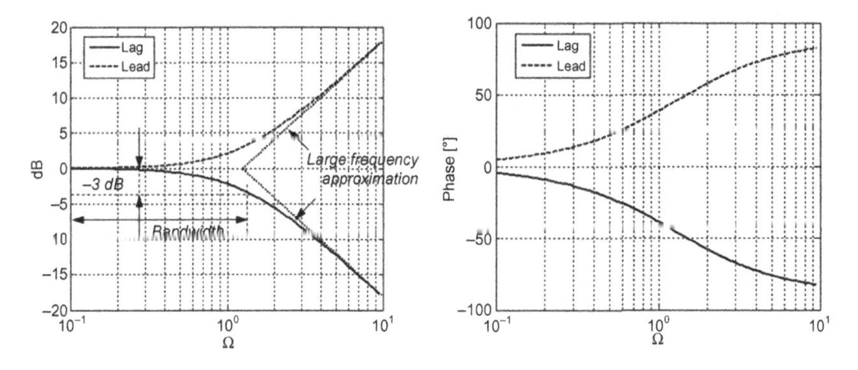

FIGURE A4.1 Bode diagrams for first-order lag and lead.

which, for a small delay time τ_R (typically in the order of 0.1–0.3 [s]) can be approximated (Taylor series extension) by a simple lag

$$G(i \cdot \Omega) = \frac{1}{i \cdot \Omega \cdot \tau_R + 1}$$

for a radial frequency (Ω) that is not too large. For larger frequencies, the major differences are the changes in gain (constant for a delay, dropping to lower values for a simple lag) and phase. As shown in Figure A4.1, the phase of a simple lag drops down to—$\pi/2$ for very large frequencies, whereas the phase of a delay, —$\Omega \cdot \tau_R$, may result into very large negative values.

iv. *A simple lead*

In case of a simple lead, the frequency transfer function is given by

$$G(i \cdot \Omega) = i \cdot \Omega \cdot \tau + 1$$

resulting in a similar low frequency behavior, and with the large frequency behavior given by

$$M_{dB} \rightarrow 20 \cdot {}^{10}\log \Omega + 20 \cdot {}^{10}\log \tau \quad \varphi \rightarrow \frac{\pi}{2} \quad \text{if } \Omega \rightarrow \infty$$

Again, both approximations are straight lines in the magnitude plot on a log scale, intersecting at $\Omega \cdot \tau = 1$; however, here, the magnitude is increasing with increasing frequency $\Omega > \tau^{-1}$. Bode plots for a simple lead (for the same value of τ) are also shown in Figure A4.1.

v. *Quadratic lag*

In case of a quadratic lag, the frequency transfer function is given by (see Eq. (A4.3))

$$G(i \cdot \Omega) = \frac{1}{((i \cdot \Omega)/\omega)^2 + 2 \cdot \zeta \cdot (i \cdot \Omega)/\omega + 1}$$

and therefore

$$M(\Omega) = \left(\left(1 - \left(\frac{\Omega}{\omega} \right)^2 \right)^2 + \left(\frac{2 \cdot \zeta \cdot \Omega}{\omega} \right)^2 \right)^{-\frac{1}{2}} \cdot \varphi(\Omega) = -\arctan \frac{2 \cdot \zeta \cdot \Omega/\omega}{1 - \Omega^2/\omega^2}$$

For small frequency (Ω), the magnitude approaches 0 dB and the phase tends to 0°, similar to the simple lag. For very large Ω, $M(\Omega)$ behaves quadratic in Ω, with

$$M_{dB}(\Omega) \rightarrow -40 \cdot {}^{10}\log \left(\frac{\Omega}{\omega} \right) \quad \varphi(\Omega) \rightarrow -\pi \quad \text{if } \Omega \rightarrow \infty$$

We depicted the quadratic lag response in Figure A4.2 for different values of ζ. Observe the resonance behavior near $\Omega = \omega$ and the behavior for large and small Ω.

We close this appendix section with the introduction of the gain margin GM and the phase margin φ_{m} for an open-loop transfer function $G(s) = G(i.\Omega) = e^{-i.\Omega.\tau_R}$.

Consider the closed-loop system, shown in Figure A4.3, which indicates the control of the vehicle by the driver, in response to feedback from vehicle performance output x (i.e., path deviation, yaw rate, lateral acceleration). The open-loop transfer function in the frequency domain is denoted by $G(i\Omega)$. According to this functional block diagram, we obtain

$$x = G(s) \cdot (u - x)$$

and therefore

$$x = T(s) \cdot u \equiv \frac{G(s)}{1 + G(s)} \cdot u \tag{A4.7}$$

The characteristic equation is given by

$$1 + G(s) = 0 \tag{A4.8}$$

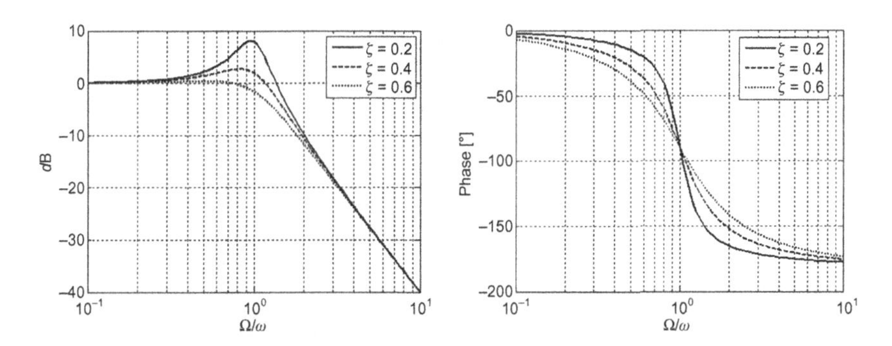

FIGURE A4.2 Bode diagrams for quadratic lag for different values of ζ.

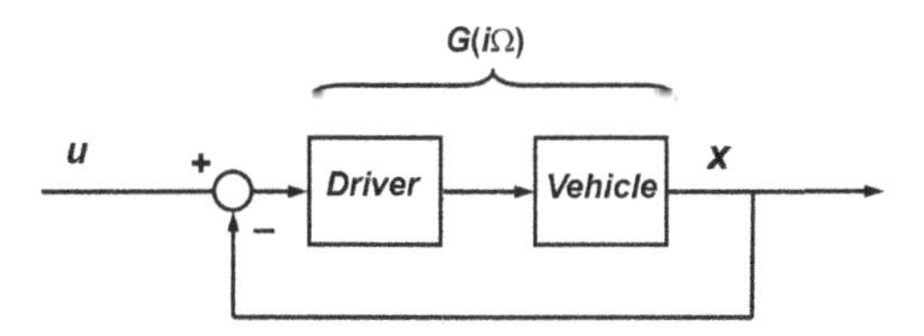

FIGURE A4.3 Closed-loop vehicle–driver system.

and stability is violated if solutions s_i of Eq. (A4.8) exist with a positive real part, i.e., lying in the right half complex s plane. In that case, a reverse transformation of Eq. (A4.7) to the time domain using partial fractions will lead to an exponential contribution $e^{\lambda t}$ in $x(t)$ with $\mathrm{Re}\cdot\lambda > 0$, i.e., yielding an unbounded solution. Letting $s = i\Omega$ follow the imaginary axis means encircling these points s_i, and therefore the origin by $1 + G(s)$, which is the same as encircling $G = -1$ by $G(i\cdot\Omega)$. This image $G(i\cdot\Omega)$ for Ω increasing along the real axis is called the polar plot. Consequently, the critical condition for stability is reached if the polar plot just passes the point where $G = -1$. This means that the denominator in the closed-loop transfer function $T(i\cdot\Omega)$ vanishes, which is identical to

$$|G(i\cdot\Omega)| = 1 \quad \arg[G(i\cdot\Omega)] = \pm\pi$$

As an example, we take

$$G(s) = \frac{1}{(s+a)^2\cdot(s+1)} \tag{A4.9}$$

for $a > 0$. We plotted $G(s)$ for $s = i\cdot\Omega$ in Figure A4.4 and the magnitude of the corresponding closed-loop transfer function in Figure A4.5 (Bode magnitude plot), for $a = 0.2$, 0.3, and 0.4. Clearly, for the polar plot passing through $G = -1$, for parameter a close to 0.3 (in fact, $a = 0.2972$), the closed-loop transfer function becomes unbounded. For lower values of a, the corresponding closed-loop system is unstable. This result is known as the simplified Nyquist criterion. This criterion states that, for an open-loop transfer system $G(s)$ with no poles in the right half s-plane, the corresponding closed-loop system is stable only if the polar plot passes the point -1 on the right-hand

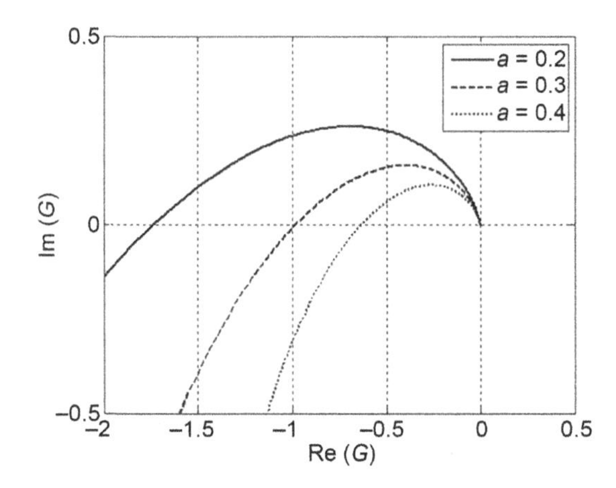

FIGURE A4.4 Polar plot of $G(i\cdot\Omega)$ according to Eq. (A.35) for various values of parameter a.

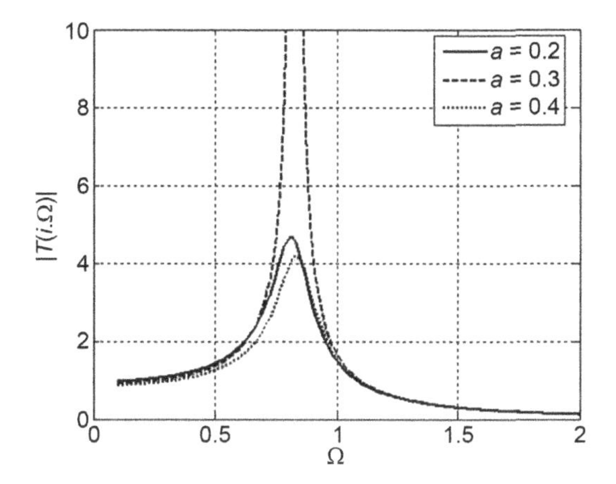

FIGURE A4.5 Magnitude of closed-loop transfer function $T(i \cdot \Omega)$.

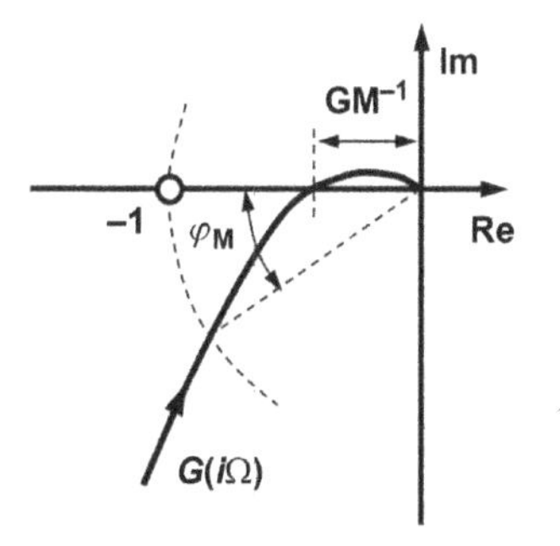

FIGURE A4.6 Polar plot, with indication of gain margin GM and phase margin φ_m.

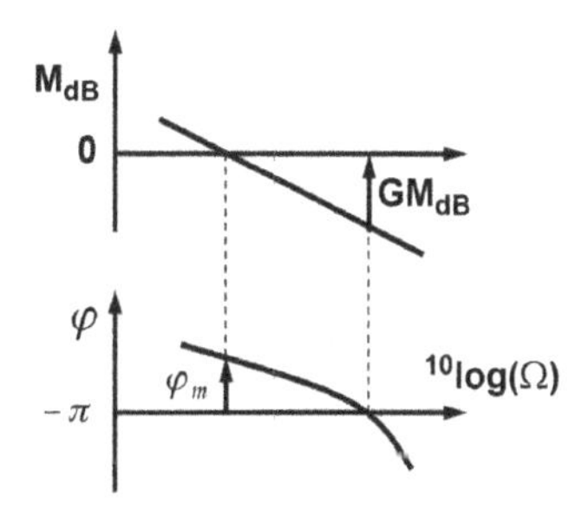

FIGURE A4.7 Indication of phase margin and gain margin in Bode plots.

side (see Ref. [54] for further reference). That means that, if the polar plot intersects the negative real axis, it will do so at a point with a magnitude less than 1. Defining the gain margin GM as the reciprocal magnitude of $G(i \cdot \Omega)$ for phase $\varphi = -\pi$, and $GM_{dB} = 20.\ {}^{10}\log(GM)$ means that at $\varphi = -\pi$, $M(\Omega) < 1$ and therefore $GM_{dB} > 0$.

Defining the phase margin φ_m as the angle on top of a phase of $\pm\pi$ for unit magnitude, $M(\Omega) = 1$, i.e., for M_{dB} $(\Omega) = 0$ (this frequency is called the crossover frequency), this phase margin will be positive for a stable system, because of the Nyquist criterion. Gain margin and phase margin have been indicated in the polar plot in Figure A4.6 and in the Bode magnitude and phase plots in Figure A4.7.

Appendix 5: Lagrange Equations

Consider a mechanical system with n generalized coordinates q_i, $i = 1, \ldots, n$. Assume that generalized nonconservative forces Q_i exist, meaning that these forces (or moments) cannot be expressed as a gradient $-\nabla U$ of a potential function $U(q_i)$ (i.e., of the potential energy), and do not depend on the generalized coordinates q_i and their time derivatives. The total transfer of energy is given by the following function:

$$L^* = T - U + \sum_i q_i \cdot Q_i \tag{A5.1}$$

with kinetic energy T, potential energy U, and work being carried out using the nonconservative forces Q_i. The first part of the right-hand side in Eq. (A5.1) is usually referred to as the Lagrangian (Lagrange function). We will follow the system for the time interval $[0, T]$ and state that the generalized coordinates correspond to the situation with an extreme total energy transfer. This means that we are searching for the extreme of the following functional (known as the *action* of the system):

$$J = \int_0^T L^*(q_i, \dot{q}_i, t)\mathrm{d}t \tag{A5.2}$$

i.e., the solution $q_i(t)$ for which the variation of J, with respect to variations in q_i and its time derivatives, vanishes. This is called the *principle of least action* and it states that a dynamical system "moves" from an initial to a final situation in a way that keeps the action minimal. The variation of J can be found as (see Chapter 4 in Ref. [21])

$$\delta J(q_i \cdot \delta q_i) = \int_0^T \sum_i \left[\frac{\partial L^*}{\partial q_i} \cdot \delta q_i + \frac{\partial L^*}{\partial \dot{q}_i} \cdot \delta \dot{q}_i \right] \mathrm{d}t = 0 \tag{A5.3}$$

for arbitrary variations δq_i and $\delta \dot{q}_i$, which all satisfy the same initial conditions.

It can be shown by partial integration that

$$\frac{\partial L^*}{\partial q_i} - \frac{\mathrm{d}}{\mathrm{d}t} \frac{\partial L^*}{\partial \dot{q}_i} = 0 \tag{A5.4}$$

where we have used the *fundamental lemma of the calculus of variations*:

$$\text{if} \int_0^T h(t)\cdot\delta q(t)\mathrm{d}t = 0 \text{ for all continuous functions } \delta q(t) \text{ then}$$
$$h(t) = 0 \text{ all along the interval } [0, T]$$

Equation (A5.4) is also known as the Euler–Lagrange equation. By substituting Eq. (A5.1) into Eq. (A5.4), for the generalized coordinates $q_i(t)$ for which expression (A5.2) has an extreme value, one finds

$$\frac{\mathrm{d}}{\mathrm{d}t}\frac{\partial T}{\partial \dot{q}_i} - \frac{\partial T}{\partial q_i} + \frac{\partial U}{\partial q_i} = Q_i \quad i = 1, \ldots, n \tag{A5.5}$$

We assumed that the potential energy only depends on q_i and not on its time derivative. In case of dissipated energy (such as for dampers), one may subtract the so-called Rayleigh function L_d from Eq. (A5.1), given by (and assumed to depend only on generalized state derivatives)

$$L_d = \tfrac{1}{2}\cdot\underline{\dot{q}}^T\cdot C\cdot\underline{\dot{q}} \tag{A5.6}$$

for matrix with damping values C (see Appendix A.5 in Ref. [11]).

Equation (A5.5) is a system of n equations that describe the behavior of the states derived from the coordinates $q_i(t)$ (e.g., yaw rate, body slip angle) for certain initial conditions and nonconservative forces Q_i (e.g., tire forces, aerodynamic loads). Setting up the equations of motion means that one must describe the kinetic and potential energy (and possibly dissipation/damping) in terms of generalized coordinates and their time derivatives and then substitute these values into Eq. (A5.5). An alternative but equivalent approach is to start from Newton's second law (impulse equals change in momentum) and likewise its rotational analogy (angular impulse equals change in angular momentum).

Appendix 6: Vehicle Data

A6.1 PASSENGER CAR DATA

This section contains arbitrary, but realistic, vehicle data for a typical sporty sedan, indicated in Figure A6.1.

We will always assume for the acceleration of gravity g that $g = 9.81$ [m/s^2].

a	1.51 [m]	Mass	1600 [kg]
b	1.25 [m]	J_x	880 [kg.m^2]
t_{front}	1.50 [m]	J_y	3110 [kg.m^2]
T_{rear}	1.51 [m]	J_z	3280 [kg.m^2]
h_{CoG}	0.57 [m]		
Steering ratio	15		
Spring stiffness front	19 [N/mm]	Spring stiffness rear	75 [N/mm]

Damper characteristics are usually described as the damper force, which depends on damper speed according to a more or less piecewise linear function that consists of three linear parts. We link three linear functions:

$$f_i(v) = a_i \cdot v + b_i \quad i = 1, 2, 3 \tag{A6.1}$$

for damper speed (v), intersecting at two speeds, $v = v_1$, $v_2 > v_1$. This can be achieved by combining these functions as follows:

$$f(v) = (1 - h_1(v)) \cdot f_1(v) + (h_1(v) - h_2(v)) \cdot f_2(v) + h_2(v) \cdot f_3(v) \tag{A6.2}$$

such that $h_1(v)$ is close to 0 for $v < v_1$ and close to 1 for $v > v_1$. Further, $h_2(v)$ is close to 0 for $v < v_2$ and close to 1 for $v > v_2$.

A function that satisfies these conditions is given by

$$h_i(v) = \tfrac{1}{2} \cdot \left[1 + \frac{2}{\pi} \cdot \arctan(D \cdot (v - v_i))\right] \tag{A6.3}$$

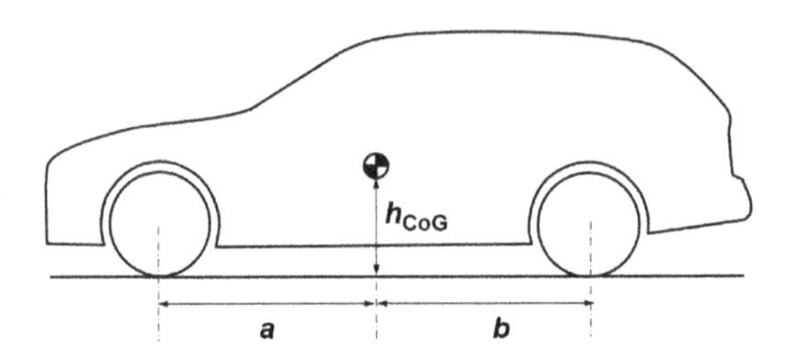

FIGURE A6.1 Passenger car with indication of CoG and axle positions.

with the parameter D tuned to make this transition either more abrupt (large D value) or smooth (small D value). For this vehicle, we selected the following parameters:

Damper Front Wheel			
a_1	250 [Ns/m]	b_1	−300 [N]
a_2	4100 [Ns/m]	b_2	0 [N]
a_3	625 [Ns/m]	b_3	500 [N]
D	50 [s/m]		

Damper Rear Wheel			
a_1	395 [Ns/m]	b_1	−400 [N]
a_2	5000 [Ns/m]	b_2	0 [N]
a_3	835 [Ns/m]	b_3	1000 [N]
D	50 [s/m]		

Note that the axle damping force amounts to twice the wheel damping force. We depicted the corresponding damper characteristics in Figure A6.2.

Further, we note that this fit approach can be used for arbitrary descriptions of the damper characteristics. This does not have to be a piecewise linear curve. Higher-order descriptions may be used in a similar way.

The axle data are expressed in terms of axle cornering stiffnesses

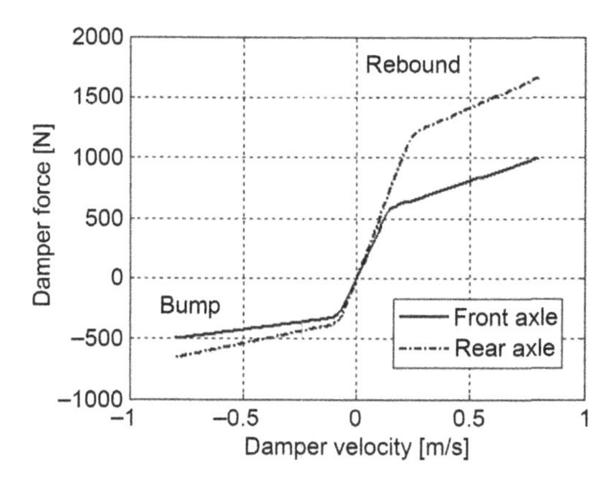

FIGURE A6.2 Damper characteristics fitted according to Eq. (A5.4).

$C_{\alpha 1}$	55,000 [N]	$C_{\alpha 2}$	98,000 [N]

resulting in an understeer gradient $\eta = 0.042$ (see Section 5.3.2).

A6.2 EMPIRICAL MODEL TIRE DATA

We use different sets of Pacejka data for tires. Some realistic data are listed here, with notation based on the Pacejka model [32]. All other Pacejka tire parameters that are not included are taken as zero.

Nominal Load	**4000 [N]**	**Vertical Stiffness**	**0.25 [MN/m]**
P_{Cx1}	1.6	P_{Dx1}	1.0
P_{Dx2}	−0.1	P_{Ex1}	0.1
P_{Ex2}	0.25	P_{Kx1}	20
P_{Kx2}	12	P_{Kx3}	−0.5
R_{Bx1}	11	R_{Bx2}	10
R_{Cx1}	1.0	R_{Ex1}	−0.5
R_{Ex2}	−0.5	P_{Cy1}	1.3
P_{Dy1}	0.9	P_{Dy2}	−0.08
P_{Ey1}	−0.8	P_{Ey2}	−0.6
P_{Ky1}	15.0	P_{Ky2}	1.8
P_{Ky3}	0.5	P_{Ky4}	2.0
P_{Ky6}	−1.0	P_{Ky7}	−0.3
R_{By1}	12.0	R_{By2}	10.0
R_{By3}	−0.01	R_{Cy1}	1.05
R_{Ey1}	0.25	Q_{Bz1}	13.0
Q_{Bz2}	−1.5	Q_{Bz9}	20
Q_{Cz1}	1.3	Q_{Dz1}	0.1
Q_{Dz6}	0.002	Q_{Dz7}	−0.002
Q_{Dz8}	−0.15	Q_{Ez1}	−1.0
Q_{Ez2}	0.8		

Appendix 7: Empirical Magic Formula Tire Model

In Subsections 2.4.2 and 2.5.2, the empirical Magic Formula is shown to be suited to describe longitudinal and lateral tire characteristics, i.e., longitudinal force F_x versus practical slip κ and lateral force F_y versus slip angle α. The general form of this formula, neglecting horizontal and vertical shifts, reads

$$F(s) = D \cdot \sin(C \cdot \arctan(B \cdot s - E \cdot (B \cdot s - \arctan(B \cdot s)))) \tag{A7.1}$$

for slip s, and factors D, C, B, and E. These factors typically depend on wheel load and/or camber within certain additional empirical relationships, as described in Sections 2.4 and 2.5. Expression (A7.1) and the additional relationships require model data to be derived from tests. There are many situations for which such data is not available, but graphical information (i.e., plots of these characteristics) can be obtained. The question then is whether such limited information can be used to derive good estimates for the factors in Eq. (A7.1). Moreover, if we can derive these estimates for different wheel loads, then one would be able to find the parameters of the additional relationships (e.g., B, C, D, or E as a function of wheel load F_z, see also Eq. (2.29) and Eq. (2.46)). The dependency on camber angle is usually less important here. Please refer to Section 2.6 where it is shown that Magic Formula data for pure slip, in turn, can be used to estimate the combined slip characteristics.

Two observations are found easily (Figure A7.1):

1. The peak value of $F(s)$ is given by D.
2. The slope of $F(s)$ at $s = 0$ is given by $K \equiv B \cdot C \cdot D$.

The values of D and K can usually be estimated quite well from available graphs. Now suppose that the value F_∞ of F for very large slip s can be estimated well, as indicated in Figure A7.1. When s is very large, Eq. (A7.1) can be expressed as

$$F_\infty = D \cdot \sin\left(\frac{\pi}{2} \cdot C\right) \tag{A7.2}$$

and thus

$$C = 1 \pm \left(1 - \frac{2}{\pi} \cdot \arcsin\left(\frac{F_\infty}{D}\right)\right) \tag{A7.3}$$

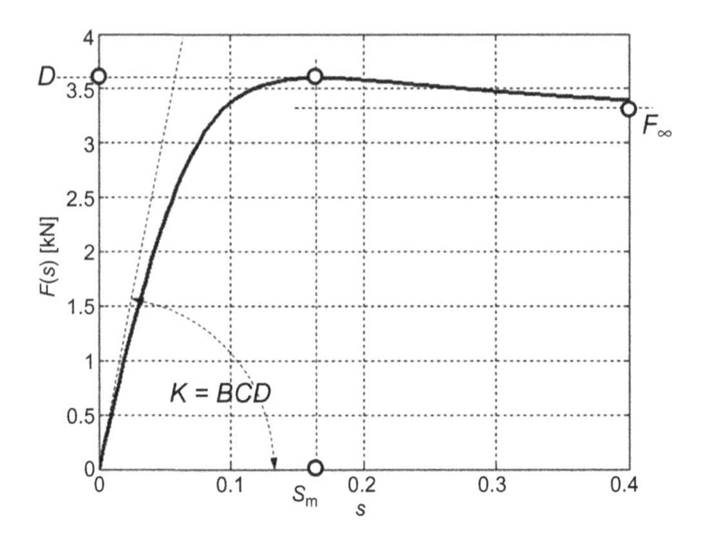

FIGURE A7.1 Tire shear force (F) versus slip (s).

Because $C > 1$, the + sign must be selected. The value for B follows now easily from K, D, and C.

Finally, if one is able to determine the point s_{m} with sufficient accuracy, for which $F(s)$ attains a maximum, one can use the relationship

$$\frac{\mathrm{d}F}{\mathrm{d}s}(s_{\mathrm{m}}) = 0$$

to show that, if $C > 1$,

$$E = \frac{B \cdot s_{\mathrm{m}} - \tan\left(\frac{\pi}{2 \cdot C}\right)}{B \cdot s_{\mathrm{m}} - \arctan(B \cdot s_{\mathrm{m}})} \tag{A7.4}$$

This estimated derivation of the Magic Formula parameters can also be found in Ref. [32].

Appendix 8: The Power Spectral Density

Let $x(t)$ be a real-valued signal derived from the behavior of a system in time. For a vehicle–driver system, $x(t)$ could be the steering angle or the vertical position of the vehicle body. The energy, related to $x(t)$, is proportional to $x(t)^2$. For the vehicle body, the square of the vertical position is proportional to the energy stored in the suspension, due to road disturbances. For a driver, the square of the steering angle corresponds to the energy that is being delivered by the driver to steer the vehicle. The corresponding average power $P(x;T)$ over a time interval with length T can be expressed as the summation of the supplied energy per unit of time:

$$P(x;T) = \frac{1}{T}\int_0^T x(t)^2 \cdot \mathrm{d}t \tag{A8.1}$$

For the example of the driver, one may be interested in the contribution to this power from the frequencies of steering, with a significant contribution for high frequencies that indicate a high workload. That means that one must discuss the system behavior in the frequency domain and consider the Fourier transform $F(\Omega;x)$, defined by

$$F(\Omega;x) = \int_{-\infty}^{+\infty} x(t) \cdot e^{-i\Omega t} \cdot \mathrm{d}t \tag{A8.2}$$

For further study, Strang [51] provides an in-depth discussion of Fourier transformation and Sneddon [52] addresses integral transforms.

Because $x(t)$ is defined for positive real time, it must be defined for negative real values as well for Eq. (A8.2) to make sense. Two different approaches are clear:

i. Take $x(t) = 0$ for $t < 0$ (one-sided description).
ii. Take $x(t) = x(-t)$ for $t < 0$ (two-sided description).

The difference between these two approaches is only a factor 2. We follow approach (ii) in this appendix.

The Fourier transform exists if the signal $x(t)$ is integrable over the entire time domain, and this condition is not always satisfied. An efficient way to deal with this, remembering that $x(t)$ is only available for a finite time span, is to replace the signal by $x_T(t)$ defined by

$$\begin{aligned} x_T(t) &= x(t), \quad 0<t<T \\ x_T(t) &= 0, \quad t>T \end{aligned} \tag{A8.3}$$

with the extensions to the negative time domain as discussed previously. We write, for the corresponding Fourier transform:

$$F_T(\Omega; x) = \int_{-\infty}^{+\infty} x_T(t) \cdot e^{-i\Omega t} \cdot \mathrm{d}t \tag{A8.4}$$

Note that Eq. (A8.1) may exist for very large T, even if the signal $x(t)$ is not integrable over the entire time domain. A clear example is when $x(t)$ is constant.

We can replace $x(t)$ with $x_T(t)$ in Eq. (A8.1). We note that a theorem exists, which states that the integral over the time domain of the square of a function is equal (except for a factor 2π) to the integral over the frequency domain of the square of the absolute value of the Fourier transform of this function. This theorem, known as Parseval's relation or Plancherel's formula [51,52] can be expressed as

$$2 \cdot \pi \int_{-\infty}^{+\infty} x_T(t)^2 \cdot \mathrm{d}t = \int_{-\infty}^{\infty} |F_T(\Omega; x)|^2 \cdot \mathrm{d}\Omega \tag{A8.5}$$

and, consequently,

$$P(x; T) = \frac{1}{T} \int_0^T x(t)^2 \mathrm{d}t = \frac{1}{2T} \int_{-\infty}^{\infty} x_T(t)^2 \mathrm{d}t = \frac{1}{2\pi} \int_{-\infty}^{\infty} \frac{|F_T(\Omega; x)|^2}{2T} \mathrm{d}\Omega \tag{A8.6}$$

This expression suggests introducing the integrand of the integral over the frequency domain as the power spectral density (PSD). More precisely, PSD is defined as this function for infinite T, i.e.,

$$\mathrm{PSD}(\Omega; x) = \lim_{T \to \infty} \frac{|F_T(\Omega; x)|^2}{2T} \tag{A8.7}$$

The average mean square (MS) of the signal $x(t)$ can now be expressed in terms of the PSD:

$$P(x; T) = \frac{1}{T} \int_0^T x(t)^2 \mathrm{d}t \approx \frac{1}{2\pi} \int_{-\infty}^{\infty} PSD(\Omega; x) \mathrm{d}\Omega \tag{A8.8}$$

for large T. The contribution of a certain frequency range (Ω_1, Ω_2) to the power (P) is derived from the integral of PSD over this frequency interval.

We treat two examples, where *x(t)* is a sine function, and when *x(t)* is the solution of a second-order equation, respectively.

Case 1. Signal $x(t) = sin(\omega_0 \cdot t)$

With

$$x(t) = \sin(\omega_0 \cdot t)$$

one finds

$$F_T(\Omega; x) = \frac{2 \cdot \omega_0}{\omega_0^2 - \Omega^2} - \frac{\cos((\omega_0 + \Omega) \cdot T)}{\omega_0 + \Omega} - \frac{\cos((\omega_0 - \Omega) \cdot T)}{\omega_0 - \Omega}$$
$$+ i \cdot \frac{\sin((\omega_0 + \Omega) \cdot T)}{\omega_0 + \Omega} - i \cdot \frac{\sin((\omega_0 - \Omega) \cdot T)}{\omega_0 - \Omega}$$

For $T \rightarrow \infty$, this function tends to zero for $\omega \neq \omega_0$. We have chosen $\omega_0 = 1$ [rad/s] and approximated PSD(Ω;*x*) for $T = 8$, 12, and 24 s. Results are shown in Figure A8.1. Observe that the PSD becomes narrower near ω_0 with increasing time interval length *T*. Increasing *T* further, the PSD tends to a Dirac function, with a finite integral, but which remains zero everywhere except for $\omega = \omega_0$.

Case 2. Signal x(t) as a solution of a second-order equation

We shall discuss the solution of the following differential equation:

$$m \cdot \ddot{x} + c \cdot \dot{x} + k \cdot x = c \cdot \dot{u} + k \cdot u \tag{A8.9}$$

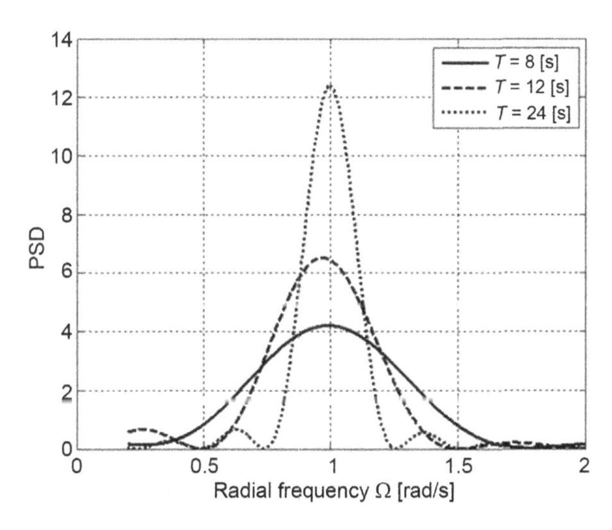

FIGURE A8.1 Approximations of the PSD for a sine function for different time intervals.

for certain values of m, c, and k, with zero initial conditions. This is the case of a single mass quarter vehicle model with vertical position $x(t)$ and road profile $u(t)$ (see Ref. [11] for more information). The absolute value of the frequency transfer function (G) can easily be obtained:

$$|G(i\cdot\Omega)|^2 = \frac{k^2 + c^2\cdot\Omega^2}{(k-m\cdot\Omega^2)^2 + c^2\cdot\Omega^2}$$

As a result, the power spectral density of the vertical body acceleration (in which one is usually interested) can be derived from PSD(Ω;u):

$$\mathrm{PSD}(\Omega;\ddot{x}) = \Omega^4\cdot\frac{k^2 + c^2\cdot\Omega^2}{(k-m\cdot\Omega^2)^2 + c^2\cdot\Omega^2}\cdot\mathrm{PSD}(\Omega;u)$$

The power spectral density of the road profile is usually expressed as

$$\mathrm{PSD}(\Omega;u) = c_u\cdot\Omega^{-2}\cdot V$$

which means that road disturbances tend to be smaller if the frequency is higher, i.e., if the road disturbance is shorter. Higher speeds correspond to a higher PSD.

We determined the power spectral density for the vertical body acceleration for parameters $m = 400$ [kg], $c = 2\times10^3$ [N/ms], $k = 3\times10^4$ [N/m], $c_u = 5\times10^{-5}$ [m^3], and $V = 25$ [m/s]. The results are shown in Figure A8.2. The maximum PSD response is found near the natural eigenfrequency ω_0 with $\omega_0^2 = k/m$, as expected.

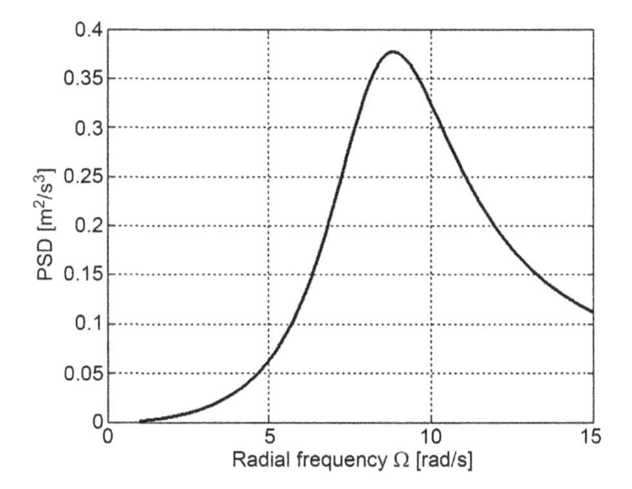

FIGURE A8.2 Power spectral density for the body acceleration with the body displacements $x(t)$ following from Eq. (A8.9).

List of Symbols

a	distance CoG—front axle
a	half tire contact length
a_e, b_e, c_e	elliptic cam parameters
a_x	longitudinal acceleration
a_y	lateral acceleration
b	distance CoG—rear axle
b	half tire contact width
c	stiffness
c_{bx}, c_{bz}	translational sidewall stiffness values
$c_{b\theta}$	rotational sidewall stiffness value
c_{cx}, c_{cy}	carcass stiffnesses per unit length
$c_{\alpha i}$	normalized cornering stiffness, axle i
CC_α	cornering compliance
C_F	lateral force coefficient
C_{Fy}	tire lateral spring stiffness
C_{Fz}	vertical tire stiffness
C_M	yaw moment coefficient
CoG	center of gravity
C_α	cornering stiffness
C_γ	camber stiffness
C_κ	longitudinal slip stiffness
d	vertical tire deflection
df_z	deviation from nominal tire load
D_p	path deviation
e	trailing arm length
f_y	normalized cornering force (lateral friction coefficient)
f_R	coefficient of rolling resistance
F_{cN}	normal contact force
F_{cT}	tangential contact force
F_R	rolling resistance force
F_x	longitudinal (brake, drive) force
F_y	lateral (cornering) force
F_{ye}	external lateral force
F_z	wheel load, axle load
F_{z0}	nominal tire load
F_{zij}	wheel load, axle i (f, r), side j (L, R)
g	acceleration of gravity
$G(s)$	transfer function

$G_d(s)$	transfer function driver
$G_v(s)$	transfer function vehicle
h_{CoG}	height vehicle CoG
J_{ay}	rim moment of inertia
J_{by}	belt moment of inertia
J_{wheel}, J_w	wheel moment of inertia
J_z	vehicle yaw moment of inertia
k	damping
k_{bx}, k_{bz}	translational sidewall damping values
$k_{b\theta}$	rotational sidewall damping value
k_x, k_y, k	tire read stiffnesses
K	gain
K_p	driver steering gain
K_s	stability factor
$K_{\varphi 1}$, $K_{\varphi 2}$	axle roll stiffnesses
L	wheelbase
L_b, L_f	parameters basic road function
L_d	Rayleigh function
L_p	driver preview length
L_s	length two-point follower
L^*	Lagrange function
m	mass (vehicle)
m_a	rim mass
m_b	belt mass
M_{dB}	$20 \cdot {}^{10}\log M(\Omega)$
M_s	static margin
M_x	overturning moment
M_y	drive, brake torque
M_z	aligning torque
M_{ze}	external yaw moment
M_{zr}	residual torque
$M(\Omega)$	magnitude transfer function
p_i	inflation pressure
P_a	accelerator pedal depression
PSD	power spectral density
q_x, q_y	integrated shear stress
q_z	integrated normal stress
r	yaw rate
r_g	radius of gyration
R	unloaded tire radius
R	curve radius
R_e	effective rolling radius
R_l	loaded tire radius
s	Laplace variable

s_x, s_y	practical slip quantities
SI	stability index
SRR	steering reversal rate
t	track width
t_{p}	pneumatic trail
T	temperature
T	kinetic energy
T_{c}	cornering kinetic energy
T_{eq}	equivalent time
THW	time headway
T_{k}	translational kinetic energy
T_{L}	lead time
T_{p}	preview time
TTC	time to contact
u, v	tire contact deflections
u, v	local vehicle speeds
u_{b}, v_{b}	tire belt deflections
u_{t}, v_{t}	tire tread deflections
U	potential energy
V	velocity
V_{gx}, V_{gy}	sliding speeds
V_{r}	rolling speed
$V_{\mathrm{s}x}$	longitudinal slip speed
V_{sy}	lateral slip speed
V_x	forward tire speed
v_y	vehicle lateral speed
w_{e}	effective road height
x_{a}, z_{a}	rim position
x_{b}, z_{b}	belt rigid ring position
x_{Guo}, z_{Guo}	vehicle states cf. Guo
x_{NS}	position neutral steer point
Y_β, Y_r	derivatives of stability
N_β, N_r	derivatives of stability
α	wheel, axle slip angle
α'	effective slip angle
β	body slip angle
β_{e}	effective road slope
δ	steering angle
γ	camber angle
γ	articulation angle
η	course angle
η	understeer gradient
φ	phase angle
φ	angular wheel position

φ	vehicle roll angle
φ_{m}	phase margin
κ	brake slip
κ'	effective brake slip
λ	eigenvalue
λ	vehicle rotating length
μ	road friction
μ_x	normalized brake force (longitudinal friction coefficient)
$\mu_{x\mathrm{p}}$	peak braking coefficient
$\mu_{x\mathrm{s}}$	sliding braking coefficient
μ_y	normalized cornering force (lateral friction coefficient)
$\mu_{y\mathrm{p}}$	peak cornering coefficient
$\mu_{y\mathrm{s}}$	sliding cornering coefficient
θ	tire parameter
θ	vehicle pitch angle
θ_{a}	rim rotational deflection
θ_{b}	belt rotational deflection
ρ	total theoretical slip
ρ_x	horizontal contact patch displacement
ρ_x, ρ_y	theoretical slip quantities
ρ_{zr}	residual tire deflection
σ_z	normal contact stress
σ_α, σ_κ	relaxation lengths
τ_{d}	delay time
τ_{lag}, τ_{L}	lag time
τ	total shear stress
τ_x, τ_y	shear stresses
ω	eigenfrequency
ω_0	undamped eigenfrequency
Ω	forced frequency
Ω	rotational speed
Ω_{bw}	bandwidth
Ω_0	rotational speed under free rolling conditions
ψ	yaw angle
ψ_{p}	path yaw orientation
ζ	damping ratio

References

[1] M. Abe.: *Vehicle Handling Dynamics. Theory and Application.* Butterworth-Heinemann, Oxford UK (2009).

[2] I.J.M. Besselink.: *Shimmy of Aircraft Main Landing Gears.* Doctoral Thesis, Delft University of Technology, Delft, The Netherlands (2000).

[3] D.J. Bickerstaff.: *The Handling Properties of Light Trucks.* SAE Technical Paper 760710 Warrendale, PA, USA (1976).

[4] W. Borgmann.: *Theoretische und Experimentelle Untersuchungen an Luftreifen bei Schräglauf.* Diss., Braunschweig (1963).

[5] J.C. Dixon.: *Tires, Suspension and Handling.* SAE International, Warrendale, PA, USA (1996).

[6] R.V. Dukkipati, J. Pang, M.S. Quatu, G. Sheng, Z. Shuguang.: *Road Vehicle Dynamics.* SAE International, Warrendale, PA, USA (2008).

[7] E. Donges.: *Supporting Drivers by Chassis Control Systems.* In: J.P. Pauwelussen, H.B. Pacejka (eds.): Smart Vehicles. Swets & Zeitlinger Publishers, Amsterdam/Lisse, The Netherlands (1995).

[8] H. Fromm.: *Kurzer Bericht über die Geschichte der Theorie des Radflatterns.* Bericht 140 der Lilienthal Gesellschaft, 1941; NACA TM 1365, 1954.

[9] W. Gengenbach.: *Experimentelle Untersuchung von Reifen auf Nasser Fahrbahn.* Automobiltechnisch Zeitschrift (ATZ), Band 70, Nr. 8 und 9 (1969).

[10] G. Genta, L. Morello.: *The Automotive Chassis, Volume 1: Components Design.* Springer, Berlin, Mechanical Engineering Series (2009).

[11] G. Genta, L. Morello.: *The Automotive Chassis, Volume 2: System Design.* Springer, Berlin, Mechanical Engineering Series (2009).

[12] R. Gnadler.: *Nassgriff und Aquaplaningverhalten von PKW-Reifen.* Verkehrsunfall und Fahrzeugtechnik (1988).

[13] J. Godthelp, P. Milgram, G.J. Blaauw.: *The development of a time-related measure to describe driving strategy*, Human Factors, Vol. 26, pp. 257–268 (1988).

[14] S. Gong.: *A Study of In-Plane Dynamics of Tires.* PhD Thesis, Delft University of Technology, The Netherlands (1993).

[15] K.-H. Guo.: *A Study of a Phase Plane Representation for Identifying Vehicle Behavior.* Proceedings of the 9th IAVSD Symposium, Linköping (1985).

[16] B. Heissing, M. Ersoy, S. Gies.: *Fahrwerkhandbuch, Grundlagen, Fahrdynamik, Komponenten, Systeme, Mechatronik, Perpektiven.* PRAXISjATZ/MTZ-Fachbuch, Vieweg & Teubner, Springer Fachmedien Wiesbaden GmBH (2011).

[17] A. Higuchi.: *Transient Response of Tyres at Large Wheel Slip and Camber.* Doctoral Thesis, Delft University of Technology, Delft, The Netherlands (1997).

[18] R. Jagacinski, J.M. Flach.: *Control Theory for Humans, Quantitative Approaches to Modeling performance.* CRC Press, Boca Raton, FL (2009).

[19] W.-D. Käppler, J. Godthelp.: *The Effects of Tire Pressure Variations on Vehicle Handling Properties and Driving Strategy.* Forschung für Antropotechnik, Bericht Nr. 72 (1986).

[20] U. Kiencke, L. Nielsen.: *Automotive Control System. For Engine, Driveline, and Vehicle.* Springer, Berlin (2005).

[21] D.E. Kirk.: *Optimal Control Theory. An Introduction.* Chapter 4. Dover Publications, Inc., Mineola, New York (2004).

[22] G.N. Lupton, T. Williams.: *Study of the Skid Resistance of Different Tire Tread Polymers on Wet Pavements with a Range of Surface Textures.* In: Skid Resistance of

Highway Pavements. American Society for Testing and Materials, West Conshohocken, PA, USA (1973).
[23] J.P. Maurice.: *Short Wavelength and Dynamic Tyre Behaviour under Lateral and Combined Slip Conditions*. PhD Thesis, Delft University of Technology, The Netherlands (2000).
[24] D.T. McRuer, E.S. Krendel.: *The human operator as a servo system element*, Journal of the Franklin Institute, Vol. 267, pp. 381–403 (1959).
[25] D.T. McRuer, H.R. Jex.: *A review of quasi-linear pilot models*, IEEE Transactions on Human Factors, Vol. 8, pp. 231–249 (1967).
[26] W.F. Milliken, D.L. Milliken.: *Race Car Vehicle Dynamics*. SAE International, Warrendale, PA, USA (1995).
[27] M. Mitschke, H. Wallentowitz.: *Dynamik der Kraftfahrzeuge*. Springer Verlag, Berlin, Heidelberg, New York (2004).
[28] S. Monsma, S. Shrey.: *Quantification of Drivers Mental Workload During Outdoor Testing Using Heart Rate Variability*. EAEC conference, Valencia (2011).
[29] D.F. Moore.: *Friction of Pneumatic Tyres*. Elsevier Scientific Publishing Company, Amsterdam, The Netherlands (1975).
[30] M. Mulder, J.J.A. Pauwelussen, M.M. van Paassen, M. Mulder, D.A. Abbink.: *Active deceleration support in car following*, IEEE Transactions on Systems, Man and Cybernetics, Part A: Systems and Humans, Vol. 40, No. 6, pp. 1271–1284 (2010).
[31] H.B. Pacejka.: *The Wheel Shimmy Phenomenon*. Doctoral Thesis, Delft University of Technology, Delft, The Netherlands (1966).
[32] H.B. Pacejka.: *Tyre and Vehicle Dynamics*. Butterworth-Heinemann, Oxford UK (2006).
[33] J.P. Pauwelussen.: *Effect of Tyre Handling Characteristics on Driver Judgement of Vehicle Directional Behaviour*. In: Pauwelussen (ed.): Driver Performance, Understanding Human Monitoring and Assessment. Swets & Zeitlinger Publishers, Amsterdam/Lisse, The Netherlands (1999).
[34] J.P. Pauwelussen, A.L.A. Andress Fernandez.: *Estimated Combined Steady State Tyre Slip Characteristics*. Haus der Technik Conference 2001, Fahrwerktechnik, Osnabrück, Germany.
[35] J.P. Pauwelussen, J.J.A. Pauwelussen.: *Exploration of Steering Wheel Angle Based Workload Measures in Relationship to Steering Feel Evaluation*. International Conference on Advanced Vehicle Control (AVEC '04), Arnhem (2004).
[36] J.P. Pauwelussen.: *The local contact between tyre and road under steady state combined slip conditions*, Journal of Vehicle System Dynamics, Vol. 41, No. 1 (2004).
[37] J.P. Pauwelussen. *Graphical Means to Analyze and Visualize Vehicle Handling Behaviour*. ECCOMAS Thematic Conference in Multibody Dynamics, Madrid, (2005).
[38] J.P. Pauwelussen.: *Dependencies of driver steering control parameters*, Journal of Vehicle System Dynamics, Vol. 50, No. 6 (2012).
[39] J. Pauwelussen, O. Patil.: *On Road Driver State Estimation*. FISITA Congress, Maastricht (2014).
[40] A.A. Popov, D.J. Cole, D. Cebon, C.B. Winkler.: *Laboratory measurement of rolling resistance in truck tyres under dynamic vertical load*, Proceedings of the IMechE, Part D, Journal of Automobile Engineering, Vol. 217, No. 12 (2003).
[41] S. Saigo, P. Raksincharoensak, M. Nagai.: *Safe Driving Advisory System Based on Integrated Modelling of Naturalistic Driving Behaviour*. International Conference on Advanced Vehicle Control (AVEC '10), Loughborough (2010).
[42] D.D. Salvucci, R. Gray.: *A two-point visual control model of steering*, Perception, Vol. 33, No. 10, pp. 1233–1248 (2004).
[43] A.W. Savage.: U.S. Patent 1203910, May 21, 1915, Vehicle Tire.
[44] M. R. Savino.: *Standardized Names and Definitions for Driving Performance Measures*. Master Thesis, Tufts University (2009).
[45] B. von Schlippe, R. Dietrich.: *Das Flattern eines bepneuten Rades*, Ber. Lilienthal Ges. 140, p.35 (1941).

[46] A.J.C. Schmeitz.: *A Semi-Empirical Three-Dimensional Model of the Pneumatic Tyre Rolling over Arbitrarily Uneven Road Surfaces.* PhD Thesis, Delft University of Technology, The Netherlands (2004).
[47] D.J. Schuring.: *The rolling loss of pneumatic tires*, Rubber Chemistry and Technology, Vol. 53, No. 3 (1980).
[48] R.S. Sharp.: *Vehicle Dynamics and the Judgement of Quality.* In: Pauwelussen (ed.): Driver Performance, Understanding Human Monitoring and Assessment. Swets & Zeitlinger Publishers, Amsterdam/Lisse, The Netherlands (1999).
[49] L. Segel.: *Force and moment response of pneumatic tyres to lateral motion inputs*, Transactions of ASME Journal of Engineering for Industry, Vol. 88B, No. 1 (1966).
[50] H.T. Smakman.: *Functional Integration of Slip Control with Active Suspension for Improved Lateral Vehicle Dynamics.* Thesis, Delft University of Technology (2000).
[51] I.N. Sneddon.: *The Use of Integral Transforms.* McGraw-Hill, New York, NY (1972).
[52] G. Strang.: *Introduction to Applied Mathematics.* Wellesley Cambridge Press, Wellesley, MA (1986).
[53] V. Urban.: *Driver's State Estimation.* Thesis European Master of Automotive Engineering, HAN University of Applied Sciences and Czech Technical University Prague (2009).
[54] J. van de Vegte.: *Feedback Control Systems.* Prentice-Hall International Editions, Englewood Cliffs, NJ (1990).
[55] P.J. Th. Venhovens, A.C.M. van der Knaap.: *Delft Active Suspension (DAS). Background Theory and Physical Realization.* In: J.P. Pauwelussen, H.B. Pacejka (eds.): Smart Vehicles. Swets & Zeitlinger Publishers, Amsterdam/Lisse (1995).
[56] D. de Waard.: *The Measurement of Drivers' Mental Workload.* PhD Thesis, University of Groningen, Traffic Research Centre (1996).
[57] D.H. Weir, R.J. Dimarco.: *Correlation and Evaluation of Driver/Vehicle Directional Handling Data.* SAE Paper 780010 (1978).
[58] X. Xia, J.N. Willis.: *The Effects of Tire Cornering Stiffness on Vehicle Linear Handling Performance.* SAE Technical Paper Series 950313 (1995).
[59] P. Zegelaar.: *The Dynamic Response of Tyres to Brake Torque Variations and Road Unevenesses.* PhD Thesis, Delft University of Technology, The Netherlands (1998).
[60] J. Zuurbier, J.H. Hogema, J.A.W.J. Brekelmans.: *Vehicle Steering by Side-Stick: Optimising Steering Characteristics.* International Conference on Advanced Vehicle Control (AVEC), Ann Arbor, MI (2000).
[61] SAE J67e (revision July 1976), Surface Vehicle Recommended Practice.
[62] ISO 3888-2: *Passenger Cars—Test Track for a Severe Lane-Change Manoeuvre—Part 2: Obstacle Avoidance*, First edition 2002-11-15. International Organization for Standardization (2002).
[63] ISO 7401: *Road-Vehicles—Lateral Transient Response Test Methods—Open Loop Test Methods*, Second edition 2003-02-15. International Organization for Standardization (2003).
[64] ISO 4138: *Passenger Cars—Steady State Circular Driving Behaviour—Open Loop Test Methods*, Third edition 2004-09-15. International Organization for Standardization (2004).
[65] *The Economic Benefits of Concrete Road Pavements.* CCANZ report TR12 (2004).

Index

Note: Page numbers followed by "*f*" and "*t*" refer to figures and tables, respectively.

H

I

J

K

L

M

S

V

W

Y